Handbook of EFFECTIVE DISASTER/ RECOVERY PLANNING

Please see pages 336–337 for an order form for additional seminar/workshop materials available from DIA*log Management, Inc.

Other McGraw-Hill Books in Software Engineering

ISBN	AUTHOR	TITLE
0-07-040235-3	Marca, McGowan	SADT: Structured Analysis and Design Technique
0-07-036948-8	Lecarme, Pellissier, Gart	Software Portability
0-07-057299-2	Shumate	Understanding Concurrency in ADA®
0-07-046536-3	Nielsen, Shumate	Designing Large Real-Time Systems with ADA®
0-07-042632-5	Modell	A Professional's Guide to Systems Analysis
0-07-016803-2	Dickinson	Developing Quality Systems
0-07-023165-6	General Electric Company Staff	Software Engineering Handbook
0-07-010646-0	Wallace et al.	A Unified Methodology for Developing Systems
0-07-010645-2	Charette	An Introduction to Software Engineering Environments
0-07-067922-3	Wallace	Practitioner's Guide to ADA®
0-07-044119-7	Musa et al.	Software Reliability
0-07-010661-X	Charette	Risk Analysis and Management

Ranade IBM Series

ISBN	AUTHOR	TITLE
0-07-065087-X	Towner	IDMS/R
0-07-002673-4	Azevedo	ISPF
0-07-050686-8	Prasad	IBM Mainframe
0-07-046263-1	McGrew, McDaniel	Online Text Management
0-07-039822-4	Malamud	DEC Networks and Architectures
0-07-071136-4	Wipfler	CICS Application Development and Programming
0-07-051144-6	Ranade	SNA Introduction to VTAM
0-07-583963-6	Ranade	VSAM: Concepts, Programming, and Design
0-07-051198-5	Ranade, Ranade	VSAM: Performance, Design, and Fine Tuning

For more information about other McGraw-Hill materials, call 1-800-2-MCGRAW in the United States. In other countries, call your nearest McGraw-Hill office.

Handbook of EFFECTIVE DISASTER/ RECOVERY PLANNING
A Seminar/Workshop Approach

Author: **ALVIN ARNELL**
Technical Editor: **DONALD G. DAVIS**

McGraw-Hill Publishing Company
New York   St. Louis   San Francisco   Auckland   Bogotá
Caracas   Hamburg   Lisbon   London   Madrid   Mexico
Milan   Montreal   New Delhi   Oklahoma City
Paris   San Juan   São Paulo   Singapore
Sydney   Tokyo   Toronto

Library of Congress Cataloging-in-Publication Data

Arnell, Alvin.
    Handbook of effective disaster/recovery planning: a
seminar/workshop approach / author, Alvin Arnell; technical editor,
Donald G. Davis.
      p. cm.
    ISBN 0-07-002394-8
    1. Electronic data processing departments—Security measures—
Planning—Handbooks, manuals, etc. 2. Disasters—Planning—
Handbooks, manuals, etc. I. Title.
HF5548.37.A75   1989
658.4'78—dc20   89-13002
                                                                               CIP

Copyright © 1990 by Alvin Arnell. All rights reserved.
Printed in the United States of America. Except as permitted
under the United States Copyright Act of 1976, no part of this
publication may be reproduced or distributed in any form or by
any means, or stored in a data base or retrieval system, without
the prior written permission of the publisher.

1234567890 HAL/HAL 8965432109

ISBN 0-07-002394-8

The editor for this book was Theron Shreve,
the designer was Elliot Epstein, and the production
supervisor was Dianne L. Walber. It was set in Times Roman
by Professional Composition, Inc. Project supervision by The Total Book.

Printed and bound by Arcata Graphics/Halliday.

The author has tried to provide herein a professional publication. However, no responsibility whatsoever is assumed by the author or McGraw-Hill Publishing Company for any errors of omission or commission or for any matter or result based upon or resulting from the use of any thing, matter, item, idea, plan, and so on presented in this publication and/or any workshop. Such responsibility belongs to, and remains solely with, the individual reader or workshop attendee (and his or her firm) when using or acting upon any of the above.

---

*For more information about other McGraw-Hill materials,
call 1-800-2-MCGRAW in the United States. In other
countries, call your nearest McGraw-Hill office.*

To Elisse, Andrew, Mathew,
Mathilde, and Nathan
whose patience, love, and confidence
never waivered.
Alvin Arnell (A^2)

To Mary Louise for
thirty-two years of love and devotion,
despite the many days and months
we've had to be apart.
Donald Davis (D^2)

CONTENTS

Preface xi
A Message to Disaster/Recovery Planners xiii
Introduction xix

PART 1   1

Section 1 Vulnerability Analysis and Recovery Requirements Definition   1

1.1	**Business Vulnerability Analysis**	1
1.1.1	Today's Dependency on Electronic Data Processing (EDP), 1	
1.1.2	Survey of Management's Viewpoint, 1	
1.1.3	Identification of EDP-Dependent Business Processes, 2	
1.1.4	Legal Requirements, 2	
1.1.5	Developing a Qualitative and Quantitative Risk Model, 2	
1.1.6	Restoration Policy for Critical Applications, 3	
1.2	**Application Recovery Priorities**	3
1.2.1	Essential Applications, 3	
1.2.2	Recovery Thresholds, 3	
1.2.3	Ranking Criteria for Creating the Applications Recovery Priority Scheme, 3	
1.2.4	Service-Level Needs, 4	
1.3	**EDP Recovery Requirements**	4
1.3.1	Computer System Sizing, 4	
1.3.2	Software Configuration, 4	
1.3.3	Network Configuration, 4	
1.3.4	End-User Computing, 5	
1.3.5	Implications of Future Applications, 5	

Section 2 Recovery Strategy Selection and Risk Management   6

2.1	**Recovery Strategy Selection**	6
2.1.1	Alternatives Analysis, 6	
2.1.2	Cost Justification, 8	
2.1.3	Strategy Selection, 8	
2.1.4	Contracts and Agreements, 9	
2.1.5	Computer and User: Backup Alternatives, 9	
2.2	**Management Considerations and Responsibilities**	10
2.2.1	Defining Application-Recovery Requirements, 11	
2.3	**Data Center Risk Analysis**	19
2.3.1	Threats Analysis, 19	
2.3.2	Physical Security Risks, 19	
2.3.3	Environmental Control Risks, 22	
2.3.4	Computer Equipment Failure Risks, 24	
2.3.5	Network Failure Risks, 24	
2.3.6	Logical Security Risks, 25	

2.3.7	Loss Expectancies, 25	
2.3.8	Quantitative Risk Model, 26	
2.4	**Disaster Countermeasures Selection**	**26**
2.4.1	Alternatives Identification, 26	
2.4.2	Cost-Benefit Analysis, Selection, and Implementation, 29	
2.5	**Insurance Coverage**	**29**
2.5.1	Coverage Analysis, 30	
2.5.2	Cost Savings Opportunity Analysis, 30	

Section 3 — Recovery Management Plan and Procedures Development — 32

3.1	**Recovery Management Plan**	**32**
3.1.1	Organization of Recovery Teams, 32	
3.1.2	Recovery Management Cycle, 35	
3.1.3	Command (Control) Center, 35	
3.1.4	Damage Assessment, 37	
3.1.5	Notifications, 38	
3.2	**EDP Teams Recovery Procedures**	**38**
3.2.1	Mobilization, 38	
3.2.2	Restart Processing, 38	
3.2.3	Interim Processing, 41	
3.2.4	Restoration, 41	
3.3	**Critical Applications**	**42**
3.3.1	Future Requirements, 42	
3.3.2	Applications Recovery Plans, 42	
3.4	**Applications Recovery Plan Requirements for Data Center Production Applications**	**42**
3.4.1	Applications Recovery Plan, 43	
3.5	**Corporate Data Center File Retention Philosophy**	**43**
3.5.1	Overview, 43	
3.5.2	File Management, 43	
3.5.3	Legal Requirements, 43	
3.5.4	Interdependencies, 43	
3.5.5	File Activity and Dump Frequency, 44	
3.5.6	Criticality, 44	
3.5.7	Cost, 44	
3.5.8	Summary of Standard Retentions, 44	
3.5.9	Conclusion, 45	
3.6	**End Users Recovery Procedures**	**45**
3.6.1	Outage Period Processing, 45	
3.6.2	Restart Processing, 51	
3.6.3	Interim Processing, 55	

Section 4 — Implementation and Testing — 56

4.1	**Contingency Plan Implementation**	**56**
4.1.1	Preparedness Standards, Policies, and Procedures, 56	
4.1.2	File Backup, 57	
4.2	**Personnel Training**	**60**
4.2.1	Awareness Training, 60	
4.2.2	Technical Training, 60	
4.3	**Validation Testing**	**61**
4.3.1	Scenarios Planning, 61	

4.3.2	Preparedness Testing, 61	
4.3.3	EDP Restart/Recovery Testing, 61	
4.3.4	End-User Restart/Recovery Testing, 63	

Section 5 — Ongoing Administration and Plan Maintenance — 64

5.1	**Ongoing Testing**	64
5.1.1	Test Plan, 64	
5.1.2	Results Analysis, 64	
5.2	**Ongoing Maintenance**	64
5.2.1	Change Management, 64	
5.2.2	Distribution, 67	

PART 2 — Effective Disaster/Recovery Plan: Model Plan—Table of Contents Review — 89

Note: The Table of Contents appearing in Part 2 represents the actual table of contents used in the workshop disaster/recovery master plan. The numerical sequence of the table of contents review in Part 2 is independent of the numbering sequence of the rest of the book.

PART 3 — Getting the Job Done — 185

1.1	**How to Get the Job Done**	185
1.1.1	The First Step, 185	
1.1.2	The Second Step, 186	
1.1.3	The Third Step, 186	
1.1.4	The Fourth Step, 186	
1.1.5	The Fifth Step, 186	
2.1	**The Plan Preparation Workshop**	191
2.1.1	Creating a Workshop Environment, 191	

PART 4 — Instructor's Guide — 195

PART 5 — Organization and Participant Profile — 203

PART 6 — Data Collection Review — 221

1.1	**Data Collection**	221
1.1.1	Information Collection, 221	
1.1.2	Personnel Assignments and Participation, 221	
1.1.3	Task Forces and Disaster/Recovery Action Teams, 222	
1.1.4	Senior Management Team, 222	
1.1.5	The Workshop Task Force Coordinator, 222	
1.1.6	The Disaster/Recovery Information-Gathering Task Force, 223	
1.1.7	Management's Participation, 223	
1.1.8	Questionnaires and Forms, 223	
1.1.9	Questionnaires, Checklists, Guidelines, and Forms, 223	

1.1.10	Limitations Concerning Questionnaires, Checklists, Guidelines, and Forms, 224	
1.1.11	Data Processing Planning Coordination: Selecting Personnel; Developing Task Forces and Plant Teams, 224	
1.1.12	Task Force Assignments, 224	
1.1.13	The Workshop Disaster/Recovery Administrator, 225	
1.1.14	The Task Force Disaster/Recovery Coordinator, 225	
1.1.15	Plan Economics and Personnel, 225	
2.1	**Data Collection: Information Requested (Definitions)**	225
2.1.1	Outline of Information Requested, 225	
2.1.2	Organizational Charts, 226	
2.1.3	Risk Assessment, 226	
2.1.4	Hardware/Auxiliary Equipment (Configurations), 226	
2.1.5	Layout of the Facility, 227	
2.1.6	List of Hardware/Auxiliary Equipment and Contacts, 227	
2.1.7	Data Center Operating Procedures and Policies, 229	
2.1.8	User Operating Procedures, 229	
2.1.9	Security Systems, 229	
2.1.10	Communications, 229	
2.1.11	Normal Communications, 229	
2.1.12	Emergency Action Plan Responsibilities, 230	
2.1.13	Emergency Response Plan, 230	
2.1.14	Pyramid Calling Tree, 231	
2.1.15	Application System Recovery—Software Data and Documentation, 231	
2.1.16	Plan Preparation Workshop, 235	
2.1.17	Testing the Plan, 235	
2.1.18	Plan Test Monitoring, 235	
2.1.19	Personnel Training, 235	
2.1.20	Maintenance of the Plan, 235	

PART 7		**237**
Section 1	**Program Profile: Corporate Business Interruption Recovery Plan**	237
Section 2	**Computer and Noncomputer Disaster/Recovery Planning Questionnaire**	247
PART 8	**Interface Disaster/Recovery Miniplan Model: Branch Office for a Financial Institution**	**265**
PART 9	**Model Disaster/Recovery Emergency Response Plan**	**273**
PART 10	**Sample Interactive Sessions**	**285**

Index 329

PREFACE

This *Handbook of Effective Disaster/Recovery Planning* was created and distilled from the thousands of pages of documentation written and used in our seminar/workshop on contingency disaster/recovery planning over the past five years. Over 160 organizations, varying in size from single-branch banks to multinational corporations, have created, implemented, and tested plans developed using the seminar/workshop methodology. The secret to a successful, implementable, practical, and effective contingency disaster/recovery plan is total organizational participation. This *Handbook of Effective Disaster/Recovery Planning* is intended to provide organizations who *need* a plan for their corporate and data center activities with a practical guide for those assigned the responsibility of designing, implementing, testing, and maintaining such plans.

The key to creating a successful plan is to understand what the scope of the plan and requirements are, how they form a cohesive internal structure within the organization, and that in the event of an adverse incident, the business services of the company can be maintained. Without ensuring an ongoing *continuity* of normal business performance, management is not fulfilling its basic responsibility to its personnel, customers, vendors, or shareholders.

Each process of planning requires a methodology. The primary objective of this handbook is to provide not only the methodology but the supporting material—the tools—to make the process work. The ultimate *product* essential for the protection of the business environment is an efficiently designed, implementable, testable, and maintainable contingency disaster/recovery plan. More important, through the effective use of this methodology, each organization will have a plan customized to its individual operational environment—a plan developed in-house and understood by the people who may have to use it, because they are the ones who have created, implemented, tested, and maintained the final product.

The planning process presented and the content of the handbook have been considerably enhanced by the author and technical editor's experience in assisting the more than one hundred and sixty corporations worldwide in the preparation of contingency disaster/recovery plans.

Those assigned the responsibility of creating a plan now have available the information on the structure and content of a viable process to prepare a plan—one that will enable an organization to cope with any form of unexpected, undesirable, or adverse incident, without having to deal with the scanty or scattered number of disconnected articles or publications.

This is not just a guideline on how to prepare an effective disaster/recovery plan. This handbook is a step-by-step, how-to-do-it program that takes into consideration the effects of a catastrophic event in the information processing environment, but it also addresses the issues relating to the entire scope of maintaining the continuity of the business and the integrity of data.

The handbook is divided into working parts starting with the concepts of effective disaster/recovery planning and including a practical, proven methodology for the development of the actual plan, section by section, paragraph by paragraph.

The reason the methodology works is that the practitioner inside the company, unlike those from the outside, understands and is more intimate on a day-to-day basis with his or her environment, organizational needs, and critical business factors. Thus, the development of the plan takes less time and less commitment of critical, usually unavailable human and financial resources, and provides a workable, effective document.

Sponsored by one of the leading computer hardware manufacturers and "Big 8" accounting firms, the author continues to perform seminar/workshops on effective disaster/recovery planning. This handbook provides the same structure and methodology presented at the public programs. Many organizations have found that unless they are outside their normal daily environment, regardless of good intentions, the job never gets done. If the methodology presented here appeals to your organization, and you prefer to save the program development time and effort, you might consider participating in one of the scheduled programs. Either way, we hope this handbook will encourage and assist more organizations to develop a contingency disaster/recovery plan to protect valuable information resources and assets.

Contingency–disaster/recovery planning is now a requirement oriented to organizational problems rather than a technological burden.

Acknowledgments

The material in this handbook draws heavily from a wide range of the authors' field experience in OSHA and the computer and disaster/recovery disciplines. Special acknowledgment is given to the more than one hundred and sixty organizations who participated in our seminar/workshops and shared their many ideas on how to make this program work, regardless of the country and cultural differences or the type of organization.

I will be forever grateful to Donald G. Davis for his continued faith and support, and the trying experience of editing and re-editing more than six previous manuscripts covering over 4000 pages of material.

Sheldon Wiederman of Grumman Aerospace provided unique guidance and special insight into planning for major corporations.

Kathy Tuzio, whose patience and fortitude in the preparation of this manuscript, deserves special commendation. Thanks for never saying "it can't be done."

Our good friends, Joseph St. Georges of Disaster Control Inc. and Robert Dever of Liberty Travel, helped us make an impact into the industry.

Last, but not least, my special gratitude to my father-in-law, Nathan Schikler, whose confidence and support never wavered and whose advice helped make the DIA*log program successful. To him I will be forever indebted.

Alvin Arnell
Merrick, New York

A MESSAGE TO DISASTER/ RECOVERY PLANNERS

The express purpose of this *Handbook of Effective Disaster/Recovery Planning*® is to provide a proven methodology to understand the concepts to design, implement, test, and maintain a business resumption plan.

This handbook prepares the planner for the tasks required to develop plans by providing a general overview of the environment relating to effective disaster/recovery planning and how the plans are used. The progression of the material in Part 1 provides the guidelines and background for the company seminar/workshop presenter to introduce the concepts, issues, and development procedures of contingency-disaster/recovery planning. This is presented in a methodical fashion, following the order of the plan's table of contents.

Part 2, the Table of Contents Review, walks the planner through the actual plan, paragraph by paragraph. The Table of Contents Review is a valuable tool for defining and describing the essential elements of the plan: the application of the background material found in Part 1.

The intent of this handbook is to enable those assigned the responsibility of effective disaster/recovery planning to develop their own in-house plan to meet the planner's unique organizational requirements. Consequently, each organization will develop their disaster/recovery strategies to "fit" specific requirements. Emphasis is placed more on plan development and less on discussions of background and justification. Risk analysis is bypassed because need is virtually mandated by organizations' total dependence on computers. The important job is to meld the plan components with the organizational environment. Some recognition that planners view their operations and environment in one manner, and the requirements for effective plans require the view from other perspectives, has necessitated the inclusion of methods to replace the emphasis from one view of the data processing environment to another.

The method of designing, implementing, and training personnel by the use of the seminar/workshop methodology is covered in Part 3. This section guides the planning task force assigned the plan responsibility in how to use the materials presented in the other parts of the handbook. The planning task force may prefer to use the material verbatim; others may wish to redesign the training and planning procedures presented. Regardless of which tact is taken, the plan is not a plan, nor will it ever become a plan, without absolute dedication, strict time constraints, and total organizational participation and support.

An effective disaster/recovery plan can be designed, implemented, tested, and maintained following the methodology of this handbook. The design aspect of the plan centers on the ability of the planners to individualize the material presented to the organization's specific functions and needs. Case histories have been avoided because they have a tendency to provide false impressions about one's specific environment. The objective of the seminar/workshop methodology is to make the plan "fit" the organization, not the organization "fit" the plan.

Do not start this project without understanding the purpose and process of the methodology presented here. This is not a one-person, one-department project. It is a *total organizational responsibility*, both user and data center.

Effective disaster/recovery plans cannot be developed in a vacuum. It is a task that encompasses and envelopes every functional unit of the organization. A single person, sequestered in a lonely room, attempting to "fill in the blanks" will not create a plan that can be effective or useful at the time of a disaster. Unless all levels of the organization, from the mandate and support of management to the janitorial services who understand the importance of protecting information by destroying printouts, are *total* participants in the design, development, implementation, execution, testing, and maintenance of the plan, *it won't work*.

The authors have proven through hundreds of seminars, workshops, and special presentations where total organizational participation exists that plans will work. Do not be deluded that once the plan is completed it can become a dust collector residing on a shelf and work when it is needed. Unless the plan will be continually maintained and tested, don't waste time in preparing it.

How to Use This Book

There is no panacea for the development of effective disaster/recovery plans for data centers, information systems, or business units. Over the past nine years, the authors have devoted their entire activity to develop one of the most unique methodologies for plan development, and the 160 plus organizations worldwide have succeeded through this effort. The seminar/workshop methodology has become a practical, cost- and time-efficient process *that gets the job done*.

Computer operation and computer safeguards are the subjects of numerous books, but the methodology that will enable a viable, productive organization that has experienced a disaster to recover from any form of catastrophic event, is not yet housed in one complete volume. This handbook provides the necessary technology on recovering from a disaster.

Planning Prerequisites

Developing an effective disaster/recovery plan for your organization requires a number of absolute disciplines. Without the prerequisites we've indicated herein, the job of completing this project will be long, tiresome, and ineffective.

1. The *entire organization* must understand the need for business continuity and disaster/recovery planning.
2. Management must totally support the project without reservation.
3. The project must be given a high priority.
4. Management must provide the human and financial resources to make the project work.
5. The proper level of management and staff must accept the responsibility to develop and maintain the plan.

Planning Necessities

Those assigned the responsibility for the development of disaster/recovery plans will need to have sufficient knowledge and understanding as to the plan requirements and components. They must also have sufficient knowledge of the workings of the organization and business performance. Finally, they must have the authority to proceed without interference.

How This Book Is Organized

Part 1

The sections in Part 1 deal with the background and applicable concepts of effective disaster/recovery planning. Those assigned the responsibility of developing disaster/recovery plans for either the data center or business unit should familiarize themselves with the basic concepts of this type of business plan.

Part 2

Rather than provide a "fill in the blank" or "cookbook" plan, the author has provided a model disaster/recovery plan in content concept. Each paragraph of the Table of Contents Review shows the planner the "what," "why," and "need" for specific documentation required to put together an effective plan.

Part 3

Making the seminar/workshop methodology work in order to develop a data center or business unit plan is not an easy task. Part 3 walks the planner through the various functions, procedures, and agendas to organize, set up, and present the material, as well as prepare a separate unit document for the disaster/recovery plan.

Part 4

Much of the material contained in Part 4 represents the actual seminar/workshop approach that has been successfully used by DIA*log Management, Inc., and the author in the presentation methodology.

Part 5

How well do you know your organization and the people in the organization? In order to document a workable plan—a maintainable and testable plan—the planners, especially those who are, or will be involved in, disaster/recovery planning, must know every aspect of the organization's operation and performance requirements. The Organization and Participant Profile shows the planners, or alerts those involved in the planning process, exactly what to anticipate at the seminar/workshop. This profile becomes the planners road map in preparing material for the seminar.

Part 6

Information provides the core of the plan. The more information you have about your organization and the more detail that is documented before a disaster, the easier it will be to develop the plan and recover at the time of a disaster. It's not what you've planned; it's what you've missed that makes a plan inadequate.

Part 6 helps the planners obtain and document, in detail, everything that will be required at the time a catastrophic event occurs. Data collection also requires a series of management decisions that assist the planner in strategy development, selection of critical applications, requirements for backup facilities, and so forth.

Part 7

What is a disaster/recovery plan, and how will the seminar/workshop methodology help each organization ensure business continuity in the event of a disaster? What is the scope of the plan and how can we get the business units to participate in the program? Each of these questions is answered in Part 7.

Regardless of whether or not your disaster/recovery plan includes the organization's business units and subbusiness units, the information necessary to assist these business functions is vital to any disaster/recovery plan. If the data center is functioning, and a "user" or business unit is affected by an adverse incident, then it is the responsibility of business unit managers to understand and document subplans in order to maintain "business continuity."

It is amazing how little managers and supervisors know about their own or other interrelated business functions, and, more important, what it actually takes to maintain the status quo in the face of a catastrophe. The questionnaire in Part 7 was designed to alleviate the pressures and provide managers and supervisors with a guide on how to recover.

Part 8

Although the data center may require a highly complex series of activities to maintain business performance in the face of disaster, all business units must understand and plan for an adverse incident. The miniplan outline in Part 8 is provided to guide the "user" in how to view disaster/recovery planning in conjunction and interaction with, or as an extension of, the data center master plan.

Part 9

The primary objectives of a disaster/recovery plan are to save lives, ensure business continuity, and have the capability of recovering in the face of disaster. Emergency response plans, emergency practices, and organizational policy are the bases for preventing minor or major incidents from paralyzing the organization. Emergency measures and procedures will be initiated to handle specific emergencies in an orderly and calm manner. The provisions to deal with emergencies will protect personnel and mitigate damage to assets and facilities. The model presented in Part 9 represents a plan devised for an actual installation and incorporates and reflects management approval.

Part 10

The development of information policies and practices that ensure security and accuracy of information systems has been relatively slow in relation to the expansion of computer activity. The definitions of computer security techniques/baseline security concepts contained in Part 10 are designed to assist data center and business unit management in incorporating prevention and security strategies without having to go through the tedious, expensive, and time-consuming risk analysis process.

Part 10 provides the planner with samples of actual interactive sessions used at public seminars and workshops. The list of interactive sessions appearing in the Organization and Participant Profile is in sequential order with regard to the plan components. These interactive sessions have to be developed by the planning leaders to reflect each organization's unique culture.

Planning Is a Step-by-Step Process

This book has been designed to guide the planner and planning groups, step by step, until the project is a complete, implementable, executable, testable, and maintainable contingency-disaster/recovery plan.

Step 1: Read the entire book and understand all the concepts, principles, processes, and procedures that are necessary to document a plan.

Step 2: Create a plan development team. Each member of this team should also complete a review of the book.

Step 3: Determine the scope of the plan.

Step 4: Review Part 3 of the book, which covers the details on how to set up and run a preplanning seminar. Create an agenda for your program that will reflect the organizational culture as well as the necessary components that are needed within your plan. This can be accomplished by carefully reviewing and abstracting the important elements from the Table of Contents Review that are applicable to your organization.

Step 5: Determine which interactive sessions are applicable to your organization.

Step 6: Create an agenda for the seminar, and review it with management. Once the agenda and plan concept have been approved, schedule a seminar date.

Step 7: Assign data collection and strategic decision responsibilities.

Step 8: Schedule a plan preparation workshop. Leave reasonable time between the seminar and workshop (about six to eight weeks) for data collection.

Step 9: Perform the plan preparation workshop.

Step 10: Review the plans produced.

Preparing the Plan for Distribution

Don't get enamoured with technology. A properly inputed plan in a simple word processing form is more than sufficient. Remember, you are only using the plan once, if ever. Plans utilizing complex databases and menus are overkill. At the time of a disaster, few people will have the time or inclination to use valuable PCs. Storing the plan in a database only means excessive time spent. Most of the information referred to, or used will be available in hard copy.

The authors will be pleased to respond in writing to plan developers regarding any specific problems that might arise prior to, or at the time of, plan development.

Note

This book contains proprietary material and effective disaster/recovery planning seminar/workshop methodologies and materials developed by Alvin Arnell and performed by DIA*log Management, Inc., worldwide. All of the material presented in this publication and other documents prepared and used during DIA*log seminar/workshops on disaster/recovery planning have been copyrighted under the laws of the United States. The purchase of this book is limited, for the purpose intended, to assist individual organizations in developing effective disaster/recovery plans within their own organization. It is being published and distributed by McGraw-Hill Publishing Company, and it is released by terms of its purchase to your organization with the understanding that the material herein contained will be used to conceive of, develop, build, and design, a company training course for the development of an internal company disaster/recovery plan. Further, by terms of its purchase you agree that this material will not be used to develop any seminar, workshop, or other form of educational program that will be offered for sale or subscription outside the purchaser's organization or operations that would directly or indirectly compete with the present seminar/workshops offered by the author and/or DIA*log Management, Inc.

INTRODUCTION

There are numerous reasons to plan for the inevitable emergencies and disasters that business, industry, and government at all levels face. Business and industry share with federal, state, and local levels of government the responsibilities of protecting people and company assets associated with natural disasters, human-made disasters, and technological accidents. Prudent management is now developing and implementing plans for the safety and security of personnel, protecting and preserving assets, and maintaining the continuity of business functions and the integrity of data.

Moreover, organizations are considering the introduction of contingency-disaster/recovery plans that go beyond the scope of data processing. Such plans would include, but would not be limited to, protecting the public from on-site incidents, as well as common area disasters, that could affect the health and safety of the entire community. This is a major undertaking. Further, consideration is being given to other critical functions of the organization. As companies perform more and more critical computer services, as networks and on-line services expand, the *human review* factor diminishes and it becomes almost impossible to sustain business performance without data processing support.

The level of protection that must be afforded critical information resources includes physical threats, natural disasters, internal vulnerabilities, personal misuse, unauthorized users, and programming errors.

Most people use the term *security* in its narrower sense, covering only protection against unauthorized disclosure of information. In this book, *security* is used to cover the integrity of data, performance assurance, continuity of operation, prevention, and preservation.

Organizational structures, businesses, locations, and sensitivity or criticality of information and dependency vary to such a degree that the effects of a disaster to one organization will impact in a completely different manner for another organization.

It's Not What You've Planned for, It's What You've Forgotten . . .

That can wreak havoc at the time of a disaster. Without a professional guide, and regardless of the number of tests you've performed—and reviews the plan has gone through—a plan is not a plan until it has been tested under fire—through an actual disaster. Even though a full recovery simulation test was performed, corporate and data processing management will come to a rude awakening when they find out what they've overlooked. Consider the panic and conditions of stress under which people will have to work at the time of a disaster. First, there is the effect of actual damage or destruction; second, the execution of the plan; third, the changes in working environment; fourth, the psychological effects on personnel who are not involved in the recovery plan and who no longer have a place of employment; fifth, the relocation of people; and so forth.

Without a highly structured, minutely detailed, and disciplined contingency disaster/recovery plan, developed by personnel in the organization under the guidance of professionals, an organization cannot be sure that the greatest percentage of infinite details has been addressed. This handbook is the first major attempt to provide organizations with a thought-provoking methodology that ensures that every aspect of the complex process of contingency-disaster/recovery planning is addressed and detailed. This is an insider plan concept, not a consultant's plan, not a cookbook plan—it is your plan, reflecting and fingerprinting your organization and its requirements.

- What about all those micros, minis- and PCs?
- When was the last time you backed up your development work or your tape library management system?
- What about an alternate communications link to a critical user?
- Who are your next door neighbors? Who is above or below your data center?
- Where do you move to at the time of a disaster?
- What do vendors require about certification after an incident?
- Where are your vital records and source documents?
- Do you have a replace and/or refurbish clause in your insurance policy?
- Can your people work well outside their home environments?
- Does your insurance require security of affected site no matter how long it is unoccupied?
- How do you hold on to employees?
- What about personalized terminal security systems?
- Do you have a list of critical tapes? Where are they stored?
- What is your communication plan?
- Do you have proper documents if you have to clear international borders?

No matter how good the plan, "business as usual" takes a long time and is costly. Stop and think of a disaster impact without a plan.

Planning, Plan Execution, Plan Testing, Plan Maintenance, and Your Organization

Contingency disaster/recovery planning is a *top-down, bottom-up* process. The planning process can be considered a series of individual exercises, with one common bond—survival of the entire organization. Each phase of planning requires a uniquely structured organization within an organization. This can be accomplished in two methods.

1. The establishment of a corporate-level contingency disaster/recovery planning task force (corporate committee) for the development of disaster/recovery plans for data processing and each functional group within the company.
2. A corporate-level contingency disaster/recovery information support group that assists data processing and each functional group in the development of their own plans.

Decisions on the structure of the planning environment are based on the organization's structure, diversity, dispersal or concentration, operating process, divisional autonomy, and so forth. Regardless of the method selected, it still must be a mandated process from senior management or it won't work.

As the process grows within the company, variations on the different planning organizations start to emerge. The initial preplanning group will vary greatly from the plan execution team in larger organizations, but it may never change in a smaller company. This is reviewed early in the handbook.

The Need for Planning Is No Longer the Question

When and *how* the job gets done must be the only criteria for effective disaster/recovery planning. Spending valuable, critical resources to confirm what you already know serves no purpose. Cost-benefit analysis may be appropriate for the determination of new equipment or hardware, the need for reduction or addition of staff, or the construction of a new building, but it is no longer a viable alternative for justifying the development and implementation of a disaster/recovery plan. The only purpose for any cost or risk analysis you will find in this handbook is the determination of the degradation of services resulting from the loss of computer or any other support service. This is not used to justify the plan, but to direct the selection of strategies and the determination of critical services and applications, which have a direct relationship in the plan development.

The methodology and content presented in this handbook guide you through all the prevention and security measures that should be employed to prevent or detect accidental or intentional disclosure, modification, abuse, or destruction of data, or the loss of the means of processing the information. Your only concern should be to provide sufficiently agreed-upon preparations—to create and implement a documented set of procedures and strategies, to respond to any level of adverse incident, and to reduce the effect on business service interruption to an acceptable level.

The concepts presented in the baseline security section define *common* or baseline security and controls that should be in effect in *every* information processing resource. The establishment of baseline requirements eliminates the necessity for costly, time-consuming analyses of already-defined cost-effective prevention and security technology. The baseline security control must be accepted by management because it effectively establishes the *standards* and environment in which sensitive, critical information processing resources *must* perform. Prudent management will recognize the importance of these security enhancements and will appreciate this approach in lieu of the prospect of expensive and long-term risk analysis studies. Although we expect organizational management to be constantly alert to the problems inherent in computer dependency, the initiation of the baseline security concept encourages rapid implementation of internal controls.

We hope the success that we have achieved in assisting organizations in developing their effective disaster/recovery plans worldwide by use of the seminar/workshop concept can be replicated within your organization. The first step is to get management's commitment; the second is to establish an *immutable schedule*; and the third is to get the job done.

The "Three-Legged Stool" of Disaster/Recovery Planning

Effective disaster/recovery planning can be viewed as a "three-legged stool." Each component of the stool serves a specific purpose to ensure a solid support resting on a solid foundation. If any component is absent, the entire stool is worthless because it will not furnish the footing upon which the organization can depend.

A sound plan can only be based on the knowledge of the performance and critical success factors unique to each organization. This leads to the identification of the critical applications, the resources required to support these applications, and the

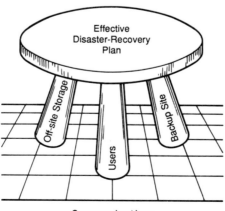

Communications

procedures and strategies that will achieve a timely recovery from any adverse incident affecting the information processing facilities and/or the user community.

A disaster/recovery plan is created to protect people and assets... not hardware.

The Legs of the Stool

Off-site Storage

Data files, software, documentation, source documents, and vital records are all subject to the same hazards and vulnerabilities of an unexpected or undesirable incident, whether they reside in the computer room, a tape library, or in the user operation. Restoration of data processing operations in the event of an adverse or catastrophic incident requires the latest version of the software, the most recent generation of the data files, associated documentation for specific application and system recovery, and the source documents properly logged by time to ensure that when the backup efforts are accomplished, the data can be brought up to the minute of the disaster. Corporate vital records, as well as other classified records, must be located in a place where they cannot be affected by the disaster. A systematic method for sending software, data files, and audit trails (daily runs) off-site is a crucial part of the organization's survival plan.

Backup Site

Backup alternative strategies have to be devised and put into place to supplant an affected site in order to maintain the continuity of operation. No two organizations are alike; each must devise its own strategies based on maximum allowable downtime (the time to recover) and critical applications based on the policy established by management. In some cases, it may be a single strategy, all or nothing; other companies may be able to perform with a number of strategic alternatives. The strategy or strategies must be based on the characteristics and merits of each individual situation. Backup strategies are not limited only to the complex array of hardware in the data center; serious consideration must be given to the impact of a disaster to one or more users.

Users

Each user department can be heavily impacted directly (by its own disaster) or indirectly (by the loss of the computer processing capability). Not only is it the responsibility and obligation of the user community to participate in the development of the data center's disaster/recovery plan, but it is imperative that each user develop a disaster/recovery plan for its department or function in order to maintain operation continuity. Every aspect of prevention, security, and disaster planning that is applicable to the data center pertains also to user functions.

The issue is not whether a disaster/recovery plan should be developed, but what the scope of the plan should be. The objective is to minimize financial losses, maintain the continuity of operations, ensure the integrity of data, and restore normal operations in a timely, cost-effective manner.

The Seat: The Disaster/Recovery Plan

The disaster/recovery plan—the solid seat which ties together the plan components—is a set of highly structured and formalized policies, systems, procedures, strategies, and action plans that will be implemented in the event of any adverse incident that prohibits the use of existing data processing or user facilities. A plan must be specific to the organization and manicured to its needs:

- A prevention plan
- A security plan

- An emergency response plan
- A disaster action plan
- A recovery plan
- A user standby provision plan

The objective of this handbook is to assist organizations in developing, implementing, testing, and maintaining a disaster/recovery plan that combines a blend of remedial strategies, emergency responses, and standby provisions to minimize the impact of a possible catastrophic event that could paralyze the organization by interruption or destruction of its data processing capabilities while maintaining the continuity of business operations.

Communications

Communications and the communications network provide the "solid floor" for maintaining the continuity of business performance. Not unlike the applications themselves, communications are subject to continuous modification and expansion as business requirements dictate. Systems grow; the requirements of the user community enlarge; new terminals and additional line capacity are needed; line speeds are upgraded; and the on-line network spreads.

Every aspect of communications—voice, data, facsimile, electronic mail, teletype—which we take for granted, is a major factor in disaster/recovery planning. The organization's dependency on time-critical functions must be understood and provided for as part of the backup and recovery. Alternate data communication topology (the re-routing of communications between sites and the backup facility) is the very structure on which the plan is developed. A solid communications plan is one where the disaster is totally transparent to the entire user network at the time of a disaster.

Effective Disaster/Recovery Planning and Your Organization

There was a time when it was reasonable to consider returning to manual operations in the event that data processing failed, for whatever reason or for whatever period of time. Today, total support is provided to a broad range of organizational activities; it is virtually impossible to conceive that any manual activities could supplant the instant response supplied by the computer. Unexpected or undesirable events, irrespective of their nature, have a unique capability of affecting the performance and/or reliability of the data processing operation, thus preventing normal operation of the business.

The first and foremost activity is to understand those adverse "security" events that would or could affect the timely performance of the support services of the data processing center. The security measures that must be employed to *prevent* or *detect* accidental or intentional acts that could disclose, modify, or destroy data or the means of processing data are products of an in-depth investigation and facilities evaluation. The need exists to develop a *plan* that would reduce to an acceptable level the consequences of any loss of data processing resources or capability. Disaster/recovery plans are not a planned response to a minor or major catastrophe. What we are concerned with is the reduction of potential damage that an unexpected or undesirable event, if left unchecked or unattended, could cause. *Magnitude* is not of primary concern. The most minor event may leave the entire facility paralyzed.

Therefore, what the plan you are about to develop should provide are methods, procedures, strategies, and action plans that will mitigate the potential destruction or effect of any destructive event.

The plan you are developing must not be directed exclusively at reaction to catastrophically destructive occurrences. During the risk assessment or evaluation aspect of the program, those responsible for identification of events that could affect the data processing operation, either at data center operations or user groups, must plan for the possibility of such catastrophic events, but more emphasis should be placed on those less-than-cataclysmic events which would seriously impede the operation and services of the data processing function and organizational performance.

The probability of an occurrence of an unexpected or undesirable event is generally inversely proportional to its magnitude. Realistically, the greater the catastrophe, the lower the probability that it will occur. The data processing operation is plagued with a far higher frequency of security breaches and smaller problems than by larger, more destructive events.

Although it is not obvious, the size or scope of a catastrophe and its effects on data processing operations are often not directly related. Without a plan, the most minor incident can cause the greatest problem; conversely, a major incident may not result in serious losses if a good plan is in effect. Plans are developed to deal with "worst case" conditions. Minor events utilize subsets of the overall plan.

Where and How to Start

During presentations of workshops throughout the country, to hundreds of organizations of every size and level of computer dependency, the first question is, "Where and how do we get started?" What this handbook intends to do is provide an orderly process for developing an effective contingency disaster/recovery plan. Starting with a qualified prevention and security program, the plan you create will limit the chances and effects of a disaster.

Relaxation of security procedures and practices increases exposure and requires a far more extensive and complex plan. Your actions in the area of prevention planning will help *prevent* an occurrence. Although these efforts will not negate the necessity for a disaster/recovery plan, they will certainly minimize the complexity and extent of the plan.

The procedures, strategies, and actions presented will save those assigned the responsibility of preparing a disaster/recovery plan the task of "reinventing the wheel." The steps presented here will assist you in expediting many of the processes required to initiate, develop, implement, test, and maintain the plan.

Giving careful attention to detail and concerning yourself with small units at a time will result in a far greater and more effective approach to the problem.

Disaster/Recovery Planning Is More Than a Data Processing or Technical Issue

Disaster/recovery planning is a business plan, a program that has to be considered a business recovery plan rather than a data processing recovery plan. Although this plan addresses data processing management, the plan development is directly applicable to the entire business activity. In fact, those organizations that do not operate any data processing operations, or those organizations who derive support from outside services, will also benefit from this approach to planning. Assuming that organizations who provide this type of support service have adequate plans to maintain the continuity of their operations and the integrity of data can be as destructive to the dependent organization as a catastrophe within its own organization. The program presented here is directly applicable to these organizations.

Plans must involve users. Although central data processing is devoted to protecting the major aspect of the computer operation, users have an equivalent responsibility. Distributed data processing only adds to the exposure of the data processing center. Therefore, because users are subject to the same unexpected or undesirable events, it is necessary that they provide for a program identical to the data processing center.

A Square Peg for a Square Hole

There are virtually no data processing facilities that are so similar in equipment configuration, application, environment, personnel, and relative sensitivity or criticality of systems that a general-purpose disaster/recovery plan of broad applicability can be prepared and applied equally well to more than the facility for which it was designed. Preparing a specific plan suitable to the particular needs of *each* data processing facility (including user groups), regardless of its size or scope, is the only viable program that should be undertaken.

Another Dimension for Planners

In addition to the viability of the plan to ensure expeditious recovery, irrespective of the magnitude of an adverse event, the plan must specifically address itself to the ultimate organizational effects. Providing support services and fulfilling user needs have a "bottom-line" benefit. Data processing management is often ignorant of the purpose of applications and utilization of its facilities and the ramifications in the event information is manipulated or data systems are abused or destroyed. This approach to security and disaster/recovery planning is discussed in detail under the assumptions section on plan development. At this point it is sufficient that the plan developers be cognizant of the implications of security failures or destruction of the data processing capability.

Understanding the Project You Are Undertaking

The disaster/recovery plan program you are undertaking has a number of organizational benefits. What you are providing for is the ability to minimize the costs resulting from losses of, or damages to, the data processing facility, its resources and capabilities, and related support services. Again, the value of the plan is based on *dependency* and the consequences in the event of an undesirable or unexpected event. The scope of computer support of your organization may be impossible to fathom, and an attempt to attach each and every element in the support services may make the project totally unmanageable and virtually impossible to develop. Therefore, the object of the plan is to compress the security and disaster/recovery measures into manageable units. The plan developers must view security and disaster/recovery planning as it relates to the support services and operations that it provides other units of the organization, allowing other units to develop their own methodology for the continuity and support. Data processing must view itself in a single-mirror reflection when determining the scope of the plan. If you attempt to view the project in two mirrors—one in front and one behind—there is no end to the project, and it will never succeed.

The single-mirror approach relates to the resources under the direct control of data processing management and includes only the people, programs, data, hardware, communications equipment and systems, power source reliability, the housing facility, and all other items related specifically to its function and operation. The survival of this entity certainly ensures the continuity of other units supported by data processing operations.

Therefore, the disaster/recovery plan for any data processing activity, regardless of its size or scope of operation, must provide for the three basic elements:

- *Emergency response.* The procedures to cover the appropriate emergency response to any form of natural or manmade disaster (fire, flood, civil disorder, bomb threat, tornado, hurricane, power failure, and so on) or any other incident or activity, to protect lives, limit damage, and minimize the impact on data processing operations.
- *Backup operations.* The procedures that ensure that essential (critical, high-priority) data processing operational tasks can be conducted after partial or

total disruption to the primary data processing facility. These would include backup of computer facilities, needed files, programs, supplies, and so on.
- *Recovery actions.* The procedures that facilitate the rapid restoration of a data processing facility following physical destruction, major damage, or loss of data.

The development of the plan, the methods and measures used to affect each of the preceding, the strategies involved, and the action plans necessary to maintain data processing continuity are based on each individual company's dependency, criticality, and sensitivity of the computer support function. These relate to and affect the critical applications required to support the business activities. The cost of the measures to maintain the support function must be balanced with the level of criticality and time.

The first consideration is the *mandatory measures*. For the most part, mandatory measures must be implemented irrespective of cost. They relate to the basic protection of people and facilities and are part of the initial prevention planning process for the *most probable* adverse incidents that can occur—fire control, alarm systems, evacuation procedures, and other emergency precautions necessary to protect the lives and well-being of personnel and protection of property.

Mandatory measures also include those needed to protect all the items that relate to maintaining the organization's viability. There is absolutely no tolerance for error in this area of activity since the very survival of the organizational entity is at stake.

If you operate a cash business it is your responsibility to hire a guard service to carry the cash to the bank to reduce exposure to robbery. How less important are the assets of the organization residing in the computer room or media vault?

When you relate mandatory measures to disaster/recovery planning, it is management's policy that establishes the criticality and ultimate applicability or mandatory measures. The costs related to their protection are mute and should be included as a normal cost of doing business.

Necessary measures are another consideration. Sound data center operation and common sense dictate that every precaution should be taken to prevent any incident from occurring that will disrupt the normal operation of the organization. Prevention, security, and disaster/recovery planning, if properly attended to, should provide a level of security and safety in data processing operations. Although mandatory measures should be identified and implemented immediately, necessary measures, which include all reasonable precautions to prevent a minor or serious disruption, should be attended to as a normal course of events.

Necessary measures are usually applicable to those areas of activity that, in the event of a disruption, although costly, will not have a "survival" implication. For example, unless a research and development contract calls for extreme penalties in the event of delays, disruption of activities in the engineering department many only result in small losses and inconvenience.

The same condition holds true in some manufacturing processes, distribution activities, sales and marketing, personnel relations, and so on. Again, measures employed must be directly related to criticality and dependency of the rest of the organization on a specific function. Here again, management policy must dictate the extent and cost of measures, and mandate the degree and speed of emergency backup required.

Desirable Measures are a third consideration. Desirable security and disaster/recovery measures relate to reasonable precautions that should be taken to *prevent* any area of the organization from inconvenience or disruption and maintain the stability of the business operation. For example, if the office activity is disrupted because of a localized fire, it is important that an alternate site be available to continue office activities. The cost of this activity is low by comparison to the disruption of personnel and the loss of morale.

Many other desirable measures require a modicum of planning, little cost, yet they provide for greater overall efficiency and productivity. Desirable measures should be implemented as circumstances allow, balancing desirability against identifiable need.

Creating High Visibility With Results

During the initial, preplanning (the plan-to-plan) phase of the program (also known as the definition stage), it is advisable to start with the high-level impact team. This team should select the plan (program) administrator or coordinator. Because the initial stages of the planning process require the majority of the input from the data processing department, it is advisable that the administrator/coordinator be a senior executive or manager from data processing.

The primary objective is to produce a plan, as quickly as possible, that will provide for the security and recovery capabilities necessary to ensure continuity of operations and data integrity. The more visible the activity, the greater attention and support it will get from management.

The preliminary planning (definition) team should include:

- Senior data processing management
- One or more of the supervisors or managers of data processing
- Security administration (OSHA administration)
- Internal risk management or a representative from the company's insurance carrier
- Senior personnel manager
- Internal auditor (or a representative from financial management)
- Representatives from critical (high-priority) user groups
- Legal department
- Building management and engineering
- Labor relations
- Public relations
- Medical services (internal or external)

This group must recognize that the plan to be developed will incorporate all the factors necessary for a complete broad-based disaster/recovery plan but in a scaled-down version. Therefore, it is important that a complete outline be prepared exhibiting the scope of the full plan, with specific notations indicating the first phase of attack. The second checklist in Part 4 provides the basis necessary for the full disaster/recovery plan. In Part 2 you will find the table of contents of a full disaster/recovery plan. In capsulized form, this table of contents is the same as you would use for your preliminary program.

Management and Data Processing Contingency Disaster/Recovery Planning

Data processing facilities generally provide a service to one or more functional areas of organizations of which they are a part. Occasionally, they provide data processing support to several organizations. Recognition that the data processing shop serves in a support role is essential to the proper conduct of many aspects of data processing management. It is no less important to the generation of realistic, cost-efficient contingency plans.

Few data processing operations and facilities are so similar in equipment configuration, applications, environment, personnel situation, and relative criticality of systems that a general-purpose contingency plan of broad applicability can be drawn and applied equally well to more than one facility.

Because the data processing facility normally provides vital and irreplaceable services to the organization, the senior management of each organization should realize the critical nature of that organization's dependency on contingency plans. These plans, if carefully prepared and executed, serve to keep within tolerable limits the consequences of losses or damage to data processing resources. Economic feasibility in contingency plans requires carefully derived decisions as to what organizational functions are deferrable and for how long. As we proceed with the program, costs can be applied to these deferrals.

It is impossible for such decisions to be reached entirely within the data processing organization. Data processing management is not usually in a position to assess accurately the relative importance to the whole organization of work done by the respective support areas. Further, the relative cost of continued support of each in the face of adversity may vary quite widely. Thus, cost of support under unusual conditions must be considered. For these reasons, it is not only appropriate but essential that senior management provide direction and support for contingency disaster/recovery planning.

Senior management should do the following:

- Demonstrate a *firm commitment* to the data processing security program by promulgating objectives and including responsibilities to attain those objectives in job descriptions and promotion plans where appropriate.
- Create a *risk assessment* task force and direct the establishment of contingency plans which are based on the results of a comprehensive risk assessment.
- Direct the *support* (both manpower and financial resources) of the planning process by all organizational units servicing and served by the data processing facility. In particular, identify those elements of the organization which are critically dependent upon the data processing facility. Of extreme importance is the assistance of supported activities in identifying those vital records and data maintained by the data processing function; that is, those that are essential to the sustained continuation of supported activities following a disruption. (Responsibility for preparation of the plan should be with the data processing facility.)
- Direct the initial and *periodic tests* of the workability of, and costs associated with, the plan.
- Direct the *periodic revision* or update of the plan as a consequence of information derived from the tests, and as a result of changing dependence of the organization on data processing. A complete review of the plan should be made upon the addition of new applications systems, reaccomplishment of a risk analysis, or a change in any of the critical dependencies.

As senior management, you recognize and understand that the capability of the data processing facility is a critical organizational resource. This recognition leads to a realization that data processing operations are subject to hazards and vulnerabilities (natural disasters and malicious attacks). Yet for management, because of the "black box mystique" of data processing operations, total strategy for dealing with potential catastrophic events is total risk acceptance.

The lack of planning by management is the result of, as Alvin Toffler states in *Future Shock*, "More and more, the specialists do not fit together into a chain-of-command system, and cannot wait for their expert advice to be approved at a higher level." Management involved in the day-to-day business operations cannot deal with "probability." The acceptance that little can be done; the absence of a universal miraculous solution; a reluctance to discuss the problem with managers and users; and a total lack of "how" to arrive at an efficient cost-effective plan leaves the company completely vulnerable.

Contingency planning can so significantly reduce the susceptability of an installation to damage, improve the ability of a business to survive outages, and reduce the duration of the outage and cost of recovery, that to ignore the importance of the process as good business is exerting poor judgment on the part of management.

Management must make a sound business decision to make a detailed review of the data processing activity with the objective of developing a contingency plan. This is a five step procedure:

1. Develop, implement, monitor, and test a contingency disaster/recovery plan.
2. Evaluate the risks and hazards to the data processing facility and the organization as a whole.
3. Develop a complete threat and vulnerabilities analysis of the data processing facility and the organization as a whole.
4. Establish a corporatewide task force team.
5. Gain complete support and application of manpower and financial resources. This should result from management's desire to:
   a. Review the accuracy, reliability, and effectiveness of the computer-based systems.
   b. Perform a reliability assessment to determine whether there is any unacceptable high risks to the entire computer environment (including data, software, hardware, and the facility in general).
   c. Determine if existing data is erroneous and whether a separate review is necessary to verify the extent of errors.
   d. Assess internal controls in financial audits (and other user areas) to provide the basis for reliance on organizational decisions.
   e. Review, in general, the security, controls, confidentiality, hazards, and vulnerabilities in data processing to ensure the continuity of operations and the timeliness of performance.

Since data processing is a continuously developing area, whatever is established today *must* be periodically revised. The continuity of protection that must be afforded to data processing operations is a continuous "link chain"—a vulnerable "link" capable of being penetrated or destroyed affects the entire length of the chain. It is management's responsibility to ensure that each "link" in the data accumulation and processing system is equally strong.

Definitions

The following definitions should be carefully reviewed and understood before embarking on the development of a disaster/recovery plan.

- *Disaster.* A disaster is any security event that can cause a significant disruption in the information services capabilities for a period of time and affect the operation of the organization. It means any situation which leaves the data processing facility in a nonproductive state. A contingency disaster/recovery plan, therefore, may be called an emergency management plan.
- *Integrity of (data).* Integrity exists when computerized data is the same as that in the source documents or data has been correctly computed from source data and has not been exposed to accidental or malicious alteration or destruction. Erroneous source data and fictitious additions to the data are considered violations of data integrity.
- *Availability of data processing services.* The state that exists when required data processing services can be obtained within an acceptable period of time.
- *Accidental and deliberate events.* Accidents, errors, and omissions generally account for more losses than deliberate acts. Thus, they should be the first focus of attention. Safeguards that can reduce the potential for harmful effects from accidents are also important for reducing the opportunities for fraud and misuse. A system tolerating frequent errors is fertile ground for criminal activity that can be masked by the errors. Interruptions in operations resulting from continuous and frequent errors and omissions are a direct source for system failure and the ability of data processing to perform tasks in a timely manner.

PART 1

Section 1
VULNERABILITY ANALYSIS AND RECOVERY REQUIREMENTS DEFINITION

1.1 BUSINESS VULNERABILITY ANALYSIS

1.1.1 Today's Dependency on Electronic Data Processing (EDP)

While it may not be readily noticeable, there has been an increased dependency on the computer since its introduction to business data processing in the mid-1950s. True, the change has been gradual, but it has also been steady. At first only the accounting department's functions were "computerized," but manufacturing, distribution, and marketing soon followed. There has been a creeping commitment by management to the new technology. Today, that commitment is total and irreversible, at least for the large corporation.

There can be no return to the "green eyeshade." The infrastructure for the old technology is no longer present. The old technology was labor intensive; it often required a severalfold increase in staffing levels to do the job. The new technology is capital intensive—computer intensive. There have been significant staff reductions with the implementation of each new application system. Oh, maybe no one was made redundant, but there was planned attrition and a slowdown in otherwise requisite hirings. If business executives think staff numbers and costs are too high today, they should ponder what those levels would be without computerization.

However, the computer did do more than reduce costs. Were it not for the computer, many business practices would just be impossible today. There would be, for instance, no telemarketing, fewer customer options, slower banking, and no on-line airline reservations system. The business executive (as well as the customer) has readily accepted these advancements, seeing an improvement in his or her competitive position.

The technology is impressive. Computers can process millions of instructions per second. The master files are maintained by sophisticated database management systems. Applications are on-line, real-time. They are developed using nonprocedural, fourth-generation languages. Access to the computer is via private networks whose connectivity is maintained by network electronics more sophisticated than many countries' telephone systems.

Such sophistication enhances risk; thus, a fail-safe system is required. Unfortunately, safeguards have not always been built in. It is now time to retrofit many safeguards; including a proper contingency backup plan. A good plan is the firm's "insurance policy" that guarantees at most a small perturbation in the financial growth of the company, come what may. And fortunately, the "premiums" are affordable.

How long can an organization go without its computers? According to the 1979 University of Minnesota report (see Figures 1-1 and 1-2), the answer, which varies by type of industry, is only a few days. Costs from downtime escalate exponentially. Note that the time frame for maximum allowable downtime to restore information capability is much shorter than that required to replace the facility or restore the operation of the affected site. Therefore, a proper contingency disaster/recovery plan calls for a standby facility, preplanned and ready to resume operations on short notice. Many organizations' maximum allowable downtime has been substantially reduced since the 1979 report was issued.

1.1.2 Survey of Management's Viewpoint

Members of the user management department will each have his or her own view as to how critical each application is and how long each application could be "down" before dire consequences ensue. Most probably, each user department manager will say that his or her application is very critical. To say otherwise would imply that the manager's department, not to mention the manager himself or herself, is of little importance. The department manager will know the relative importance of each job within an application. He or she will be able to predict the consequences within the department if the application was not available. He or she may also be able to quantify costs in terms of overtime (for catching up) or lost sales.

Only top management can assess the companywide impact of computer outages or the loss of a critical user func-

MAXIMUM DOWNTIME ALLOWED BY INDUSTRY		
		TODAY*
Financial	2.0 days	24 hours or less
Distribution	3.3 days	48 hours or less
Miscellaneous	4.8 days	48 hours or less
Manufacturing	4.9 days	24 to 48 hours
Insurance	5.6 days	24 to 48 hours
Average	4.8 days	24 to 48 hours
Maximum allowable downtime of computer support services before the organization will be heavily impacted with regard to finances, competitive position, contractual implications and penalties, and so on.		

Figure 1-1 (*University of Minnesota, 1979.*)
(*Author's observation)

DOLLAR LOSS OF AN ORGANIZATION WHOSE ANNUAL GROSS SALES IS $240 MILLION FOLLOWING DATA CENTER DISASTER		
		TODAY*
5 days	$94,200	First day
10 days	$879,000	2 to 3 days
15 days	$2,449,200	5 to 7 days

Figure 1-2 (*University of Minnesota, 1979.*)
(*Author's observation)

tion, and thus fix the maximum allowable downtime. Notwithstanding, upper management will still have a tough time with the exercise because it is hard to imagine what corporate life would be like without the computer—the scenario is so overwhelmingly bleak that the mind represses any imaginings. Impacts should be quantified in dollars. The maximum allowable downtime is the period after which costs skyrocket and/or the viability of the organization is suspect. Many organizations fix maximum allowable downtime at one or two days. Small, less dependent organizations may estimate one or two weeks, or perhaps longer.

A disaster will cause corporate amnesia and operational paralysis, yet management may need input from information systems (IS) management to help them remember all the ties to various information systems. Top management may know what "numbers" they use to manage the business, but not where they come from. The chief information officer (CIO, who is the head of information systems) should be an individual who is capable of defining these concepts.

1.1.3 Identification of EDP-Dependent Business Processes

What applications are likely to be EDP dependent? For the large organization there is likely to be almost universal dependency. Figure 1-3 shows a list of applications (information flows) which are likely to be substantially computerized. For most of the elements listed, it is impossible to perform the operation without the aid of a computer.

1.1.4 Legal Requirements

In evaluating the business risks of computer outages, it is necessary to look at various legal requirements that are not necessarily forgiven just because the firm's data processing capability is suspended. There are federal, state, and local reporting requirements which must be filed before a deadline. Contractual requirements to employees, contractors, and customers may require substantial computer support. Finally, the Foreign Corrupt Practices Act of 1977 requires protection of the assets of the company; this is difficult without the aid of data processing.

The federal government requires many reports which are conveniently generated from the company's databases. These include:

1. Income tax reporting
2. Sales tax reporting
3. SEC reporting
4. EEO reporting
5. Government contract reporting
6. Labor statistics reporting
7. Customs reporting

Employees are, of course, entitled to a regular payment of wages. At retirement, the company pension fund must calculate accrued benefits and issue cash payments. Health insurance claims and other benefits require substantial record keeping. Performance reviews are to be kept accurately and completely. This includes attendance and tardiness records. Stockholders require regular dividends and faithful registration of stock certificates. Contractors require statements of cost and revenue.

The Foreign Corrupt Practices Act, in effect, requires a good management accounting system to control all assets: inventory, labor, capital, and so on. Information itself (e.g., the customer database) may be regarded as an asset. Common law, as well as various statutes (worldwide), requires that a corporation be run by the officers and directors in a prudent (i.e., not reckless) fashion, thus protecting the interests of the stockholders. What could be more reckless than top management's not providing for the continuity of data processing and business functions?

1.1.5 Developing a Qualitative and Quantitative Risk Model

Simply put, the risk envisioned here is that if disaster should strike the data center, then several business processes would be crippled until adequate alternate facilities could be provided. This loss of capability would increase costs, reduce revenue, and invoke several additional liabilities. Furthermore, there would be opportunity costs associated with the

EDP-DEPENDENT BUSINESS PROCESSES

1. General ledger
2. Order entry
3. Invoicing
4. Accounts receivable
5. Finished inventory
6. Manufacturing parts inventory and WIP (work-in-process) inventory
7. Purchase order generation
8. Accounts payable
9. Tax accounting (sales, income, asset)
10. SEC reporting
11. Management accounting (including cost)
12. Cash management (treasury function)
13. Spare parts inventory
14. Customer service dispatch
15. Customer database
16. Marketing research
17. Marketing support
18. Point-of-sale transactions
19. Automatic teller machines (ATMs)
20. Process control (including assembly)
21. Personnel (including EEO reporting)
22. Payroll and benefits
23. Financial analysis (spread sheet)
24. Telecommunications
25. Stockholders
26. Word processing

Figure 1-3

disaster, because all new projects and programs would undoubtedly be suspended. The pressure on cash flow and profit would be substantial and exponentially increasing. In order to arrive at the risk exposure, and ultimately a maximum downtime parameter, it is necessary to outline the losses associated with a computer interruption.

For selected time intervals (e.g., one day, two days, one week), estimate the consequences and consider every business process that cannot return to manual operations or that would be substantially affected. Quantify each loss or potential exposure and aggregate the effects. This analysis should directly yield a maximum downtime assumption and help to determine the amount of expenditure justified to protect income.

1.1.6 Restoration Policy for Critical Applications

The maximum downtime parameter should be officially recognized within the firm. This can be accomplished by creating a policy. It is then the responsibility of data processing management to see that the necessary backup facilities are secured for each type of computer involved. This parameter will apply to all critical applications (i.e., those that must run to keep the business viable).

1.2 APPLICATION RECOVERY PRIORITIES

1.2.1 Essential Applications

Section 1.1.3 identified what business processes (i.e., applications) were EDP-dependent. These applications can be further categorized on the dimension of priority. The dimension is continuous, but three terms are convenient to describe certain relative priorities: *mandatory, necessary*, and *desirable*. Mandatory applications are those that are required for survival, including those needed to meet inflexible legal requirements. They are usually associated with the basic *raison d'être* functions of the organization (e.g., taking orders and shipping product). Necessary applications are those inexorably linked to mandatory functions. Finally, desirable applications are the remaining applications which, by reason of their existence, can be assumed to provide some positive yet noncritical function.

1.2.2 Recovery Thresholds

Although an application may be judged essential, it is not true that the entire application must be recovered in order to keep the business running. For instance, some reporting could undoubtedly be curtailed. Certain transactions could be eliminated or postponed. Certain file maintenance functions might well be delayed until after full recovery of computing capacity. Perhaps certain functions could be handled manually. What is left is only a core subsystem which is indeed critical.

An example is a plan developed for a major university medical center data processing center. With a database of over 23 gigabyte, it would be impossible to obtain sufficient backup capability to support the system in the event of a disaster. After careful review of critical support requirements and the necessary information to maintain operational continuity, a "skinny database," dropping historical data and only requiring the system to process current patient history, was conceived, and the plan was developed to support this activity.

Sometimes even the core system may not be critical for several workdays, owing to the cyclic nature of the work load and various imposed deadlines. For example, the general ledger application is not critical until the end of an accounting period, specifically a quarter. Interim reports are for management's use and are thus not mandatory. However, management will not want to fly in "autopilot" for too long. Thus, the general ledger application will rise in priority the longer it is delayed.

1.2.3 Ranking Criteria for Creating the Applications Recovery Priority Scheme

If an application is essential (i.e., mandatory or necessary), there is no need to rank it; the set of all such essential applications defines the minimum size for the backup computer. However, other applications can be ranked. It is important to separate optional features from the core application. It would be best to have a steering committee decide the rankings given the survey input discussed in Section 1.1.2. Do not forget to take into account the processing cycle of the application involved. It will be necessary to revise this general ranking each day during the disaster/recovery period.

The steering committee should take a broad view of criticality and assign priorities based on a cost-benefit analysis, not on the political strengths of the departments involved. Input for the analysis will come from the cost estimates discussed in Section 1.1.5. Attention should be paid to the ability to do the main functions of the company, not the work of support departments.

1.2.4 Service-Level Needs

Each application will have particular service-level requirements. For batch systems this is a certain number of jobs each day, each with certain computer requirements (such as processor time). For on-line systems a certain number of transactions per hour is required with a maximum average response time. A table of requirements should be prepared for each application (see Figure 1-4 and see Figures 3-24, 3-25, and 3-26 in Part 1, Section 3 for a sample worksheet).

1.3 EDP RECOVERY REQUIREMENTS

1.3.1 Computer System Sizing

The computers at the backup sites must be large enough (in aggregate) to handle all the essential applications. They must meet this criteria in every dimension: processor, I/O processor, I/O paths, disk tape, print, and data communications. If a computer of equal or greater capacity is chosen, then all applications can be run and the user should notice no degradation in service level. This is ideal, but it is not cost effective for the average firm unless the firm shares the backup site with many co-subscribers.

The backup computer must be of a like architecture to the computer at the home site. If there are multiple architectures at the home site, then there must be multiple architectures at the backup site.

1.3.2 Software Configuration

If possible, run the software of the original site. This will not be strictly feasible if the configuration varies, since the configuration file must be modified and the appropriate software generated. However, the same level operating system and environmental software should be used. In many cases, it can be brought with the recovery team and directly loaded. This process must be tested. Application software should be able to be directly executed in object form. The job control language may have to be modified. This also must be thoroughly tested. All changes should be noted, and copies of the

---

APPLICATION REQUIREMENTS WORKSHEET*

1. Application name
2. Responsible user manager
3. Responsible IS manager
4. List of batch job names
5. Required batch processing schedule (attach schedule): job name/frequency/resource requirements†
6. List of on-line transaction processors
7. Required on-line processing schedule (attach schedule): Teleprocessing name/required schedule/Transfer-Volume/response time/resource requirements†
8. Job classes used
9. Teleprocessing support programs used
10. User codes used
11. Disk (pack) families required
12. Datacomm lines used
13. Terminals Required: kind and number

* Also see Figures 3–24 through 3–26 in Part 1, Section 3.
† Resource requirements include (a) processor time, (b) I/O time, (c) number of tape drives, and (d) work space on disk.

Figure 1-4

---

changes should be stored with the recovery documentation, ideally on tape. If problems arise, it may be necessary to recompile the software using source code which has been safely preserved off-site.

1.3.3 Network Configuration

Chances are the new network configuration will be substantially different from the original. Often times dial-up lines are substituted for leased lines. A new configuration file must necessarily be built. Using those new specifications, new data communications software will be generated. A complete test of the new topology is necessary. When dial-up, different line speeds, various black boxes, and so on, are introduced, the odds are that something will fail to work or will work only erratically. An emergency is no time to work out the bugs.

Leased lines may be replaced by dial-up lines. Dial-up lines are usually at lower baud rates; 9600 baud is the usual maximum for dial-up lines. It is advisable only to purchase modems that can handle dial-up connections and which can drop down in speed if necessary. If such modems were not purchased, then new standby modems will be required for the backup environment. Since it may be too expensive to purchase these modems in advance, you may wish to line up an emergency supplier and only purchase the equipment required to test the principle.

Direct-connect (two wires, unloaded, no modem) lines may be replaced by dial-up lines as well. If the terminal cannot handle an RS232 connection, it will be necessary to purchase a protocol conversion box for each line.

In either case the replacement dial-up line, because of its slower speed, will be capable of supporting less terminals. This may dictate the splitting of current lines into multiple substitutes, which is inconvenient and will add substantially to the cost. Remember that for each dial-up line there must be a port on the host computer.

Dial-up lines are often considered for the backup situation because they incur virtually no cost until the time of the disaster. Although charges would mount quickly during backup operations, this should be of little consequence if extra expense insurance was purchased.

Leased lines may also be handled by reterminating the lines at the backup site. This of course makes reconfiguration easy. The trouble is that the telephone company would take several weeks to reterminate the lines if standby local loops were not leased in advance. If the backup site is close (when it is serviced by the same exchange), the cost of the standby circuits are minimal. However, if the site is interstate, the cost for idle circuits is substantial. The telephone company will install at your expense a switch to make the cutover almost immediate. If desired, the switch can be activated by dialing a phone number.

Very large companies have private, dynamically reconfigurable networks. The internode links are leased from one of the long line companies. These networks are by definition very flexible and thus provide the ultimate backup protection.

1.3.4 End-User Computing

End user computing is usually defined as either of two environments: the personal (micro) computer or do-it-yourself mainframe computing (i.e., database inquiries).

Applications run on the microcomputer may or may not be essential. If they are not, they can be suspended until the emergency is resolved. If they are, they must be planned for with a backup strategy. Quite often, microcomputers can be replaced very quickly by a computer vendor. If not, or if "quick" is not fast enough, then a backup "site" must be made available. You can have your own microcomputer in a second location or it can be rented or borrowed. It is important to remember that when all the secretarial staff is on word processing, you cannot even send a letter without a microcomputer. It would be difficult to get the office staff to return to manual "processing."

Database inquiries by end users, although important to profit maximization, are not usually core activities. This is also true of many time-sharing services. But if such activities are essential, they must be planned for at the backup site.

1.3.5 Implications of Future Applications

The backup computer is configured based on current requirements. If those requirements should change over time because of the addition of new applications or expansion of the old, it becomes necessary to revisit the capacity issue. Accordingly, new applications should have resource requirements noted in their specifications. They should include requirements for the primary site as well as for the backup site (essential processing only).

If you have an agreement with another site for backup, it is necessary to be cognizant of the fact that the backup site may expand or contract in capacity without regard to the consequences for the other site. Notification of such changes should be a requisite feature of the backup agreement.

Section 2
RECOVERY STRATEGY SELECTION AND RISK MANAGEMENT

2.1 RECOVERY STRATEGY SELECTION
2.1.1 Alternatives Analysis

Depending on the degree of business dependency and the maximum downtime allowed, it is now possible to choose an appropriate recovery strategy. Here, *recovery strategy* means the interim ability to process data while a full recovery of the primary computer site is underway. It might require a multiphase approach, having both a short- and medium-range strategy. Also there may be cause for a different strategy for each essential application (see Figure 2-1).

The fallback strategy of *manual processing* cannot be directly dismissed. It holds promise, particularly for those processes that have little dependency on the computer. Furthermore, backup manual operations may already be in place for the short-term "disaster." For example, banks always have a procedure for making deposits and handling withdrawals during the short periods of time that their computers are invariably down each month. It is only necessary to extend the duration of this strategy. Of course, it will be necessary to provide other safeguards. An unscrupulous client may take advantage of the situation by cashing checks beyond the balance of an account once the word is out that the bank's computer is no longer in operation.

If manual operations is the strategy to be employed, it must be thought out completely. Manual forms must be ready. Provision for temporary staffing might also be necessary.

One client in the grocery distribution business decided on this strategy as part of its plan, and made arrangements to double the size of its warehouse staff by adding temporaries. The cost would be handled by extra expense insurance.

Another special case recovery strategy is the *fortress strategy*. Here it is assumed that a super-secure center can be constructed, such that the probability of an outage is negligible. First of all, let it be said that this strategy is not inexpensive. Lowering the probability of failure to this degree is very costly. It involves first-rate security on every dimension: hardware, power, environment, physical, logical, and so on. Such a fail-proof system design was done for the Apollo space program. Undoubtedly every system component (e.g., the mainframe) would be duplicated. Because of its cost and uncertainty, this strategy is only adopted in situations where the center is necessarily one of a kind.

The fortress strategy is not to be followed by default. The true default strategy is the *null strategy*, which is no plan at all. As a result of lead times for replacements and a relatively short maximum downtime postulate, the null strategy means the loss of the business.

Other strategies fall on two categories. First there is the choice of seeking a backup site within the firm, amongst other "sister" firms, or through a commercial service. Second, to what degree will the site be operationally ready? The backup site can have bare flooring, complete with environmentals or with computers ready to process, or it can be an ongoing data center operation offering "time."

Providing capability within the firm is best, in one sense, because the most control can be manifested. On the other hand, the expense will be high unless the firm is large and can spread the overhead cost among many internal data centers. The backup site would require adequate administration to keep it in a state of readiness.

A corporate backup capability strategy was selected by a Fortune 500 multiproduct manufacturer with over 90 component organizations. A backup center was constructed to support virtually every computer environment within the corporate structure. Each component was required to pay its "fair share" for the operation of the backup site. It was eventually determined that the charge for most components far exceeded the cost for each component if each had subscribed to a commercial backup facility. The present company backup site has been abandoned.

Yet, other organizations, owing to the critical nature of their operations and their total dependency on computer support, cannot depend on outside backup. Such companies include those involved in aspects of **national security and defense production**.

An agreement amongst sister firms is, on the surface, very practical. It spreads out cost. However, the logistics of setting up such a center as a joint venture are oftentimes

MULTIPLE STRATEGIES ACROSS TIME AND APPLICATIONS		
	Short Term	**Medium Term**
Application 1	Strategy A	Strategy B
Application 2	Strategy C	Strategy D
Application 3	Strategy E	Strategy F

Figure 2-1

insurmountable. The staff usually assigned to plan such an activity have no experience in setting up a business. Quite often, they are not at a level high enough in the corporation to be able to interact expeditiously with legal, treasury, and executive management. Also, the more firms that are involved, the more logistical troubles there are.

The commercial service is already established and offers a known product. However, such a firm might not exist, and if it does it may be a monopoly. The cost, although much less than going it alone, might well seem too expensive to the cost-conscious executive.

The *cold site* or *empty shell* is usually fully capacitated except for a computer. All environmentals are present (i.e., power, air-conditioning, and telecommunications). The assumption is that a computer can be delivered and installed in a short period of time (perhaps two weeks).

This contrasts with a *hot site*, which would already have the computer installed. Service bureaus, of course, are a special case. They already have a fully functioning center with many active clients.

For completeness, each possible backup site is summarized as follows (a summary comparison chart is provided in Figure 2-2).

- *Private cold site*. A company builds its own empty computer room at a site remote from all other data centers. The empty site need not be unproductive space because it can be used for the environmentally protected storage of backup media. It can also be used as a warehouse for supplies and equipment.
- *Cooperative cold site*. A consortium of sister firms builds an empty computer room remote from all the member firms.
- *Commercial cold site*. An independent business offers an empty computer room on a subscription basis.
- *Private hot site*. A firm builds its own disaster/recovery data center, complete and immediately functional. The center could be used for low-priority work load, such as development. The center should be remote from all existing data centers.
- *Cooperative hot site*. A consortium builds and maintains a disaster/recovery data center. Barring the security problems of multiple firms, the center could be used for low-priority tasks, such as development. Time would have to be set aside for the testing of disaster/recovery plans, however. The members would have to agree on software levels if they share the center for development. If other users subscribe to the center, this is really a commercial venture (discussed below).
- *Commercial hot site*. A commercial enterprise is formed and offers its service to clients. There may be multiple classes of subscribers, each with different preemption rights. A professional staff is usually available for consultation. Almost all of the available computer time is reserved for the testing of disaster/recovery plans. The typical contract is discussed in Section 2.1.4.2.
- *Cooperative agreement*. Sister sites pledge mutual backup for some or all applications. There are many problems with the cooperative agreement. The most looming problem is the fact that if one party declares a disaster, then there is, in effect, a disaster for the other party, as that party must degrade its service in order to handle the operations of the other. There is the issue of security. There is also the problem of incompatible system and environmental software. When can testing time be allocated?
- *Service bureaus*. Service bureaus are operational, commercial data centers which offer "shared" use of a computer system. The disadvantages of using service bureaus are similar to those of the cooperative agreement. However, they are somewhat mitigated by the fact that the commercial enterprise is organized to handle the multiple-user environment. Management of such a center would never, for example, let capacity become an issue. Service bureaus would normally offer backup services to established clients only.
- *Time brokers*. Time brokers keep a list of firms with available computer time who would be willing to provide such time to third parties on a temporary basis. This strategy should be considered a last resort since the firms involved often have no experience with foreign users and would provide, at best, second-class services.

For most firms, the commercial hot site is the appropriate solution. Exceptions are very large firms and those not having heavy computer dependencies. Sometimes a commercial service is not available for the subject mainframe. If so, the firm may wish to opt to try and form a consortium. This is a difficult route.

The commercial hot site is the preferred, nearly universal solution. Typical features include:

- Variable lead time requirements (24, 48, or 72 hours)
- Disaster declaration fee
- Hourly usage charge
- Two test periods per year
- Large mainframe for class of computer
- Variety of peripherals
- Dial-up terminations
- On-site terminals
- Consulting services

COMPARISON OF BACKUP STRATEGIES		
Strategy	Advantages	Disadvantages
Manual processing	No capital outlays	Labor intensive Requires forms A lost art No computer reports
Fortress	Only one site to manage	No redundancy Very expensive
Private cold site	Moderate cost Good as secondary strategy	Two-week delay
Cooperative cold site	Inexpensive Good as secondary strategy	Two-week delay Difficult to start
Commercial cold site	Very inexpensive Good as secondary strategy	Two-week delay Remote location possible
Private hot site	The best Excellent control	Very expensive Must be managed
Cooperative hot site	Excellent backup Good control	Expensive Difficult to start Must be managed
Commercial hot site	Excellent backup Very cost-effective	As good as management
Cooperative agreement	Almost no ongoing cost OK for selected applications	Creates second disaster Difficult to test May not be enforceable Not recognized by regulating bodies
Service bureaus	Low cost until disaster Probably good for selected application	Only for permanent clients Probably not a total solution
Time broker	No cost until disaster	Last resort only Can a fit be found?

Figure 2-2

2.1.2 Cost Justification

Cost justification can be a long and arduous process. Thus, it should be "avoided" if possible.

For instance, there is no need to argue if any regulatory body has ruled that a certain type of backup capability is required. Banking, to take one industry, is heavily regulated. The Comptroller of the Currency (Commercial Banks) in BC177 requires all banks to have a disaster/recovery plan which includes provisions for backup. The Federal Home Loan Board (Savings and Loans) has released a regulation (Memorandum 67) which also mandates a disaster/recovery plan including backup. In this instance, cooperative agreements are specifically declared unacceptable; a hot site of some kind is required.

If no regulations are operative, the next step is to show that the proposed strategy is "baseline" for data centers in general, or for one industry in particular. *Baseline* refers to a prudent and generally accepted practice. No justification should be required. (*Computer Crime/Computer Security Techniques*, U.S. Department of Justice.)

If the use of the baseline security techniques fails, a formal cost justification will probably be required. The baseline technique has been successfully applied to virtually every type of data processing facility, in a multitude of environments.

First, an annual loss expectancy (ALE) must be calculated. The following is a highly simplified calculation using global averages. Statistics show about a 2 percent annual probability that a disaster will strike. This probability should be multiplied times the cost of a fallback to manual operations while an alternate site would be readied on an emergency basis—about six weeks. If the loss is $10 million, $10 million $\times$ 2% = $200,000 (note that this is the ALE). This number must be compared to the cost of capital times the cost of acquiring a backup site plus the incremental annual operating expenses. For example, 10% $\times$ $1,000,000 + $50,000 or $150,000. In this case the site is justified since $200,000 is greater than $150,000. The reason this analysis is to be avoided is because the 2 percent assumption is not firm and incremental annual costs are hard to estimate. Furthermore, upper management has a propensity for using such studies as a delaying or avoidance tactic and remaining at a risk posture.

2.1.3 Strategy Selection

The main determinant in strategy selection is maximum allowed downtime. Another is, of course, cost. For most companies the choice will be a commercial hot site, if available.

Some sites, because of costs, may be forced to reject hot and cold sites as too expensive. The cooperative agreement is the only feasible option left. Some of the negative features of this choice can be eliminated by partitioning the work load across several computers. In this way the additional loading is small and, accordingly, the target computers will not be forced to substantially degrade their service. Although this is not ideal, it is marginally acceptable.

2.1.4 Contracts and Agreements

2.1.4.1 Cooperative Agreement

In the cooperative agreement, it is important to preserve the rights of the owner of the "backup" computer without making it impossible to run the transferred application(s).

The following points should be addressed:

- A definition of *disaster*
- Resources that may be used: (1) N spindles or M megabytes, (2) processor & memory requirements, (3) priority and response time requirements, and (4) time of day scheduling
- A list of job and program names
- A determination of whose operators will control the jobs
- An understanding of what system and environmental software is to be used
- An agreement on whether block or shared time be granted
- Site rules
- Maximum duration of disaster allowed
- Mutuality of backup
- Costs
- Testing time allotted
- Formal notice when configurations are changed

2.1.4.2 Hot Site Contract

For a hot site, many of the same issues and some new ones should be addressed:

- A definition of what is a disaster
- An agreement as to what happens when more than one client experiences a disaster
- Costs
- Testing
- Maximum duration of a disaster allowed
- Lead time notice requirements
- Minimum configuration
- Communications
- Technical support
- Provision of supplies
- Site rules

2.1.4.3 Cold Site Contract

The issues for a cold site contract are as follows:

- A determination of what equipment will be provided
- The maximum power load
- The maximum heat load
- Costs including refurbishment charges
- Duration limits

2.1.5 Computer and User: Backup Alternatives

The circumstances and influences affecting the development and operation of a computer installation system, application, or user community within an organization represent its environment. The paradox of this environment is that it can perform continuously for its entire lifecycle, or it can be affected by unexpected and undesirable incidents that impact on the performance of the entire organization. Computer *dependency* is the yardstick by which survival of catastrophic events is measured.

If your organization can drastically scale down computer-supported business operations, or revert to manual operations during a disaster, then your organization is not computer-dependent, and the need for data processing backup, fortunately, is a subject you will not have to address. Unfortunately, for most organizations and financial institutions, this is not the case.

Computer security risks threaten the very existence of an organization. Critical decisions regarding the adequacy of prevention and security safeguards in data centers and sensitive applications must be made by qualified managers based on reliable technical information. Beyond this, management must realize that systems fail. In the event of a failure, they must have in place appropriate, tested disaster/recovery plans and strategies and action plans, and they must be able to maintain the data processing operation and ensure the integrity of data. Those responsible for the "business" must provide the mechanisms necessary to ensure continued performance; in a computer-dependent organization, this means being capable of supporting the organization if the data processing operation, or *any* functional unit of the organization, experiences a disaster. This requires an alternate processing capability—one leg of the three-legged stool of effective disaster recovery.

2.1.5.1 Disaster/Recovery Planning

Disaster/recovery planning or contingency planning, which should be part of the normal business process, involves the preparation of strategies, procedures, and action plans that will mitigate potential damage in the event of an adverse incident and facilitate timely recovery from events that will disrupt data processing or other critical services of the organization.

Disaster/recovery planning requires that management provide appropriate technical and administrative procedures to ensure that these vital functions of the organization remain viable after an adverse incident. The plans *must* be accepted by management as a cost of doing business, and *must* have management's total support if they are to be developed, implemented, tested, and maintained. Only under these conditions can plans succeed.

There are five aspects to disaster/recovery planning:

1. Prevention and security
2. Emergency-response plan
3. Backup plan
4. Recovery plan
5. Post-disaster evaluation and requirements

This section focuses on the backup plans.

2.1.5.2 Alternate Processing Strategy

Following a disaster, limited resources may require curtailed processing. Identification of the critical processing requirements is the first step in planning for alternate processing strategies. The determination of critical processing requirements is a user responsibility requiring management approval. Once these critical applications have been identified, the proper computer plans can begin. The following descriptions provide the guidelines for disaster/recovery planners to define and select one or more alternatives. These alternatives include:

- Service bureaus
- Dedicated disaster/recovery centers (commercial)
- Empty shells (commercial or organizational)
- Reciprocal agreements
- Time brokers
- Membership in shared disaster/recovery facilities
- Multiple data centers under the same organization
- Fortress concept with full redundancy
- Use of microcomputers

Other considerations include:

- Portable sites
- Empty buildings
- Reversion to manual operations

See also Figure 2-3.

The specific backup strategy for each organization will vary depending, first, on the size and complexity of the data-processing support operation, and, second, on the assumptions as to the severity and longevity of the outage. For example, a strategy could be developed that would provide temporary use of a service bureau, while a shell site is outfitted with hardware. The decision to utilize this strategy would depend on the commitment of the hardware vendor to provide an affected company with equipment. Such a strategy could provide recovery within two to three weeks.

Obviously, the selection of any one strategy or a combination of strategies must be planned for in advance and incorporated into a well-documented, thoroughly tested, and workable disaster/recovery plan. The objectives of any strategy are time-sensitive, and based on a number of legal, regulatory, and/or organizational requirements.

2.2 MANAGEMENT CONSIDERATIONS AND RESPONSIBILITIES

Computer dependency has placed most organizations in a vulnerable position. If the interruption of data processing operations threatens the very existence of an organization, management has to make critical decisions regarding the adequacy of disaster/recovery planning and processing alternatives. Otherwise, management assumes legal liability to stockholders and regulatory agencies for not taking prudent steps to protect the assets of the organization. Senior management must take every precaution to protect the organization and themselves from exposure by not providing for alternate processing resources that would mitigate the losses to their organization resulting from the loss of data-processing support.

Planning, funding implementation, testing, accreditation, and certification of a qualified disaster/recovery plan that includes appropriate alternate (backup) processing strategies are the direct responsibilities of management. In viewing the planning process, management must do the following:

- Establish organization policies concerning the management of information, operation, security, and disaster recovery of the data-processing activities.
- Assign the responsibility to design, develop, implement, test, and maintain a disaster/recovery plan that includes alternate processing strategies.
- Demonstrate to all functional units, and at all levels, the organization's commitment to, and support of, the disaster/recovery planning activity.
- Provide the resources (financial and personnel) to develop the plan.
- Monitor development of the plan.
- Require and verify periodic testing of the plan.
- Require and verify periodic review, updating, and certification of the emergency procedures, strategies, and action plans.
- Determine whether existing strategies are sufficient to accommodate the growth of the organization.

The development, implementation, testing, and maintenance of the plan involves the entire organization, and is not limited to the planners or data processing personnel. The scope of the program should be carefully defined by management and include users, support offices, branches, divisions, and so on.

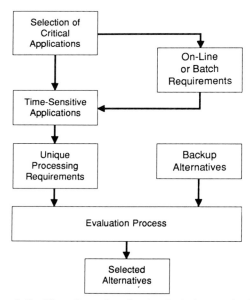

Figure 2-3 The alternative (backup) strategy selection process.

2.2.1 Defining Application-Recovery Requirements

In order to support critical applications, the necessary alternate capability must be defined. Once these requirements have been documented, the backup facility alternative(s) can be determined. The following criteria should be used to define the site's specific needs and determine the most appropriate processing alternatives:

- Availability and reliability of the alternate facility
- Compatibility of hardware
- Compatibility of software
- Physical capacity of the alternate facility
- Environmental support
- Communications support
- Personnel support
- Logistical considerations
- Sufficiency of test periods
- Security
- Cost-effectiveness
- Contractual requirements

Availability refers to the amount of lead time the particular alternative site requires in order to provide backup services. A commercial site offers immediate backup, whereas a cooperative (or mutual) agreement may require days before resources can be available. Management must determine the extent of interruption (maximum allowable downtime) that can be tolerated before the organization is exposed to significant losses. The acceptable delay period is the primary criterion in selecting the alternate processing site.

Another consideration is the length of time the alternate site will provide backup processing support. The backup site should allow on-site operations to process critical work load while its data-processing facility is being restored.

Reliability is yet another factor that planners must take into consideration. Promises of availability of equipment in a backup site are not sufficient; in determining the viability of the facility, a personal review of the site should be made. Other factors to consider concerning reliability are:

- Reputation of vendor
- Length of time in business
- Solvency
- Verification of support from other customers
- Contractual stipulations

2.2.1.1 Compatibility

Obtaining equipment at the time of a disaster to ensure compatibility will not work. The planners have to determine beforehand whether the backup facility has, or will provide (at the execution of a contract), the necessary equipment to support critical applications.

Defining the minimum hardware and software support needed to process critical (mandatory) applications before seeking a backup site will ensure that the system configuration, minimum hardware requirements (including operating system), compilers, utilities, database management, peripherals, and communications are available.

Periodically, planners must review with the backup site any changes in equipment or software, either at the data center or at the backup site, that may result in incompatibility.

2.2.1.2 Physical Capacity

It is important that there be sufficient work space to accommodate the affected site. Although many organizations require only a few of its key personnel to be at the backup site, others may need enough space to accommodate, with relative permanence, a large number of people from the affected site. The facility should be capable of providing space not only to perform computer operations, but for office and administrative functions, such as I/O control, data entry, programming, and scheduling.

Conversely, if the backup facility cannot provide sufficient "ready-room" space, the planners must consider alternate office space. In some organizations, a cohesive work environment may be lost if tasks normally performed together are separated. This may not be true when it comes to administrative activities.

The plan must take into account and provide the items required to perform administrative and office functions. This includes business furniture, equipment, expendable supplies, typewriters, terminals, forms, checks, and general operating supplies (computer and noncomputer).

2.2.1.3 Environment

Normally, commercial sites have been efficiently planned and provide all the environmental support necessary to operate a computer system. If consideration is being given to a "shell" or unequipped site, it is essential to ensure the adequacy of the following:

- Raised floor
- Electric power
- Emergency power backup
- Lighting
- Air-conditioning and humidity control
- Security systems
- Fire prevention, suppression, and detection
- Water detection
- Intrusion alarms (physical security)
- Communications

2.2.1.4 Communications

Owing to the lead time required to establish a communications network, once the alternate(s) to backup processing have been selected, communications must be installed. This key element in disaster recovery is used to determine whether on-line communication is vital to continue processing. Commercial sites usually have communications professionals who can assist you in designing the backup communications network. This plan must be in place before any processing can be performed.

Lease lines are preferable, but they are costly to maintain in an unused condition, and they usually will not be available when they are needed. With only a limited number of lines available, the telephone companies are constantly seeking dormant lines. Even though you have leased lines, they will, if unused, be given to someone else.

Dial-up lines, although expensive to maintain when unused, will surely be available at the time of a disaster, since by definition dial service is normally dormant. During a backup situation, the use of dial-up can be expensive, but if the proper insurance coverage is available, these costs will be borne by the insurance carrier. Preplanning of a dial-up network can substantially reduce the impact of a disaster.

2.2.1.5 Location

The location of the alternate site, in today's operating environment, is immaterial. Although there are several considerations that should be reviewed in making a backup site selection, location is quite low on the priority scale. The only real consideration in terms of location is that the alternate facility should not be located where it might be affected by the same adverse incident.

A sound disaster/recovery plan requires that only a limited number of people be used at the backup site. The majority of personnel, users, data entry, operations, and so on, if located on the site of the affected facility, should remain closer to home. Therefore, the more important plan is to provide adequate backup facilities for these people.

If a severe disaster (such as a flood, tornado, hurricane, or earthquake) occurs, key personnel might be reluctant to leave their families. During the preplanning phases, all employees who will be involved in the backup operations should be polled to determine their commitments. Identification of key job functions should be made, and cross-training of vital operations should be performed to ensure sufficient alternates.

2.2.1.6 Testing

A plan that has not been tested *is not a plan*. No matter in how much detail each of the activities has been designed, unless a series of tests are performed to determine the viability of the actions and strategies, the plan will usually not work at the time of a disaster.

In selecting an alternate data processing site, it is imperative that adequate time be available in which to test backup operations procedures. Plan deficiencies will usually occur despite careful planning. In many instances it takes at least three tests before the plan is debugged. Failure at the first test is to be expected.

Thus, the most useful procedure is to test the backup strategy and certify it at the alternate site at regular, prearranged intervals. Many commercial sites will not continue contractual arrangements unless testing has been performed.

When an alternate site, other than a commercial facility, has been selected, you will find that testing is difficult to perform. When a test is performed, it will disrupt a normal processing operation, and many facilities are reluctant to relinquish valuable computer time for the performance of a test.

If a shell site is selected, testing of operations cannot be performed. Few companies will expend the resources necessary to move computer equipment into a shell site in order to test a plan.

Testing should follow a prescribed set of criteria, beginning with the most fundamental steps of the plan, and at least one major, full-scale test that requires complete, on-line processing at the alternate facility.

The foreign environment of an alternate facility may result in numerous changes to data processing operations, which will result in necessary modifications to equipment, programs, and documentation. Testing will ensure the

- Adequacy of the plan
- Compatibility of hardware/software at the alternate site
- Recovery of critical applications using backup files and software stored off-site
- Communications
- Adequacy of training for personnel

Testing assures familiarity and confidence in the plan during a crisis, and it serves to reduce confusion and anxiety.

2.2.1.7 Security

Processing at an alternate site is no excuse for relaxation of security—both physical and logical. Although most commercial sites provide some level of physical security, they do not assume any responsibility or liability for losses the user might incur owing to theft or manipulation of data. Thus, the company that uses a backup facility takes possession of the site and operates it independently of its personnel. When a facility other than a commercial site is used, the security problem becomes more critical. Unless the organization involved in a disaster can partition its processing, it is virtually impossible to protect the integrity of data in a mixed-processing environment.

If the selected alternate facility is unable or unwilling to support security measures, it may be necessary to select another site.

2.2.1.8 Cost Considerations

Many risk analysis specialists evaluate backup capabilities with the formula that an alternate processing method should not exceed the potential estimated loss that could occur in the event of a disaster. This is called *annual loss expectancy* (*ALE*). First, a backup facility should be considered when an organization's dependency on data processing could have an adverse impact on the mission of the company. Second, the backup alternative should depend on the time frame necessary to support processing. Both of these elements will determine the annual cost expenditure necessary to maintain the continuity of data-processing operations and the organization's functional units.

The planners have to evaluate the number of applications to be supported, the equipment configuration required to support the applications, and the necessity to provide data processing support.

Commercial facilities have a series of standards on which the annual cost is based. These include membership (subscription), testing, notification fee, installation assistance, daily disaster operation costs, and other services.

There are three basic costs associated with subscription in a commercial or other type of backup facility. The first two costs are predisaster expenses, the third is a postdisaster expense.

1. *Initial cost.* The cost of initial setup, which includes membership, construction, or other fees.
2. *Recurring operating costs.* The recurring costs for maintaining and operating the facility, including rent, utilities, repair, and ongoing backup operations.
3. *Activation costs.* The costs involved in the actual use of the backup facility which include disaster notification fees, facility usage charges, overtime, transportation, and other costs.

The issue of cost has two functions in the consideration of backup processing support. The first is as a tool for comparing various alternatives. The second is in the justification of any backup capability. It is important to be careful that costs alone do not dictate the choice of an alternate method of processing. As has been stated before, the most vital consideration is providing the requirements necessary to continue critical processing and the time necessary to recover from an adverse incident. The maintenance of, or the costs involved in, maintaining the backup strategy should be treated as a normal business operating expense.

2.2.1.9 Contracts

Contracts vary among vendors, even though intent of services to be supplied is the same. Nothing should be taken for granted, and all agreements should spell out exactly what the organization's requirements are, and what the vendor will supply. This applies not only to commercial facilities, but also to reciprocal agreements, service bureaus, and so on.

Contractual requirements, in writing, hold true even for those within the same organization who have committed to support other groups' data processing operations.

2.2.1.10 Evaluating Alternatives

2.2.1.10.1 No Backup Strategy

No backup strategy means exactly that. Here management has made a decision to ignore the potential threats, hazards, and vulnerabilities that exist, regardless of the risk assessment or any other risk evaluation, and will not fund any strategy to back up the data processing facility or any other operation that is critical to the performance of the business.

Plus or Minus

In a computer-dependent organization, the effects of a disaster to either the data center or a major user may spell complete organization failure. Management should be encouraged to heed the results of disasters on other organizations, or at least perform a *cost-of-downtime* evaluation.

In those smaller organizations, the justification for redundant facilities, commercial backup sites, and so on may far outclass the dependency on the computer.

Cost

There is no cost.

2.2.1.10.2 Service Bureaus

Some service bureaus will contract to provide disaster backup capabilities for a fee. This may be either a monthly maintenance fee alone or it may include a disaster-mode (time-utilization) fee. Time-shared activities can support processing by batch and/or interactive systems. Advance planning is necessary to ensure proper communications.

Plus or Minus

Service bureaus limit services to current subscribers. Unless previously contracted, service bureaus will not accept a facility that has encountered a disaster. Serious delays will occur unless the service bureau's system has been previously tested.

Service bureaus can, at the time of an emergency, provide limited support, if their work load permits.

Costs

Contracts are usually by annual subscription (paid monthly). Additional daily time-share fees are paid at the time of a disaster.

2.2.1.10.3 Dedicated Disaster/Recovery Centers

Dedicated disaster/recovery centers (commercial hot sites) are fully equipped computer centers which provide one or more computer models (usually from the same vendor) and all the necessary peripheral equipment. Some hot sites are large enough to accommodate several users. Most of these facilities have all the security and environmental conditions found in the most sophisticated data centers.

A hot site will usually have backup power supplies, redundant environmental conditioning, abundant communications lines, fire protection and warning devices, intrusion-detection devices, and physical security.

Preplanning with the hot site ensures that its systems are either compatible or can be quickly converted to those of a large number of subscribing organizations. These facilities can provide short- or long-term system backup and are most advantageous in serving organizations that have experienced a total loss of data processing facilities and that cannot defer computer support.

Plus or Minus

Fully capacitized facilities, dedicated to serving organizations that have had a disaster, are available for occupancy immediately upon notification. They will not provide services for any organization that has not been a continuous subscriber up to the time of a disaster. Most facilities limit the number of subscribers from a specific area, particularly when the subscribers are in an area that might experience a common disaster (floods, earthquakes, tornadoes, hurricanes).

The length of utilization is usually limited to a period of six weeks. Some will allow companies to continue operations for a short period beyond six weeks, with the understanding that if another organization has a disaster, the occupying company will immediately vacate the premises.

Many have shell sites, where the company can temporarily move its facilities (new hardware) until its site is recovered. If several subscribers have concurrent needs for the facility, attempts are made to share existing capabilities.

Some centers work on a first-come, first-serve basis, and if the disaster/recovery backup facility has more than one center, the second organization may be required to use another location.

Compatibility

Backup sites provide a basic hardware configuration that should be compatible with subscribers' configurations. If additional equipment or special peripherals are required to support the system, the backup facility will either provide the capability for an added fee or have sufficient influence with vendors to obtain equipment quickly. Vendors will attempt to spread costs among subscribers who require additional equipment to reduce any one organization's expense.

Costs

Customers contract with commercial backup sites for a specific period of time (usually two years) and pay a monthly membership fee. The fee is based on the configuration of equipment required to support the organization and how quickly it is necessary to occupy the site.

A disaster-notification fee is imposed on subscribers. This fee is used to discourage companies who might use the backup site to handle overload situations. Once the backup site is occupied, the subscriber pays a usage fee for the number of hours (shifts) the facility is used. The monthly fee usually covers a specified number of tests per year; additional tests will cost more. Some facilities also charge fees for use of communication equipment.

If an insurance program is properly designed, all the expenses incurred in using a commercial backup facility, extra expenses, and so on will be covered.

Other Services

Some backup facilities provide space for temporary location of operations, data input, or other administrative personnel. Prior to occupancy of these facilities, documentation will have to be provided to indicate how these facilities will be used. The space used by a company not in a disaster mode will be reclaimed in the event another subscriber has suffered a disaster.

Floor plans of office space to be used have to be provided, as does an inventory of furniture and other auxiliary equipment (typewriters, terminals, copy equipment, an so on). The backup facility will usually work with local rental companies to provide such equipment. Again, proper insurance coverage should pay for any and all requirements and equipment necessary to support the affected site.

2.2.1.10.4 Empty Shells

Empty shells are large, unequipped, unfurnished spaces which are prepared to accept hardware and communications equipment, and which, like commercially facilitated sites, can be leased. They are equipped with raised flooring, utilities, and communications.

The client must supply the hardware and prepare the shell for processing (including arrangements to wire equipment, plumbing, and so on). The affected organization is responsible for performing environmental testing at the empty shell to ensure that communications lines, air conditioners, chillers, and power and other systems perform properly. The company occupying the shell must restore the facility to its original state when it vacates the premises.

Commercial empty shell sites should be considered when processing can be deferred for a period in excess of two to three weeks—the time it will take to obtain all the necessary hardware, peripherals, and communication system.

Many companies consider the use of shell facilities as an in-house backup alternative.

Plus and Minus

Extensive preplanning is required to prepare the shell for backup processing. Organizations evaluating the use of a shell (commercial or in-house) must consider the methods to support the organization's data processing needs while the shell is being prepared. The time necessary to prepare the shell may far exceed the time elements necessary to support critical applications. A short-term alternative may be required to meet the critical processing needs if the shell is considered as part of the long-term plan.

Compatibility

Hardware compatibility is not a problem when using a shell site, since the user must provide the computer equipment and hardware. Time is the only problem that presents an inconvenience to the user, because the vendor will have to ensure that the equipment supplied in an emergency is compatible with the version the customer has used. Delays arise if the destroyed equipment is of a previous generation. During the preplanning and decision stages of developing the backup alternative strategy, it might be appropriate to consider upgrading the equipment at the time of a disaster. To ensure that the new version will be compatible, systems and programming (new equipment team) will have to investigate modifying software to suit the upgraded system. Arrangements will have to be made to test the new programs and assure the auditing department that the existing database will perform with the modified software.

Costs

In a commercial site shell, subscribers pay a monthly membership fee for the availability of the facility. In some cases, the shell owner imposes a notification fee. A daily usage fee will be paid while the user occupies the site.

Testing

Even though it is impossible to test the backup procedures at a shell site, there are a number of tests that can be performed at the data center. Simulations of events ranging from a total loss to a minor interruption can be performed. It is recommended that these tests be performed on off days, such as holidays and weekends, so as not to disrupt normal operations. Static tests on-site provide useful information and training and highlight obvious plan deficiencies, errors, and omissions in the backup strategy.

Other Problems

Shell sites built by companies or commercial sites require a wide range of preparatory activities. These may have to be performed by licensed tradespeople and must conform to the area building codes. These plans should be in place, and contracts for their services should be prepared. It is important to discuss electrical (power), communications, and environmental conditions with the vendor to ensure conformity to the requirements of equipment upgrading. This will certainly expedite installation activities in a disaster.

Security

Developing a company shell site requires the transfer of security to the shell site. If a commercial facility is used as backup and is to be shared with other subscribers, physical security measures will have to be discussed with both the center's management and the user. Both logical and physical security must be developed.

Special Concerns

If an empty shell is selected as a viable alternative, it is important to detail a plan for the maintenance of source documents reverting back to the last backup places off-site.

2.2.1.10.5 Reciprocal Agreements

Reciprocal agreements, sometimes called mutual aid agreements or partner agreements, are agreements between two or more computer facilities and must be formally signed documents. During an emergency those organizations that are unaffected allow the affected company to use their computer resources. Both organizations accept the fact that at the time of an emergency, each company will only be able to operate at a reduced capacity if the facility is shared simultaneously. Management of all organizations involved in a reciprocal agreement must identify only high-priority critical applications.

Plus and Minus

Reciprocal agreements are not recommended, especially when both companies must run simultaneously. If one facility uses the unaffected site facility in off hours, it will be necessary to determine if reciprocal agreements will work. The host company will in the latter case have to clear the entire system, since neither organization should intermix data and software. Reloading of the system by the host company will be a time-consuming daily project.

Beyond this, many companies that have had reciprocal agreements have found that at the time of a disaster the host facility has denied the affected company access. Don't depend on any legal remedies; they are unenforceable.

Compatibility

Software and hardware must be constantly monitored for compatibility. Slight modifications of equipment that go unreported will automatically invalidate any potential use of the facility.

Reciprocal or mutual aid agreements are frowned upon by EDP auditors, who recognize that such agreements will not work. Although costs are minimal, you get what you pay for.

2.2.1.10.6 Time Brokers

Time brokers provide a unique service, not only to those who have experienced a disaster, but to many companies who need extra computer time to develop programs or test them. Processing arrangements are made through this third party. Where time can be obtained through a broker, there is no assurance that the facility has excess capacity for testing programs or for running specialty jobs. There is also no guarantee that a time broker can obtain processing time or equipment configurations to accommodate critical applications at the time of a disaster. Preplanning in obtaining facilities that will conform to a required configuration does not ensure availability when needed in an emergency.

2.2.1.10.7 Membership in Shared Contingency Facilities

A shared contingency facility is essentially the same as a dedicated contingency center, except that membership as a prime subscriber (*A* subscriber) is limited to those organizations who originally formed the group. The membership comprises organizations that use the identical hardware. Each organization proportionally funds the facility and configures it to satisfy its critical processing requirements. This approach has been successful primarily in the financial industry.

A secondary level membership (*B* subscriber) is offered to other organizations, usually with a contractual stipulation that if an *A* subscriber experiences a disaster, the *B* subscriber can be preempted from the facility at a moment's notice.

Shared facilities will usually offer other organizations nondisaster capabilities for development work or overloads. This is also done with the understanding that if an *A* or *B* subscriber experiences a disaster, it can be preempted.

Plus and Minus

One of the primary drawbacks of a shared facility for *B* subscribers is that it can be preempted from the facility at a moment's notice. Another is that most of the membership is made up of organizations located in the same city. Exposure to a common area disaster will create an overload and excessively diminish processing. Owing to the requirements of two or more companies, each organization will have limited resources available. When financial institutions use the same hardware and software, processing can be accomplished by mixing databases, which could result in slow response time, an overload status of the equipment, or disclosure of the proprietary equipment.

The virtue of this backup alternative is that cost sharing and further acceptance of *B* subscribers reduce the budget impact on each of the primary members. Operating expenses are usually lower, since it is established as a nonprofit operation. The addition of *B* subscribers may make it profitable and further reduce *A* member costs.

Other Problems

Shared contingency facilities take a long time to develop. Attempting to form any consortium is difficult, and history has proven that more fail ultimately to materialize than succeed.

The more successful situation, especially for large corporations, is to have dedicated facility professionals develop a backup facility in which the primary member invests in the facility and the professional group is allowed to sell subscriptions to other organizations. The preemptory rights for the single *A* subscriber still exist. Yet *B* subscribers have better longevity, since they are only concerned with one *A* subscriber.

The costs for *B* subscribers are about the same as those found in dedicated facilities.

2.2.1.10.8 Multiple Data Centers under the Same Organization

Some organizations have two or more data centers located geographically apart so that an areawide disaster might affect only one of these facilities. Identical or redundant hardware may not be completely the same, but in order to support the critical work load there must be sufficient capacity to handle the affected and unaffected site. A detailed disaster/recovery plan is required for the same reason as for other backup alternatives.

Major considerations are:

1. Sufficient capacity to support an affected center.
2. Some compatibility of hardware to match the requirements of an affected site.
3. The ability of the unaffected center to reduce processing requirements.
4. Transportation of people, materials, and software data to the backup facility.
5. A written agreement between sites stating that backup services will be provided.
6. Continuous monitoring of equipment configuration changes, processing capacity, software, communications, and so on—anything that would affect compatibility.

This backup alternative has many of the same disadvantages as reciprocal agreements. Only a constant monitoring program of each facility gives this program some chance of working.

EDP auditors, either from regulatory agencies or major accounting firms, do not consider any form of reciprocal agreement a viable backup alternative. Before extensive planning is done in these areas, they should be consulted.

2.2.1.10.9 Fortress Concept with Full Redundancy

For organizations that have to satisfy high security (DOD Tempest) levels, it may be economically feasible to house a primary and secondary redundant equipment backup within the same facility.

Under the fortress concept, two complete, separated centers are built within one highly secure facility. Special considerations are:

1. The area in which the facility is to be located.
2. Potential threats or vulnerabilities from floods, tornadoes, hurricanes, earth faults, or highway dangers from transportation of hazardous waste or explosive materials located within a company's industrial complex and subject to fire or other hazards.
3. Alternate communications networks available.
4. Continuous uninterrupted power resources, which may need to be supported internally.

Those organizations that consider the fortress concept must look to the types of organization that are presently using such an alternative. These include highly sensitive military installations, critical funds transfer groups (who build redundant sites as well as the fortress facility), air-transportation command and control, air-traffic control, and electronic mail centers of the federal government, to name only a few.

This alternative is expensive, not only because of the redundancy of hardware, but because of the housing of the facility and the security that must be installed to protect the environment.

2.2.1.10.10 Use of Microcomputers

A primary objective of any backup alternative is to ensure the support of the organization's information requirements, maintain the continuity of operations, and capture current data to ensure timely recovery from any adverse event.

The use of high-power microcomputers, which can in certain applications revert on a local support basis to a host computer, is a viable alternative. The present technology in hardware and software makes this alternative functional for many varied types of organizations. As a primary function, microcomputers can replace dumb terminals as the user's equipment is required to interface and interact with the host.

If the user and/or host or communications experiences a disaster, the reversion to microcomputers to perform local processing, storage, data entry, and query and word processing can be performed without any significant interruption. Replacing microcomputers and finding the locations in which to operate them are usually easier than with large mainframes, since microcomputers do not require all the amenities or extensive environmental conditioning necessary for mainframes. Vendors can usually supply microcomputers out of stock, or they can be quickly obtained from other sections of the company where they are not needed to perform critical applications.

Equipment planners must be extremely careful to prevent incompatibility, incompleteness, or inconsistency of databases. Databases between the host and microcomputer versions must remain compatible. Software that is equipment-configuration sensitive may be difficult to use in this type of backup alternative.

To utilize this backup alternative, the following considerations must be reviewed by the planners:

- Review all the applications, establish their criticality, and determine whether these operations can be accomplished by a microcomputer.
- Determine whether commercially available software will provide the support of critical applications processing.
- Review current and future needs and consider the applicable hardware that will support these requirements.

Planners should be seeking a method viable within the context of organizational (user) requirements that can be supported by this backup alternative. Careful review of each application requirement with systems people will aid in developing this backup support and provide a unique flexibility in processing before, during, and after any adverse incident.

2.2.1.10.11 Vendor Facility

The vendor facility backup strategy involves an agreement with the equipment vendor. This alternative provides for the use by a customer of existing data center facilities (vendor data center), demonstration area equipment, or a test center in the event of a disaster. Compatibility is one of the considerations in selecting this strategy. This would include hardware, software, operating system, and so forth.

Plus and Minus

Any facility of the vendor is subject to continuous change. Although the system may be conceivably compatible at the time of the agreement, it is necessary to continuously monitor changes. Testing of plans are considerably difficult, since

- Data center facilities of the vendor have the same capacity problems as other data centers.
- Test facilities may be scheduled for client demonstrations, or long-term tests are being performed that would be costly to interrupt.
- Demonstration areas are far from being stable.
- Communications may not exist to the extent required by the customer.

Other Problems

Unless the vendor is dedicated to support customers and creates a true backup environment and facility, detailed agreements are hard to come by. Systems are generally batch-oriented rather than real-time services, especially in demonstration and test facilities. They have limited disk and tape capabilities and usually have little or no room to accommodate outsiders.

Security

Security is limited or nonexistent in many of the facilities, as centers are used by in-house personnel as well as other customers.

2.2.1.10.12 Distributed Network

The distributed network does offer a possibility in some cases where other organizations with large-size computer systems and networks have excess capability that can be used as a backup alternative. In fact, in the financial industry, a number of institutions with either primary sites and/or their own backup capabilities (redundant facilities) can and are willing to sublease facilities for backup services.

The distributed network can function in the following manner:

1. Arrangements can be made to utilize large computer system networks to provide backup capability for on-line/database systems through the switching capabilities that will transfer the load from the affected site to the new network configuration.
2. For the backup facility that has a number of data centers in different locations, the network can be designed to transfer the operation of the affected site to one or more of these facilities. Usually these networks are quite intricate and sensitive. Substantial planning has to go into this activity.

Where there are restricted communication lines and

limited switching capabilities (usually government-owned telecommunications found in smaller countries), additional expenses may be incurred to install company-owned networks.

3. Local area networks (LANs), in conjunction with microcomputers, offer yet another alternative. Extensive planning is necessary to provide software that is compatible with mainframe operation, and at the time of a disaster, can defer certain activities to micro- or minicomputers.

Plus or Minus

There are specific limitations concerning this type of arrangement. Primarily, if the facility with which the agreement is made has a limited capacity and is not truly a redundant or dedicated backup facility, you are in the same position as you would be in making a cooperative agreement.

If the distributed network is an in-house situation, the use of other facilities within the same organization is a good possibility, if the other centers can accommodate the critical, high-priority processing.

Local area networks provide a modicum of security that processing can continue if one facility fails. A local area disaster will usually close down every facility within the area.

Security

In all cases, special consideration must be given to the physical and transmission security of communication lines.

Cost

Unless absolutely necessary as part of the normal processing activity, an overall system can be expensive, and if installed as a backup capability alone, it cannot be justified. If the plan can include other operational needs, then there may be a chance to justify the cost.

2.2.1.10.13 Manual Backup

The level of dependence on computer operations as well as the ability of an organization to revert to manual procedures will determine the extent of manual operations that can replace computerized functions at the time of a disaster. Obviously, there will be a certain amount of manual processing. In fact, there are many operations that should be planned to convert to manual at the time of a disaster in order to relieve the limited processing capability that will be available to support critical services.

If the organization's dependency is a low-level one, then reversion to manual procedures is a viable alternative. The manual system will be sustained while the affected facility is being reconstructed or refurbished.

A history of short-term outages may necessitate manual backup, since the cost of providing for alternate processing capabilities may far exceed the necessity.

There are a number of preparatory considerations relating to manual backup. There may be the need for additional personnel. Manual systems have to be detailed, and personnel have to be trained on the procedures. The system may radically change from original manual activities, because it is important that all documents be designed for easy transferability to computer input once the disaster is normalized.

You must carefully plan to eliminate subsequent problems and the process to recapture manually from prepared source documents to update computer files/data.

1. Transaction recording must follow the same field formats used in the computer entry system.
2. Planning to capture data on micros or minis will speed up the process of updating files.

User departments must be involved in the disaster/recovery planning activities. It is their responsibility to provide the mechanisms, standards, and procedures to assure the continuity and integrity of information.

Cost

Reverting to manual operations can be a costly backup alternative in terms of manpower, hours to keep the business systems performing, alternate locations for users if the building has been damaged, and so on. Insurance coverage can pay for most, if not all, of these costs if properly addressed.

2.2.1.10.14 Other Considerations

In the event of an unexpected or undesirable event, the data center, a critical user, a special application, or a combination of these may affect the ability of an organization to fulfill its business obligations. A strategy or combination of strategies, running either parallel or one after another, can and usually will result in the successful backup of an affected site.

Commercial ventures in backup have considered numerous alternatives, which include:

- *Portable sites.* Airplanes or trailers that are equipped with a reasonable configuration of equipment and basic dial-up communication capability, and are complete with security and environmental conditioning. These units could be moved to an area close to the affected site(s) within a short period of time, to be used as a short-term support. Although such ventures have not "gotten off the ground" commercially, those organizations that can afford the luxury of redundant systems, have expensive transportation equipment standing by, and that require virtually immediate support may consider this alternative.

  There are rental companies that provide office trailers on a moment's notice. The trailers provide limited space on-site for important clerical and administrative functions; they also serve as a control center for the disaster/recovery administration team.

- *Unoccupied, unrented, surplus office buildings.* In many urban and suburban areas, there exists a large volume of unrented (unoccupied) office space. Many of these facilities have air-conditioning and built-in communications equipment, are reasonably easy to secure and have owners who will consider contingency leasing for nominal annual fees. The owner has the right to rent, but if under contract, is required to notify the contingency lessee when such negotiations are entered into and when contracts are completed. Alternate space is usually provided in either the same building or in others of the same owner.

The organization experiencing a disaster may consider establishing a shell site in its own surplus empty buildings and provide the communications required to bring up on-line users on a dial-up mode. The facilities provide a less expensive alternative and accomplish a more successful recovery than does dependence on rented space.

Using conference and banquet facilities of local hotels and motels may, in some cases, provide a viable office and communications center, as well as an excellent facility for users.

Neither of these alternatives can be established within the processing schedules for critical applications unless properly planned and converted to accommodate computer and user operations. Consideration should be given to these alternatives only if the organization faces a long-term outage.

2.2.1.11 Insurance Is not an Alternative

If the corporation's management has decided to remain at an "at-risk" posture by not doing advance planning, and if they depend solely on insurance reimbursement to recover from a disaster, they may be courting disaster. Insurance only provides the dollars to recover and covers the losses of buildings, hardware, software, data and so on if the proper coverage exists. *It does not put you back in business* serving your customers, or provide the necessary organizational computer support.

Only alternative processing facilities, software, and data backup, and a detailed, tested disaster/recovery plan, will mitigate the potential business losses, lessen business service interruption, and provide the strategies and action plans needed to recover rapidly from an adverse, unexpected, or undesirable event. There is no other choice.

In today's high-cost insurance environment, many insurance underwriters will negotiate lower premiums when disaster/recovery plans are in place—and will not provide coverage if plans are not in place.

2.2.1.12 The Positive Approach

There is realistically only one successful approach for organizations which are computer-dependent: the development, implementation, and testing of policies, standards, procedures, and viable alternatives for backup and disaster/recovery plans that will reduce the risk of loss.

Alternative processing facilities that do not meet the "time" requirements for continued support of critical applications are not viable alternatives. Cost not withstanding, a corporation should not rely on insurance to extricate it from a crisis; adequate insurance protection is essential. Don't underestimate the importance of your data processing operations, because a disaster could threaten your company's very existence.

Professional, commercial, dedicated contingency centers provide the most viable, timely recovery of most alternatives available. These facilities are operationally ready for immediate occupancy. They provide the broadest range of equipment configurations to suit your requirements. They are environmentally controlled and secure. Communications professionals at these organizations can design the necessary alternate communications to accommodate the user community. Although after-the-disaster costs may be high, these costs will be borne by the insurance companies.

Other than an organization's ability to develop a totally redundant site and have it sit idle, no other alternative or combination of alternative strategies has proven more effective.

Only extensive risk analysis of your entire organization's computer dependency and operations will determine what the probable threats, hazards, and vulnerabilities are that will affect your organization. The bottom line is to provide the potential ALE that a company will experience if it were to lose computer support. The important consideration that management must face is that whatever the loss, the loss is forever—the dollars, the customers, and perhaps the entire business. There are no trade-offs.

2.3 DATA CENTER RISK ANALYSIS
2.3.1 Threats Analysis

There are many threats which can have serious consequences on data center operations. Everyone thinks of fire, power failures, and equipment failures, but a strike where employees cannot or will not cross the picket lines is potentially just as disruptive. Figure 2-4 should stimulate one's imagination as to what could befall his or her site.

Another way of addressing the issue is to list all the assets one is trying to protect and then consider what negative forces could be brought to bear. (Refer to Figure 2-5 for a list of assets.) The point is that a complete list of all threats needs to be compiled.

After compiling the list, assign probabilities of occurrence. By only selecting those with a probability greater than 1 percent per annum and coalescing things which have only slightly different variations on a common theme, a list of perhaps 10 can be constructed.

2.3.2 Physical Security Risks

Both nature (acts of God) and people pose threats to the data center. Natural risks vary considerably by geographic location. Cataclysmic events, such as nuclear war, are not covered

ALPHABETICAL LISTING OF POSSIBLE THREATS

Accident	Embezzlement	Misappropriation
Act of God	Environmental control failure	Mischief
Agent access (physical, communication)	Environmental spillage	Municipal services deterioration
	EPA/FEMA evacuation	Neighboring hazard
Alcohol abuse	Error or omission	Nuclear accident (evacuation)
Area evacuation	Falling aircraft	Picketing (demonstrations)
Arson	Falling object	Political upheaval
Asbestos	Fire	Postal strike
Blackout	Flood (including a tidal wave)	Power failure
Boiler explosion	Food poisoning	Power fluctuation
Bomb threat	Forest fire	Program alteration
Building inaccessibility	Fraud	Program error
Brownout	Hacker	Radar interference (emission)
Chemical spill	Hardware failure	Radon emission
Civil disorder (demonstrations of any manner or magnitude)	Heat/humidity	Riot
	High crime area	Sabotage
Cold weather	Hostage condition	Salami techniques
Communications failure	Hot weather	Sanctions
Computer virus	Hurricane	Sand storm
Construction disturbance	Ice storm	Service delay
Crime (computer related)	Illicit code	Snow storm
Data diddling	Industrial espionage	Software failure
Delivery service interruption	Industrial spy	Strike/Labor disputes
Demonstrations	Job-related effects	Superzapping
Dirty buildings	Labor dispute	Terrorist activity
Disgruntled employee	Lack of fire suppression	Theft
Disinvestment	Legionnaires disease	Tornado
Drug abuse	Local area disaster	Trash picking
Dust storm	Logic bomb	Trojan horse
Earthquake	Magnetic fields (accidental)	Utility failure
Eavesdropping (wiretapping)	Malicious damage	Volcano
Electronic bulletin board	Malicious software	Water (sprinklers, high-humidity condensation)
Elevator failure	Microwave tap	
Embargo		

Figure 2-4

by a contingency plan since all commerce would cease and employees would be worried about personal, not company, survival. Physical security implies protection against intruders—whether vandals, thieves, or people determined to commit a violent act.

This risk category includes the following major risks:

- Fire
- Flood (all water-related damage)
- Earthquake
- Wind
- Cold and hot weather
- Lightning
- Theft
- Vandalism
- Rioting
- Terrorism
- Labor disputes

2.3.2.1 Fire

Fire is probably the first type of data center disaster one thinks of, and for good reason. Most data centers suffer a fire at one time or another. The computer itself is a significant fire hazard. There are a million connections, any one of which might short-circuit. A large data center uses several hundred kilovolts of electricity, the equivalent of several thousand 100-watt lamps burning within a confined area. The paper used for the printer is a ready fuel, as are the tapes in the tape library. Solvents used to clean tape drives and printers will quickly spread the fire. If a fire starts inside electronics cabinets, smoke from burning insulation is likely to be more threatening than the flames or heat.

Cleanup from a fire is a dirty business: ash and water will be everywhere. Salvage of disk and tape drives, although sometimes feasible, is a long and arduous job. The tape library is very vulnerable to smoke and water damage. Imag-

EXAMPLES OF ASSETS

Software
1. Operating system
2. Programs
   a. Application
      (1) Source
      (2) Nonsource
   b. Contract programs and packages
   c. System utilities
   d. Test programs
   e. Communications

Information
1. Operations
2. Tactical
3. Planning
4. Defense
5. Financial
6. Statistical
7. Payroll
8. Personnel
9. Trade secrets
10. Other

Hardware
1. Central machine
   a. CPU (central processing unit)
   b. Main memory
   c. I/O channels
   d. Operator's console
2. Storage medium
   a. Magnetic media
      (1) Disk (pack)
      (2) Magnetic tapes
      (3) Diskettes (floppies)
      (4) Cassettes
      (5) Other
   b. Nonmagnetic media
      (1) Punched cards
      (2) Paper tape
      (3) Paper printout
      (4) Other
3. Special interface equipment
   a. Network front ends
   b. Database machines
   c. Intelligent controllers
4. I/O devices
   a. User-directed I/O devices
      (1) Printer
      (2) Card reader
      (3) Card punch
      (4) Paper tape reader
      (5) Terminals
         a) Local terminals
         b) Remote terminals
      (6) Modems
   b. Storage I/O device
      (1) Disk drives
      (2) Tape drives

Administrative
1. Documentation
   a. Software documentation
      (1) File
      (2) Program
      (3) Job control language
      (4) System
   b. Hardware documentation
   c. Operations
      (1) Schedules
      (2) Operations guidelines and manuals
      (3) Audit documents
2. Procedures (written documentation)
   a. Emergency plans
   b. Security procedures
   c. I/O procedures
   d. Integrity controls
3. Inventory records
4. Other records
5. Operational procedures
   a. Vital records
   b. Priority run schedule
   c. Production procedures

Physical
1. Resources supply system
   a. Air-conditioning
   b. Power
   c. Water
   d. Lighting
2. Building
   a. Structure
   b. Computer operations
      (1) Computer room
      (2) Data reception
      (3) Tape and disk library
      (4) Field engineer room
      (5) I/O area
   c. Data preparation area
   d. Physical plant room
   e. Stationery storage
3. Backup Equipment
   a. Auxiliary power
   b. Auxiliary environmental controls
   c. Auxiliary supplies
4. Waste materials (to be considered for disclosure)
   a. Magnetic media
   b. Paper
   c. Ribbons
   d. Hardware

Communications
1. Communications equipment
   a. Communications lines
   b. Communications processor
   c. Multiplexer
   d. Switching devices
   e. Telephone

Personnel
1. Computer personnel
   a. Supervisory personnel
   b. Systems analysts
   c. Programmers
      (1) Applications programmers
      (2) Systems programmers
   d. Operators
   e. Librarians
   f. Security officers
   g. Maintenance personnel
   h. Temporary employees and consultants
   i. System evaluators and auditors
   j. Clerical personnel
2. Building personnel
   a. Janitors
   b. Guards
   c. Facility engineers
3. Installation management
4. Other personnel

Figure 2-5

ine cleaning 5000 reels of tape. The tape vault will physically survive a major fire, but after it is retrieved from the flooded basement to which it fell, the tapes inside will not be in a readable condition. A computer will survive being sprayed with water, provided it is not powered and if it is promptly dried out. However, water and ash create an acidic solution which will damage circuits if left in contact.

Fire must be prevented; if not, then it must be contained. A data center manager will want his or her staff to fight the fire while taking care to escape its danger.

2.3.2.2 Flood

For data centers built on flood plains, it is just a matter of time. The Federal Emergency Management Agency can furnish 30- and 100-year histories of the area. A basement data center anywhere is at risk. Sewers back up; water lines internal to the building break sending all the water to the basement. If no one is present, water leakage will go on for hours. The quantity of water used to fight a major blaze is incredible, and it, too, will settle in the basement. Roofs will leak from heavy rains and accumulated snow.

2.3.2.3 Earthquakes

Common fault lines, like the San Andreas Fault in California, represent an "accepted management risk." Few organizations, however, are aware of the numerous fault lines present throughout the country until an adverse condition arises (for example, the recent earthquakes in New York State). Although most earthquakes do not cause substantial damage, they continually interrupt power and communications.

2.3.2.4 Wind

Wind generally strikes by blowing down power lines. Hurricanes and tornadoes are much more direct. The National Weather Service can tell you the incidence of these storms in your area.

2.3.2.5 Cold and Hot Weather

Cold weather can cause pipes to break, thus creating flooding. In the winter, the indoor relative humidity falls; moisture must be replaced by a humidifier.

Hot weather can make the air-conditioning run at its rated capacity. If the air-conditioning was not configured correctly for midsummer conditions, then it will fail to cool the room adequately or will break down. In either case, the computer must be shut down. Hot weather also puts a maximum load on the electric utility. When overloaded, brownouts and blackouts will result.

2.3.2.6 Lightning

Lightning can cause power outages and, with a direct hit, fire. Some data centers, in an area where electrical storms are common, have their computers go down whenever a storm is in the area. Considerable time can be lost this way during the summer months.

2.3.2.7 Theft

The data center represents a high concentration of capital. However, most of the equipment is bolted down and has a limited "secondhand" market. Microcomputers are the major exception to the rule. Test equipment is also very "portable."

Theft of data is primarily covered in Section 2.3.6, "Logical Security Risks"; however, in the case of diskettes, the data is physically stolen along with the diskette. Cleaning people can easily sweep floppies into a dust bin and retrieve them later. Data on diskettes is often summary information, not raw data, which increases its value greatly.

2.3.2.8 Vandalism

The data processing activity is not exempt from the senseless destruction of property. The perpetrator could be a neighborhood youth or more likely a disgruntled employee with easy access to the center.

2.3.2.9 Rioting

Rioting is usually a civil protest which has gotten out of hand. Rioters damage any convenient property; looting is often part of the scenario. This is why the location of the data center is so important.

2.3.2.10 Terrorism

Large corporations are frequently the target of terrorist organizations. Disruption of business activities and computer operations is a favorite objective. These firms are primary players in the so-called military/industrial complex. Often they support unpopular (i.e., right-wing) regimes and resist popular or unpopular causes. A partial list of terrorist organizations is shown in Figure 2-6. Bombings or threats of bombings are their main weapon.

2.3.2.11 Labor Disputes

Labor actions may become violent, looking more like a riot. Even an orderly picket line may convince employees not to cross. Incidents of vandalism may be very costly and disruptive. Bomb threats are a tactical weapon. Take the recent fire (started by an incendiary device) in the Dupont Plaza in Puerto Rico. A labor dispute is assumed to have spawned the many bomb threats and the subsequent fire.

2.3.3 Environmental Control Risks

Computers have stringent specifications for input power and environmental conditions. Failing to provide the correct environment will lead to substantial downtime.

This risk category includes the following major risks:

- Power failure
- Air-conditioning failure (temperature, humidity, particulates)

2.3.3.1 Power Failure

A power failure can be caused by a number of factors. Storms (ice and wind) can cause downed power lines. Lightning can destroy any part of the distribution system. Overloads can cause blackouts directly or indirectly when controllers act to protect the grid through instigation of brownouts or selected

\multicolumn{4}{c}{**TERRORIST ORGANIZATIONS**}			
Group Abbreviation	**Group Name**	**Group Abbreviation**	**Group Name**
PFOC	Prairie Fire Organizing Committee	ANC	African National Congress
DGI	Cuban General Directorate of Intelligence	SWAPO	South West African Peoples Organization
WUO	Weather Underground Organization	ZANU	Zanu Popular Front Zimbabwe
M19 CO	May 19th Communist Organization	MLN	Movement for National Liberation
UFF	United Freedom Front	CDP	Committee for Popular Defense
JBAKC	John Brown Anti-Klan Committee	23 SEPT	23rd September Communist League
BGF	Black Guerrilla Family	ORPA	People in Arms
BLA	Black Liberation Army	EGP	Guerrilla Army of the Poor
CPUSA	Communist Party USA	CISPES	Committee for Solidarity with the People of El Salvador
WAG	Women Against Genocide	FMLN	Farabundo Marti National Liberation Front
CDOFF	Committee to Defend the October 20 Freedom Fight	CFJ	Crusade for Justice
AIM	American Indian Movement	NLF	New World Liberation Front
ARU	Armed Resistance Unit	RGF	Red Guerrilla Family
RFG	Revolutionary Fighting Group	GJB	George Jackson Brigade
RNA	Republic of New Africa	EZ	Emmillo Zapata Unit
FALN	Armed Forces for the National Liberation of Puerto Rico	TT	Tribal Thumb
SMJO	Sam Melville/Jonathan Jackson Unit	WW/AA	Willie Wolfe/Angela Atwood Unit
FLQ	Front for the Liberation of Quebec	LB	Ludlow Brigade
TPIA	Turkish Peoples Liberation Army	TH	Tom Hicks Unit
RB	Red Brigades	CLF	Chicano Liberation Front
ASAIA	Armenian Secret Army for the Liberation of Armenia	PF#9	People Forces #9
JRA	Japanese Red Army	SLA	Symbionese Liberation Army
IRA	Irish Republican Army	CRP	Revolutionary Commandos of the People
ETA	Basque Separatists	EPB	Puerto Rican Popular Army
PLO	Palestine Liberation Organization	FARP	Armed Forces of Popular Resistance
PFLP	Popular Front for the Liberation of Palestine	MAR	Revolutionary Action Movement
PMLA	Popular Movement for the Liberation of Angola	RGRO	Red Guerrilla Resistance Unit

Figure 2-6

blackouts. A construction worker can accidentally cut through a power line.

Poor-quality electricity is not infrequently delivered. Voltages can be permanently low. Surges and sags can suddenly strike. Power quality can be degraded within the building if the computer is not isolated from heavy electric equipment such as elevators.

Input power outside of specifications not only can cause immediate system halts but can also lead to time-lagged component failure or, worse yet, flaky, intermittent problems which are difficult to diagnose.

The quality of electricity and the incidence of outages vary considerably from location to location. The consequences of an outage is proportional to the degree of dependence on the computer. If uptime is critical and power outages are frequent, then a study of the input power should be made to determine what remedial actions can be taken. A special meter with a printer can be rented for this purpose. An electrical contractor should be able to do the study if local staff cannot.

2.3.3.2 Air-Conditioning Failure

The air-conditioning system controls temperature (along with the heating system) and humidity and provides air filtration. The compressor is the most likely unit to cause a long-term outage. However, there could also be a breakdown in the humidifier, a leak in the refrigerant line, or a burned-out fan. A failure in the air-conditioning system will cause the immediate cessation of processing, because the room temperature would immediately rise. Computers generate enormous quantities of heat. Air flow is just as important as air temperature (cooling).

The proper environmental specifications for each computer are given by the manufacturer, but, in general, computers should be run in a room with an ambient temperature of 70 degrees, ±5 degrees and a relative humidity of 50 percent, ±10 percent. Excess humidity causes hydrolysis (condensation) which affects all circuits and read/write heads. Lack of humidity causes static discharge. The air should be filtered to remove paper dust and other particulates from the building. Smoking, drinking, and eating should be prohibited in the computer room: Smoke causes tape and disk errors, food particles gum up keyboards, and spilled drinks short-circuit any electronics.

2.3.4 Computer Equipment Failure Risks

Although the reliability of computer systems is improving, they still fail frequently. System interruptions at even a well-managed site may be one or two per week. Again, as with any system, the more "moving parts," the more system failures. Even at the macro level, a large computer has a lot of components in both the hardware and software. A dynamic environment where hardware and/or software is constantly in the state of flux, increases the odds.

This risk category includes the following major risks:

- Mainframe failure (hardware and software)
- Peripheral failure

2.3.4.1 Mainframe Failure

A system interruption on the mainframe can result for several reasons. The system may halt because hardware or software faults. Or, the system can be halted and restarted by the operator who notices that the computer is in a failing condition but has not yet reached a fatal halt. The environmental software or an application may fail. These interruptions may not require the computer to be restarted, but the computer will be just as useless to the users involved. If most of the computer's users are on a single application, then no useful work is being done.

The trouble with interruptions is that they disrupt the work flow of the users. At first the user waits for a response. If none comes, he or she tries to guess the magnitude of the failure and how long it might last. A restart of the computer and its software may take 15 or more minutes for full recovery. Usually no firm estimate of downtime can be given. Should the user wait or try to do another task? What if the end user is a customer, on-line?

Long-term failures have effects much greater than productivity issues or goodwill issues. The question soon becomes whether the backup plan should be invoked. Is this a disaster? First, estimate the downtime. The field engineer can help in this appraisal. There is a disaster if the estimated downtime plus its estimated variance is greater than the maximum allowed downtime parameter. Estimates may be hard to figure, but a chart like the sample shown in Figure 2-7 can be made in advance and used once the problem is diagnosed.

2.3.4.2 Peripheral Failures

Peripherals will fail frequently. The failure may cause a system interruption for all users or perhaps only those using that peripheral. Either way, recovery is as already noted. However, any one peripheral should not be critical to recovery and should not cause the invocation of the backup plan.

Disk drives will fail, perhaps once every few months; tape drives fail a bit more often. Printers are in need of continuous maintenance.

2.3.5 Network Failure Risks

Today almost all large business computers are run in an on-line mode. Networks connect remote terminals to the mainframe. The number of terminals on-line may run into the thousands. If the network does not keep the connection, the

TIME TO REPLACE EQUIPMENT (Assuming Availability and Emergency Expedite Delivery of Order)			
	Lead Time to Locate	Shipping Time	Installation Time
Processor	less than 3 days	1 day	2 days
I/O processor	less than 3 days	1 day	2 days
Memory	less than 3 days	1 day	2 days
Disk controller	less than 3 days	1 day	1 day
Disk unit	less than 3 days	1 day	1 day
Tape controller	less than 3 days	1 day	1 day
Tape drive	less than 3 days	1 day	1 day
Printer	less than 3 days	1 day	1 day
Console	less than 3 days	1 day	1 day
Boards for any of the above	less than 3 hrs	less than eight hrs	1 hr

Figure 2-7

mainframe might as well be down. In fact, the probability of network failure may well be greater than the failure of the mainframe and its software. This need not be cause to fire the network manager, since a higher failure rate may be inherent in such a complex environment.

This risk category includes the following major risks:

- Equipment failure
- Line failure

2.3.5.1 Equipment Failure

Networks today are very complex in terms of the number of devices and their sophistication. For example, modems, which are required at each end of a line, are subject to occasional failure. Operating speeds from 110 to 19,200 baud are possible in the United States. The modems for speeds above 9.6 K are very sensitive to line noise. Drop backs to slower speed are frequent.

Time-division multiplexers are built into some high-speed modems to allow for several links at lower speeds. Sometimes these multiplexers are stand-alone. These devices, although usually reliable, do introduce additional variables. Statistical multiplexers take advantage of the fact that computer data is full of redundancy. Multiple blanks in the data stream can be compressed, for instance. These devices may cost tens of thousands of dollars and are, in effect, small computers. Hardware or software errors may occur several times per month.

Terminals and terminal controllers can fail, thus causing a whole line to be unusable. Front-end computers are very sophisticated and have a greater probability of failure than does the mainframe.

One's actual experience with the failure of these devices can give a good idea of expected failures and their duration (and, thus, the consequences to the business).

2.3.5.2 Line Failures

Most networks of any sophistication are implemented on leased lines. Several vendors offer long line services, including AT&T, MCI, US Sprint, and RCA. All vendors, while providing good service on the majority of their circuits, do have a significant number of failures and a certain number of problem circuits. Those users who want above 99 percent uptime are well advised to have alternate paths.

Multidrop circuits (as opposed to point-to-point circuits) incur more problems. This undoubtedly stems from the fact that there are more links involved and more hardware (both user and vendor) on the circuit. Digital links are more reliable than analog links. However, the user surrenders most diagnostic capabilities to the vendor.

A study was made a few years ago of the causes for line outages. It was found that the number one cause of outages was the "backhoe": construction people digging up and breaking buried cable. Other causes included weather conditions, equipment failure, and human error. Network managers will tell you that many trouble reports ("trouble tickets") are mysteriously resolved with only the annotation "came clear in testing" or "no problem found." These are usually nondisclosed human errors such as leaving the line "open."

Dial-up lines involve the creation of a temporary circuit each time a call is made. Sometimes good links are chosen, and sometimes bad. If they are bad, the problem is fixed by redialing and getting a new circuit. A maximum speed of 4800 baud is the norm.

2.3.6 Logical Security Risks

Logical security is every bit as important as physical security. The computer can be illegally programmed to steal both the physical and informational assets of the company. This risk category includes the following major risks:

- Theft of confidential information (personal privacy issue)
- Theft of secret information (trade secrets, proprietary information)
- Theft of assets (embezzlement)
- Hacking and malicious damage

The various types of computer crime are explained in Figure 2-8.

2.3.6.1 Theft of Confidential Information

Congress and most state legislatures have passed privacy and computer crime legislation. Companies with personal and confidential data, such as credit bureaus, are obliged to keep it that way—confidential.

2.3.6.2 Theft of Secret Information

Theft of information is a special type of crime. Rarely does the thief physically steal the object of his or her crime; he or she only takes a copy of a file; leaving the victim with the original and thus its presumed utility. As a result, it is virtually impossible to convict such a thief using common law principles. Today we have a federal computer crime law which clearly defines copying as a crime. Most states have a similar law. However, the mere existence of a law does not guarantee protection. A low conviction rate is a poor deterrent.

2.3.6.3 Theft of Assets

Embezzlement is clearly covered in the criminal code. Today, embezzlement is often a computer-assisted crime. Exposures are significant, often in the hundreds of thousands of dollars. Losses in the millions are not that rare.

2.3.6.4 Hackers and Malicious Damage

Whether done for amusement or vengeance, a knowledgeable programmer can do devastating damage. Errors introduced in databases can disrupt normal information processing oper-

```
┌─────────────────────────────────────────────┐
│           CLASSIFYING THE CRIME             │
├─────────────────────────────────────────────┤
│ Computer-related Crime Methods              │
│ Data diddling                               │
│   The changing of data before or after      │
│   input to the computer.                    │
│ Trojan horse                                │
│   Covert insertion of code to handle        │
│   extraneous commands.                      │
│ Salami techniques                           │
│   For example, penny-rounding.              │
│ Computer virus                              │
│   Illegal code that can affect software or  │
│   data, or cause systems to do repetitive   │
│   operations, thus prohibiting normal       │
│   operation, occupying space, or            │
│   destroying data.                          │
│ Computer worm                               │
│   Code placed into a system that worms its  │
│   way through programs and data either      │
│   delaying functions or destroying data.    │
│ Superzapping                                │
│   A program which cuts through all          │
│   security and can "fix" anything.          │
│ Trap doors                                  │
│   Similar to the Trojan horse.              │
│ Logic bombs                                 │
│   Code that causes negative consequences    │
│   when triggered by a future event          │
│   (i.e., date).                             │
│ Scavenging                                  │
│   Retrieval of discarded data, programs,    │
│   and so on.                                │
│ Data leakage                                │
│   Theft of data.                            │
│ Piggybacking                                │
│   Pretending to be another so that one has  │
│   another's security clearances.            │
│ Wiretapping                                 │
│   Overhearing of data communications.       │
│ Unauthorized use of computer time           │
│   Theft of service; includes a sale to an   │
│   innocent third party.                     │
└─────────────────────────────────────────────┘
```

Figure 2-8

ations. The seeds for future disruption can be planted in computers as time bombs. If the databases are not backed up, and the integrity of application software protected, an irreversible disaster can happen.

2.3.7 Loss Expectancies

An ALE can be computed for each of the risk categories listed in Section 2.3.6. The expected number of occurrences of a particular outcome (loss) is multiplied by the dollar loss; all such results are summed giving an ALE.

Example: The expected number of power failures of short duration (1 to 15 minutes) is 8 per year. The expected number of medium-duration failures (15 minutes to 2 hours) is 1 per year; and the expected number of long-duration outages (more than 2 hours) is 0.1 per year. The cost of each type is $2000, $10,000, and $100,000, respectively. The ALE is calculated as follows:

$$\begin{array}{r} 8 \times 2{,}000 \\ 1 \times 10{,}000 \\ + \; 0.1 \times 100{,}000 \\ \hline \$36{,}000 \end{array}$$

The ALE can be used to compute the break-even point of any preventive measures that might be proposed. A proposed scheme would eliminate some of the risk. In the example given, all short-term risks are eliminated; that is, 8 events at $2000 per event or $16,000. This is an annual amount, so the $16,000 must be capitalized at the firm's cost of capital, say 16.67%. The most one would spend would be $96,000 ($16,000 × 1/16.67%).

2.3.8 Quantitative Risk Model

An ALE can be derived for each risk. However, you must be careful not to double count. For instance, power failures have a certain ALE. When calculating an ALE for lightning, you cannot consider lightning's effect on power because it has already been accounted for under the power-failure calculation.

An aggregate ALE can be calculated for all risks. This is the total exposure. Some of that risk can be eliminated with proper preventive and security measures. Of the remaining risk, only long-range outages are strictly the object of contingency planning. A proper contingency plan, including off-site storage, a backup site, and adequate procedures, can, in effect, cap that exposure. The capped exposure can then be assigned to an insurance carrier in exchange for a moderate premium.

Figure 2-9 shows a dual probability table, one for all events and one for long-term outages. *Long term* is defined as an outage lasting longer than the maximum downtime criteria. In the sample it is two days.

2.4 DISASTER COUNTERMEASURES SELECTION

2.4.1 Alternatives Identification

2.4.1.1 Countermeasures against Fire

Fire is one of the more significant threats to data-processing operations. Fortunately there is much technology to apply to this threat.

- *Good housekeeping.* This is one of the basics, but it is often overlooked. Clutter, especially paper goods, provides a ready fuel for any fire. It also is a cause of accidents. Paper should be stored in boxes in neat stacks

PROBABILITIES OF DISASTER		
Disaster Type	All Events	Effects Lasting More than 2 Days
Data corruption	Once/month	Once/year
Power failure	Once/month	Once/year
Equipment failure	Once/day	Once/year
Flood	Once/3 months	Once/5 years
Fire	Once/year	Once/20 years
Vandalism	Once/year	Once/10 years
Ice storm	Once/year	Once/10 years
Tornado or windstorm	Once/year	Once/15 years
Earthquake	Once/5 years	Once/30 years
Labor action	Once/5 years	Once/30 years
Civil disturbance	Once/5 years	Once/30 years

Figure 2-9

away from the computer and any paths to exits. Scrap paper, such as discarded printouts, should be placed in fire-retardant waste containers. These containers have special tops which prevent the spread of fire.

There are many solvents used in cleaning tape drives, disk packs, and printers. These must be kept in closed, approved containers. Rags treated with solvents should be disposed of immediately. The vapor from these solvents could easily be set aflame by a spark from any of the electrical circuits within the computer equipment.

The air plenum under the floor should be kept clean—free from dust.

- *Proper construction.* The data center should be constructed with fire walls, floor to ceiling, on all sides. Even double-thickness drywall provides substantial protection. Metal doors are appropriate. Metal flooring is preferable to wood. All electrical wiring should be in metal conduits. The tape library, storage rooms, and maintenance rooms should be isolated to prevent the spread of fires. Windows should be kept small or eliminated. Wire mesh will mitigate the damage of an explosion. There should be at least two clearly marked exits.
- *Emergency power off and lighting.* At each exit there should be an emergency power-down switch. Powering off the computer and the air-conditioning is the single most important thing that personnel on site can do, with the exception of a safe evacuation. The power to the computer, if not turned off, would continue to provide the energy (heat) necessary for combustion. Many electrical fires go out after the electrical source is turned off. The air-conditioning, if left on, would fan the fire, rapidly increasing its intensity.

Since lighting often fails (intentionally or as a result of fire), emergency battery-powered lighting must be provided to aid in evacuation and firefighting. Battery-powered lights must be tested periodically to ensure their viability.

- *Hand-held fire extinguishers.* Many fires can be extinguished quickly by trained personnel on-site using a hand-held fire extinguisher. Fire extinguishers are classified as A, B, or C (or a combination thereof) to show which type of fire they should be used on. The three main types of fires are (A) wood, paper, and other cellulose-based material; (B) oil and gas; and (C) electrical. Water-based fire extinguishers are only good on type A fires. Dry chemical fire extinguishers are appropriate for types A, B, and C, but they have only limited ability to put out type A fires; when used on type C fires, they are effective but leave a residue which is ruinous to sensitive electrical equipment such as computers. Although CO_2 extinguishers are only very effective on type C fires, because they may cause an acidic reaction, they are inferior to Halon or other nontoxic gas type fire extinguishers.

A fire hose is the best weapon against large cellulose-based fires, such as paper. Just a few boxes of paper can generate an overwhelming volume of heat requiring hundreds of gallons of water to extinguish. Be sure that a fire hose is on every level of the building and that it can reach every corner. Partitions make the path to the fire site even longer; so be sure to check each area with a rope the same length as the fire hose.

- *Halon flooding.* Halon flooding will extinguish electrical fires automatically with no damage to the equipment and no harm to operations staff, provided they leave the room within a few minutes. The room must be relatively airtight to keep the Halon concentration high until the fire is extinguished. Halon is not effective on cellulose-based fires because they smolder and would reignite once the Halon concentration drops. Halon flooding can be installed above the floor via nozzles in the ceiling or in air plenums. There should be a delay after detection and before discharge to allow for manual fire fighting. Electricity should be automatically turned off to eliminate a source of the fire. Halon should be triggered by smoke and/or heat sensors. By requiring two sensors to detect alarm conditions, false alarms are reduced. A Halon discharge costs several thousand dollars.
- *Sprinklers.* Sprinklers have some advantages over Halon: they require no electronic sensor to activate, they do not shut off, and they are much less expensive. But the independent heat-activated heads can be a disadvantage as well. Many data center managers do not like the idea of a wet pipe which could leak or activate because of physical damage. Therefore, they keep the pipes dry and activate them with electronic sensors similar to those used for Halon extinguishers.

Sprinklers are superior on paper fires and are preferable when the fire threatens to take the entire building. If the electricity has been turned off, electronic equipment can survive being watered. Fans can be used to dry equipment before repowering. A flow detector should be installed to power down the equipment should the sprinklers be activated.

A scheme whereby the computer room is protected by both Halon and sprinklers is preferred. The sprinkler heads are set to activate only on high temperatures. By this time the Halon would have dumped. If it is unsuccessful, the sprinkler follows, protecting the building. If the heads cannot be protected, a dry-pipe system can be installed."

2.4.1.2 Countermeasures against Water Damage

Water may enter the data center because of flooding in the area. Entry of water may occur because of a high water table combined with the failure of the sump pump. Drains can back up. Pipes can burst. A fire can be fought with large quantities of water. No matter what the source, water is very damaging to electrical equipment. Countermeasures can be effective however.

- *Site selection.* Data centers should not be situated in an area prone to flooding, like a flood plain, lakeshore, swamp, or subdivision with poor storm sewers. Historical records can be consulted before site selection is made.
- *Building construction.* The data center should be located out of the basement and in a location with no pipes passing overhead. The floor should have drains to remove any water coming from leaking pipes or condensation (e.g., from the air conditioner). Holes in the ceiling should be plugged to prevent water from entering from the floor(s) above.
- *Water detectors.* Water detectors should be installed on the floor. Activation of any two sensors should cause the automatic power down of equipment.
- *Water covers.* Plastic sheeting should be available for the covering of computer equipment. Fitted covers are also available. Of course, covers should only be used once the power is shut down.

2.4.1.3 Countermeasures against Intrusion

Intrusion is the precursor to several other problems such as robbery, assault, and vandalism. When applying countermeasures in this area be careful not to put "steel doors in paper walls."

- *Site selection.* A good building site is one in a good neighborhood surrounded by open spaces for good surveillance. A good, responsive police department is a definite plus. If this is not the case for your firm, consider moving. It may be cheaper than the consequences or the cost of other countermeasures.
- *Locks.* Conventional locks should be present on all doors and windows. Security limitations include lost keys, staff turnover, and ease of duplication.
- *Combination locks.* An electronic combination lock solves the problems of lost keys; the combination is simply changed. There should be enough different combinations to prohibit a 100 percent search of all combinations. Four-decimal digits (10,000 combinations) is satisfactory, whereas a mechanical lock with five pins (32 combinations) is not.
- *Card-key access.* The card key allows for multiple keys so that each individual can be assigned a unique key. This uniqueness allows the actions of each keyholder to be tracked. Also, access privileges can change by hour of the day or by day of the week. An attached microcomputer can generate all sorts of management reports. This system may not be justified for the data center alone, but it is quite justifiable for the facility as a whole.
- *Man trap.* A man trap allows only one person to enter at one time. The man trap has two doors separated by a narrow chamber. The doors, only one of which can be open simultaneously, are operated by card key. While in the inner chamber, the entrant can be examined for weapons, including bombs.
- *Low-profile building.* The theory is that if they do not know where you are, they cannot intrude. *Low profile* means that there is no sign as to a company's name or what is going on inside, no large glass windows, and no unusual architecture.
- *Lighting.* Good lighting provides some deterrence to half-hearted intruders, such as vandals.
- *Fencing.* Fencing is also a deterrent against would-be intruders. The fence can be augmented by height, barbed wire, razor wire, or electrification.
- *Guards.* Guards, whether or not armed, provide a good measure of deterrence and also provide immediate assistance in aiding victims and apprehending perpetrators. Unarmed, the guard will be limited in his or her response and will depend heavily on police backup. However, unarmed guards may enhance the positive image of the company and avoid any accidents.
- *Cameras.* Surveillance cameras provide some degree of deterrence. Do not imagine, however, that a guard can be alert to all problems using the camera. It is physically impossible for a guard to stare at monitors 100 percent of the time. Cameras are best used to investigate problems which have been detected by more passive means, such as intrusion detectors. Surveillance cameras are available that will detect movement or changes in area density (i.e., a person enters an empty room) and sound an alarm to alert guards to look at the monitor.
- *Intrusion detectors.* Intrusion detectors vary in price and can detect the following conditions: open doors, open windows, broken glass, vibration, movement, and presence. The technology is electrical (open switch), seismic (vibration), light (broken beam), infrared, and microwave (distorted reflection). Once an adverse condition is detected, it can be relayed to the guard or police department.

2.4.1.4 Countermeasures against Power Problems

Poor computer power can lead to recurring system outages and component failure. Input power quality can be measured using a sophisticated meter which measures voltage sags and surges and their duration. The following is a list of devices which alleviate some or all of the input power deficiencies. They are ordered from least to most expensive. You must determine how sophisticated the device should be in order to eliminate the problems experienced.

- *Isolation transformer.* The power leading to the computer should be on a separate or isolated transformer. In this way there is partial protection against other devices on the same power source. Such devices as elevators or anything with a large electric motor will induce voltage fluctuations on the line. By having the computer on an isolated transformer, these fluctuations are somewhat masked.
- *Constant voltage transformer (CVT).* This device will produce a steady voltage on output regardless of input voltage. This works well in an area where the power company frequently reduces load by dropping the voltage of the power supplied to its customers. This device does not handle fluctuations of short duration or large magnitude.
- *Motor generator (MG).* This device takes the line power and, via an electric motor, converts it to mechanical energy, the turning of a large flywheel. A shaft from the flywheel then turns an electric generator which produces electricity. So, in net, electric energy has been turned into electric energy. But owing to the momentum of the flywheel, the fluctuations in the input power have been smoothed. In fact, the motor generator will ride out an outage of approximately one second. This may be enough to remove most fatal aberrations in input power. The by-products of the double conversion are heat and noise.
- *Uninterruptible power system (UPS).* A UPS is similar to the MG except the flywheel is replaced by a bank of batteries. Alternating current (ac) input power is turned into direct current (dc), some of which is stored in the batteries. The dc power is reconverted into ac power. Because of the presence of the batteries, the voltage generated is constant. The number of batteries in the system determines the length of time the load can be supported without input power. Fifteen minutes is typical. If power is not restored, in five minutes, for example, the computer system can be shut down gracefully. If the UPS is intended to keep operations running for more than a few seconds, the air-conditioning must be powered by the UPS as well. Constant air flow is necessary to keep the units cool.

  Since a UPS is roughly comparable in cost to an MG and filters out many more problems, it has virtually replaced the MG in all applications.
- *Diesel generator.* The diesel generator can be used to create a mini-power plant right on the company's site. Power can be generated as long as diesel fuel is supplied. Combined with a UPS, the data center would not even notice a blackout. The line power would be instantaneously replaced by battery power, and a few seconds later the diesel generator would assume the load. The few seconds are required for the diesel to gain operating speed.

An airline reservations service would want the UPS/diesel combination. Many banks opt for this same level of power protection.

2.4.1.5 Countermeasures against Computer Equipment Failure

Computers are more reliable today than every before. Still, failures in the hardware produce several mini disasters each week. There are, however, countermeasures.

- *Purchase of Reliable Equipment.* Some vendors' equipment is more reliable than others. Certain models have better maintenance records as well. You should choose a vendor's equipment with these facts in mind.
- *Maintenance agreements.* A formal maintenance agreement means that competent service people will work on the equipment. There is the opportunity for a continuity of care from the same service people. For mechanical equipment, preventive maintenance can be scheduled and completed on time.
- *Redundancy.* Every part of the configuration can be overdesigned with some redundant units added (e.g., an extra drive, modem, channel, and so on). With switch gear, these units can be quickly called into service. Thus, outages are either prevented or expeditiously alleviated. If necessary, the whole computer system can be duplicated.

2.4.2 Cost-Benefit Analysis, Selection and Implementation

Many of the countermeasures noted already should be regarded as baseline, that is, standard for the data processing industry. If so, management should be informed of the standard and asked to comply.

Where there are alternatives, such as for power protection, a cost-benefit analysis should be made. An ALE can be calculated before the implementation of a countermeasure and after implementation. The difference is the annual benefit. This can be compared to the annual cost of the countermeasure (i.e., the amortized capital outlay, annual carrying charges, and operating costs). The option with the greatest net benefit should be chosen.

After implementation, the countermeasure should be reviewed for actual effectiveness. Possible weaknesses should be resolved to give the results anticipated. When analyzing future countermeasures, past measures must be regarded as sunk costs.

2.5 INSURANCE COVERAGE

This section is an informed commentary, but it should not be confused with quality insurance and legal advice which should only be obtained from a licensed insurance agent and attorney.

2.5.1 Coverage Analysis

2.5.1.1 Types of Insurance

Insurance companies offer special coverage for EDP equipment and related assets, expenses, and profits. You should not rely on standard property insurance since that policy often excludes data processing equipment or many of the perils to which it is subject.

The types of EDP-related coverage normally offered include the following:

- *Equipment (all risk).* This protects data-processing equipment against everyday perils such as theft, vandalism, wind, lightning, and fire. Naturally the risks of war, insurrection, and falling meteors are excluded. Watch out for additional exclusions such as flood and electrical malfunction. These are significant risks and do require coverage. The insurer may choose to eliminate the exclusions if the deductible is raised or if an additional premium is paid. The federal government provides flood insurance when other carriers do not.
- *Media.* This will replace all damaged media with new (blank) media. It does not address the data stored on the media.
- *Data (includes both data and software).* This protects the data on the media. It will pay for the reconstruction of data and programs lost. For software program products or purchased databases, it will pay for replacement copies.
- *Extra expense.* This coverage pays for all extra expenses associated with recovery from the emergency condition. For instance, transportation to the backup site, housing at the backup site, and other travel expenses would be covered. The normal level of expenditure (nonemergency) is not covered and is subtracted from expenses incurred if they cannot be clearly identified as extra. The insured cannot be extravagant; he or she must act as if he or she were not covered by insurance.
- *Business interruption.* This protects the revenue flow of the company. Thus, added to extra expense coverage, the profits of the firm are protected. The loss incurred is calculated by comparing predisaster sales to postdisaster sales.
- *Accounts receivable.* This is protection for the accounts receivable asset. Should records be lost and corresponding payments not be received, this coverage will make up the difference.
- *Valuable papers and records.* This coverage provides money to replace valuable documents.

All coverages are subject to specific and global deductibles. There are also dollar limits to each form of coverage. You must carefully add up the potential losses and provide enough coverage for replacement. Inflation and configuration modifications will constantly change this amount. Since coverage becomes increasingly less expensive per thousand, you should not err on the low side. There is frequently a different form for microcomputers.

Additionally, you should also consider honesty insurance (fidelity bonding) for employees engaged in data-processing activities. This protects a firm, and thus its clients, against any dishonest (i.e., criminal) acts by its employees. Liability insurance extends that protection to intentional acts and negligence. If the firm is a financial organization or a service bureau, for instance, there is a clear fiduciary responsibility, and liability clearly rests with the firm.

2.5.1.2 Valuation Method

Policies generally value equipment in one of two ways:

1. *Replacement cost.* This is the cost of a replacement unit that is functionally like the original.
2. *Actual cash value.* This is the original cost of the unit less an allowance for physical depreciation and functional or economic obsolescence.

Firms should normally opt for the first method since it obviates the need for additional cash outlays.

2.5.2 Cost Savings Opportunity Analysis

The cost of insurance can be controlled in several ways. First of all, you should make sure that you have the correct amount of insurance coverage. Make an exact inventory of assets and their current replacement value. (Be sure to update these costs yearly, since costs for equipment in the computer industry are declining with respect to capability.) For items with a subjective value, such as data, be sure to do a thorough job of cost estimation. When assets are sold or become obsolete, remove them from the policy.

When the choice of site location presents itself, make sure the data center is located in a building, which minimizes the risk of fire, flood, and so on. Never put the center in the basement. Do not locate in a multitenant building, especially next to tenants handling dangerous materials. Position the data center away from restaurants or cafeterias. Do not have steam or water lines passing overhead.

Install various safety features if not already present:

- Sprinkler system (buildingwide)
- Halon system
- Hand-held fire extinguishers
- Master power-down switch
- Fire doors and walls

Store duplicate copies of data and software off-site. This may lessen or eliminate the need of insurance in this area. Conduct business in a safe fashion:

- Keep solvents in approved, covered containers of minimum volume.
- Store paper outside the computer room and not along an exit path.
- Turn off the machine and cover it when not in use.
- Keep the area generally neat and tidy.
- Have someone on-site during the entire period the computer is turned on.

Have a tested disaster/recovery plan in place to lessen the period of outage. This will lessen the risk in the areas of extra expense and profit protection.

Deductibles can be raised. In general, insurance should be for major disasters. Minor interruptions can be self-insured, since they will have little effect on the company's cash flow or profitability. However, since the small losses are much more frequent, they have a disproportionate effect on costs to the insurance company and, thus, the cost of coverage to the firm.

It has become common practice for insurance companies to review existing disaster/recovery plans, test analysis, and plan maintenance programs. Up-to-date disaster/recovery plans plus sound off-site storage programs entitle organizations to premium reductions.

Section 3
RECOVERY MANAGEMENT PLAN AND PROCEDURES DEVELOPMENT

3.1 RECOVERY MANAGEMENT PLAN

3.1.1 Organization of Recovery Teams

In order to do the many tasks associated with the disaster/recovery process, it is advisable to form teams, each with a specific area of responsibility and corresponding benchmarks. This division of labor will optimize the process by assigning experts to each task and by performing many tasks in parallel. A list of possible teams is shown in Figure 3-1.

Teams are required during the planning process, for the emergency response, and for recovery operations both at the backup site and the primary site. The number of teams formed will probably be proportional to the size of the organization. For example, in a small firm, the hardware, software, database administration, and data communications teams could be combined to form a backup site operations initiation team.

3.1.1.1 Disaster/Recovery Planning Team

The planning phase will be directed by the Disaster/Recovery Planning Team. The responsibilities of this team are outlined in Figure 3-2. As suggested in Figure 3-1, this team may charter other teams to do parts of the planning. However, for our purposes here, we shall assume a single unified team. Members should be recruited from many areas of the firm, not just data processing. Users should bear much of the responsibility. Specialists from risk management, security, building maintenance, finance, audit, and so on are good additions. The support and direction of top management are essential. A senior executive member would enhance that support.

Within the planning process there will be a data collection phase. This involves collecting those documents and policies which are already in existence. For example, you should find and critically examine maintenance agreements, job descriptions, insurance policies, and the like. A summary list is included in Figure 3-3. Some documents or policies will be missing. These should be regarded as deficiencies. For instance, a site may not have a Data Center Users Guide. This is an excellent control mechanism which enhances the performance of the center and those who use it. Although not directly part of the disaster/recovery scheme, it is part of the prevention aspect that is outlined in the plan.

The planning team needs to then turn its focus to the strategic decisions that must be made. These are outlined in Figure 3-4. The most important decision is the choice of backup strategy. In general, all the issues in this book must be addressed. Various prevention, security, and disaster/recovery strategies and tactics must be put in place. These will cost substantial amounts of money in the aggregate. The planning team must be prepared to take on the selling aspect as well.

Of course there is the task of writing the plan. The document should be concise yet comprehensive. A sample outline is given on pages 90–97. Throughout the project, but particularly at this time, discipline is important. It is critical not to let the project drag out. Writing, if not disciplined, will stretch on and on. The enthusiasm built by the team cannot be expected to last forever.

After the draft is written, it must be reviewed for completeness and correctness. This is the process of certification or validation. Next a series of dynamic tests should be scheduled. The planning team will direct these efforts as well.

Ongoing maintenance, although not overly time consuming, is ever-present. The team will do this task as well. From viewing the charter of this team you would expect one or more people to be assigned full-time to maintenance, at least during the development period.

3.1.1.2 Emergency Response Team

The emergency response team (outlined in Figure 3-5) handles the immediate consequences of an emergency. They put out the fire, administer first aid, and work with all the authorities, like the fire department and FBI. Usually the director of security heads this team. Of course, the medical department and others would participate. All emergency response actions should be documented in the Emergency Response Procedures Manual.

POSSIBLE DISASTER/RECOVERY TEAMS	
Teams Active During the Planning and Preparation Phase Disaster/Recovery Planning Team (overall coordination) Off-site Storage Site Selection Team Backup Strategy and Site Selection Team Network Design Team Capacity Planning Team File Backup Strategy Team Retention Policy Team Executive Management Liaison Team User Disaster/Recovery Planning Team User Liaison Team Physical Security Design Team Software Security Design Team Application Prioritizing Team Audit Team Risk Management Team (insurance) Alternative Office Space Team Office Recovery Team Training Team Public Relations Team Emergency Response Team **Teams Active During the Emergency Response Phase** Emergency Response Team (overall coordination) Fire Fighting Team Bomb Threat Team Security Team Medical Team Evacuation Team Building Shutdown Team	**Teams Active During the Recovery Phases** Disaster/Recovery Management Team (overall coordination) Disaster Assessment and Declaration Team Off-site Storage Team Hardware Recovery Team Software Recovery Team Database Administration Team Data Communications Team Microcomputer Team Office Automation Team Operations Team Salvage Team New Facility Team New Hardware Team User Teams Data Input Team User Liaison Team Security Team Audit Team Administrative Team Supplies Team Transportation Team Voice Communications Team Public Relations Team Customer Service Team Labor Relations Team (personnel) Legal Team Risk Management Team

Figure 3-1

3.1.1.3 Disaster/Recovery Management Team

The Disaster Recovery Management Team (DRMT) is responsible for all activities after the emergency response to a disaster is complete. This team, in effect, supersedes the normal management structure until recovery is complete. All other recovery teams report to this coordinating body. The number of teams necessary will depend on the size of the firm. Less than 15 teams is envisioned. Even if the teams are coalesced, all the functions described will have to be performed by someone for every firm.

Two special positions are contemplated. The first is the coordinator; he or she actively drives the recovery process. The second is the administrator; he or she sees to it that the necessary paperwork is completed, makes sure funds are available, and keeps upper management informed. Alternates for each position are required. A smaller firm may wish to combine the two positions. A charter for this team is outlined in Figure 3-6.

3.1.1.4 Backup Site Activation Teams

If the disaster will result in downtime exceeding the maximum downtime limit, it will be necessary to commence operations at the backup site. To accomplish this changeover, the following teams are activated:

1. Off-site Storage Team (see Figure 3-7)
2. Hardware/Software Team (see Figure 3-8)
3. Applications Team (see Figure 3-9)
4. Communications Team (see Figure 3-10)
5. Operations Team (see Figure 3-11)

Experts will be chosen from the staff to fill these teams.

SAMPLE CHARTER

The Disaster/Recovery Planning Team

Team Charter

This team has responsibility for producing the Disaster/Recovery Plan, as well as testing and maintaining it over its lifetime. The planning cycle consists of data collection, strategy selection, justification, drafting, promotion, approval, accreditation, training, testing, and implementation. The team must have the full confidence of top management. The team will address issues cutting across all the divisions of the company. It must have the cooperation of all.

Recommended Members*

Coordinator:	Director of Information Systems (the Chief Information Officer)
Alternate:	Deputy Director Information Systems
Members:	Vice President of Administration
	Director of Security
	Building Manager
	Audit Staff Member
	User Department Managers
	Data Center Manager
	Controller
	Director of Risk Management
	Staff Attorney
	Staff (as required)

*Lower-level personnel can be substituted in larger firms; however, sufficient top and middle management should be on the team to provide the necessary credentials and decision-making authority when the disaster mode is invoked.

Figure 3-2

DATA COLLECTION REQUIREMENTS

1. Current status checklists and questionnaires (These are available in texts and from consultants and cover all aspects of prevention, security, and disaster recovery.)
2. Manuals, handbooks (hopefully already in existence)
   a. Personnel policies
   b. Data center and user job descriptions
   c. Operator's manual
   d. Data center users guide
   e. Security policy
   f. Application development standards
   g. User department policy and procedures
3. Emergency response plan
   a. Evacuation procedures
   b. Evacuation routes
   c. Fire fighting guidelines
   d. Procedures for each type of emergency
   e. Emergency phone numbers
4. Organization charts (top, data center, user)
5. Vital record retentions and storage policy
6. Layouts (data center, user, building, area—show all utility lines and security features)
7. Power distribution system diagram or description
8. Air-conditioning system diagram or description
9. Employee telephone and address (home and office) list
10. Vendor list (including contacts and products)
11. Supply list (one month's usage)
12. Mainframe configuration
13. Mainframe data flow diagram
14. Pack family layout and usage
15. Datacomm configuration
16. Maintenance contracts
17. Insurance policies
18. Software escrow agreements
19. Off-site storage agreement
20. Backup site agreement (including configuration)
21. Tape library system sample reports
22. Forms catalog (list, samples, vendors, leadtimes)

Figure 3-3

3.1.1.5 Primary Site Recovery Teams

After backup operations have been established, attention can be shifted to repairing, rebuilding, or replacing the primary site. These functions are performed by the following teams:

1. Salvage Team (see Figure 3-12)
2. New Hardware Team (see Figure 3-13)
3. Facilities Team (see Figure 3-14)

3.1.1.6 User Teams

The user departments may have been affected directly by the disaster conditions or indirectly from damage to the data center. To recover the user operations, the following teams will be activated:

1. User Operations Teams (see Figure 3-15)
2. Data Preparation Teams (see Figure 3-16)
3. User Liaison Team (see Figure 3-17)

3.1.1.7 Administrative and Special Teams

To handle administrative details for the foregoing teams, the following teams are activated:

1. Administrative Team (see Figure 3-18)
2. Supplies Team (see Figure 3-19)
3. Transportation Team (see Figure 3-20)

4. Public Relations Team (see Figure 3-21)
5. Audit Team (see Figure 3-22)

3.1.2 Recovery Management Cycle

The disaster/recovery process consists of four phases. They are:

Phase 1. Emergency response
Phase 2. Backup site activation
Phase 3. Primary site recovery (with continued operations at the backup site)
Phase 4. Primary site reactivation/normalization of disaster mode

The emergency response phase will last from the inception of the disaster until the disaster is under control. Cooperation with community emergency services is vital. Firefighters, for example, should be allowed to direct the firefighting efforts. However, before their arrival, it is up to the members of the firm's fire brigade to take whatever actions they can. First aid and CPR by staff members may save many lives. When the crisis is over, phase 2 begins.

The DRMT will take over control from the Emergency Response Team. They will assemble at the command (control) center and call upon other recovery teams to do their assigned tasks. The first order of business is to move operations to the backup site and to restore operations as was preplanned.

THE EMERGENCY RESPONSE TEAM

Team Charter

This team is responsible for handling an emergency situation. The long-term recovery operations are handled by the Disaster/Recovery Management Team. The particular emergency situations addressed will be dependent on the particular firm, but the following are typical:

1. Fire
2. Flood
3. Medical emergency
4. Intrusion
5. Storm
6. Bomb
7. Chemical spill
8. Building evacuation
9. Building shutdown

All procedures are documented in the Emergency Response Procedures Manual.

Recommended Members

Coordinator: Director of Security
Alternate: Head Guard
Members: Head of Medical Staff
Building Manager
Fire Wardens

Figure 3-5

Phase 3 involves the recovery of the primary site. If the site was completely destroyed, a new location will be utilized. The primary site must be reinstated before the contract for services at the backup site expires. If this is impossible, a second backup site must be acquired.

Phase 4 is the reactivation of the primary site at the original or perhaps new location. The activation of this site need not be as rushed as at the backup site; however, it could prove more challenging since all applications will be restored, not just the essential ones. There may be substantial processing to bring all databases up to date.

At the end of the four phases there will be an opportunity to review what went well and what did not. Deficiencies can be rectified for the future by augmenting the plan accordingly.

3.1.3 Command (Control) Center

Upon the declaration of a disaster, a command or control center must be established. This provides a focal point for all disaster/recovery operations, as well as temporary office space for team members. The command center should include separate meeting rooms and work areas. Telephone lines must be quickly installed. It would be advisable to have some telephone lines preinstalled. This can easily be accom-

---

STRATEGIC ISSUES

1. Members of the Disaster/Recovery Planning Team (and appoint coordinator)
2. Members of the Disaster/Recovery Management Team (coordinator, administrator, alternates)
3. Other teams and their membership
4. Scope of plan
5. Risk Assessment
6. Maximum downtime requirements
7. Cost of downtime
8. File Backup Strategy (source, object, job, database, transactions, and so on)
9. Off-site storage strategy (including level of recovery possible)
10. Prioritization of applications and reports
11. Vital records strategy (retention schedule)
12. Available backup site options (including final choices)
13. Available office space and strategy
14. Employee commitment in time of disaster
15. User recovery plans

Figure 3-4

DISASTER/RECOVERY MANAGEMENT TEAM

Team Charter

The Disaster/Recovery Management Team (DRMT) is responsible for the overall coordination of the disaster/recovery process. All other teams report to this team in time of disaster. In effect, this team succeeds the Disaster/Recovery Planning Team which heads efforts in the predisaster mode. There will be a large carryover in membership.

Typical Duties

Coordinate teams
Secure financial backing from the president and board
Approve all actions not preplanned
Give strategic direction

Role of the Coordinator

To be the working head of the team
To execute all disaster/recovery team decisions via this and other teams

Role of the Administrator

To approve or disapprove all actions with respect to the strategic direction of the firm
To be the liaison to upper management
To finance all operations
To expedite matters through all bureaucracy

Recommended Members

Coordinator:	Director of Information Systems
Alternate Coordinator:	Deputy Director of Information Systems
Administrator:	Vice President of Administration
Alternate Administrator:	Controller
Members:	Director of Security
	Building Manager
	Auditor
	Key User Activity Managers
	Data Center Manager
	Director of Risk Management
	Chief Attorney
Expanded Membership:	Heads of all other teams

Figure 3-6

OFF-SITE STORAGE TEAM

Team Charter

The Off-site Storage Team is responsible for the management of all backup tapes, off- or on-site. At the time of the disaster, the team must secure the correct tapes for transport to the backup site. An alternative off-site service will have to be set up at the backup site.

Typical Duties by Phase

Phase 2
- Inventory and select the correct tapes
- Transport to the backup site

Phase 3
- Establish off-site storage at the backup site

Phase 4
- Inventory all tapes at the backup site
- Transport the final dumps and other tapes back to the primary site

Recommended Membership

Coordinator:	Tape Librarian
Members:	Database Administrator
	Systems Programmer
	Application Analyst

Figure 3-7

HARDWARE/SOFTWARE TEAM

Team Charter

The responsibility of the Hardware/Software Team is to reconstruct the system software environment at the backup site and to do the same upon returning to the primary site.

Typical Duties by Phase

Phase 2
- Confirm the system file backups
- Load the system files to disks
- Load the configuration file to disks
- Bring up the operating system
- Test the hardware and software

Phase 3
- Support
- Reload the operating system at the primary site
- Test the hardware and software

Phase 4
- Support

Recommended Membership

Coordinator:	Chief Systems Programmer
Alternate:	Systems Programmer
Members:	Systems Programmer
	Field Engineer

Figure 3-8

plished if other facilities of the same organization are selected as the control center. This would incur only a nominal charge and would allow use of designated rooms for normal office overflow. These areas can be completely furnished if the disaster is prolonged.

A firm may choose to locate the command center in the conference room of its building. If space is not plentiful or if

Section 3 Recovery Management Plan and Procedures Development **37**

APPLICATIONS TEAM

Team Charter
The Applications Team is responsible for restoring the applications singly or all together at the backup site and then again upon return to the primary site.

Typical Duties by Phase

Phase 2
- Restore the database to as current a version as possible
- Restore object and source code to current
- Work with users and auditors to verify the database and functionality of system

Phase 3
- Monitor processing
- Help restore tape backup and off-site storage operations
- Prepare return to the primary site
- Set up parallel environment

Phase 4
- Support cutover

Recommended Members

Coordinator: Chief Applications Analyst
Members: Database Administrator
               Applications Analyst
               Applications Programmer
               User Analyst

Figure 3-9

COMMUNICATIONS TEAM

Team Charter
It is the responsibility of the Communications Team to prepare communications to the backup site and to restore communications to the primary site. Both voice and data are included.

Typical Duties by Phase

Phase 2
- Determine the requirements for both voice and data
- Install the network including lines, modems, and other communications gear
- Test the network

Phase 3
- Operate the backup network
- Determine damage to the primary site network
- Order replacements
- Install the primary site network

Phase 4
- Support the primary site network
- Dismantle the backup site network

Recommended Members

Coordinator: Chief Telecommunications Analyst
Members: Voice Communications Analyst
               Data Communications Analyst

Figure 3-10

OPERATIONS TEAM

Team Charter
The Operations Team is responsible for restoring an operational environment and for processing the scheduled work load.

Typical Duties by Phase

Phase 2
- Assist Off-site Storage, Software, Communications, and Applications Teams in restoring the processing environment

Phase 3
- Establish a schedule with assistance from the users
- Run the daily schedule
- Perform backups
- Assist other teams in preparation of the primary site
- Perform parallel operations at the primary site

Phase 4
- Return to normal operations

Recommended Members

Coordinator: Operations Manager
Alternator: Operations Supervisor
Members: Operators

Figure 3-11

---

much office space was destroyed, then a remote location must be identified. This may involve renting new office space or securing space at a local hotel. An alternate command center should be designated in case the first choice is not occupiable.

3.1.4 Damage Assessment

The first task of the DRMT is to assess damage. The main purpose of such an assessment is to determine the expected length of downtime. The actual recovery process will be carried out by one of the recovery teams, perhaps the salvage team. However, right now the decision must be made as to whether the events constitute a true disaster: one involving a move to the backup site.

Experts should be called in if they are not part of the DRMT. Their best guess as to the likely duration of the problem should be sought. If their estimate is that it is likely the time required will exceed the maximum allowable downtime, then there is a disaster. Do not second guess; invoke the plan.

It is important to factor in all the lead times to restoration. There is the time to diagnose the problem and what is required to remedy that problem. Additionally, some time is

---

SALVAGE TEAM

Team Charter

It is the responsibility of the Salvage Team to mitigate damage at the primary site and to negotiate with the insurance company for settlement of all claims. This depends on a prompt realization of what is salvageable and what is not. Repair and replacement orders will be filed for what is not in operational condition.

Typical Duties by Phase

Phase 2
- Assist in the immediate salvage operation
- Contact the insurance agent
- Inventory damaged and undamaged items

Phase 3
- Salvage equipment and supplies
- Settle property claims with the insurance company

Phase 4
- Settle extra expense and lost profit claims with the insurance company

Recommended Members

Coordinator: Director of Risk Management
Members: Field Engineer
Public Adjuster
Building Manager

Figure 3-12

required to secure required resources like spare parts. Then there is the time to implement the solution, and finally, to test the integrity of the system.

3.1.5 Notifications

When time allows, but certainly within the first few hours, there are a number of parties inside and outside the organization who must be informed of the disaster. A partial list follows:

Party	Who Breaks the News
Executive management	DRMT (administrator)
Board of directors	President
Customers	Public Relations Team
Public	Public Relations Team
Employees	Administrative Team
Insurance carrier	Salvage Team
Regulators	DRMT
Suppliers	Supplies Team
Backup site	DRMT
Off-site Storage	Off-site Storage Team

3.2 EDP TEAMS RECOVERY PROCEDURES

3.2.1 Mobilization

The DRMT will decide, for the subject disaster, the appropriate teams to mobilize. They will assemble at the command center and immediately assess the applicability of the previously completed contingency plans.

If applicable, the plan is immediately invoked. Interteam coordination is provided by the DRMT. The immediate goal is to get the backup site into operation as quickly as possible and within the maximum downtime parameter. The backup site activation teams (i.e., Off-site Storage, Software, and Operations) will perform these tasks.

3.2.2 Restart Processing

Assuming that the backup computer is to be completely assigned to backup operations, the computer will be configured to match requirements as closely as possible.

NEW HARDWARE TEAM

Team Charter

It is the responsibility of the New Hardware Team to order replacement hardware for the equipment damaged in the disaster. Hardware ordered may not be a one-for-one replacement, since this may be the best time for an upgrade, consolidation, and so on. All areas of hardware are to be dealt with, including:

- Mainframe
- Peripherals
- Data communications
- Terminals
- Micros
- Environmental control equipment

Typical Duties by Phase

Phase 2
- Obtain a list of damaged and destroyed equipment

Phase 3
- Decide on new hardware
- Order new hardware
- Install and test the new hardware

Phase 4
- Evaluate its performance

Recommended Members

Coordinator: Data Center Manager
Alternate: Information Systems Manager
Members: Field Engineer
Chief Systems Programmer

Figure 3-13

FACILITIES TEAM

Team Charter

The Facilities Team is responsible for reconstructing the partially destroyed site or for choosing and constructing a new site. The purview of this team includes the site in general, and in specific, the data center.

Typical Duties by Phase

Phase 2
- Assess the damage

Phase 3
- Decide on in-place replacement or a new site
- Assist in site preparation

Phase 4
- Assist in the move back to the primary site

Recommended Members

Coordinator:	Vice President of Administration
Members:	Building Manager
	Data Center Manager
	Staff Attorney
	Controller
	Contracted Architect
	Real Estate Agent

Figure 3-14

DATA PREPARATION TEAM

Team Charter

The Data Preparation Team is responsible for all activities related to the input of data. This will include restoring the data preparation area complete with data entry terminals. There may be a shift from on-line operation to off-line if complete recovery is delayed. Depending on how the firm is organized, this function may be carried out by the user or as an adjunct to data center operations.

Typical Duties by Phase

Phase 2
- Set up the data preparation environment
- Switch to off-line if necessary
- Handle data preparation

Phase 3
- Continue operations

Phase 4
- Continue operations

Recommended Members

Coordinator:	Data Preparation Supervisor
Members:	Data Preparation Employees

Figure 3-16

USER OPERATIONS TEAMS

Team Charter

User Operations Teams will be responsible for recovering the user departments. This includes recovery of source documents and other departmental records. The user team must restore office operations. This may include new office space and furnishings. The user will verify the database when it is declared recovered by the Applications Team.

Typical Duties

Phase 2
- Secure office accommodations
- Recover departmental records
- Identify source documents as processed or not processed
- Confirm the status of the database

Phase 3
- Ongoing degraded operations using backup site

Phase 4
- Assist in cutover to the restored primary site

Recommended Members

Coordinator:	Department Manager
Members:	Department Members

Figure 3-15

USER LIAISON TEAM

Team Charter

The User Liaison Team is responsible for coordinating all user activities with respect to the data center. Particularly, all priority issues will be resolved. This team will serve as a conduit for all communications to and from the backup site, freeing the operations staff.

Typical Duties

All Phases
- Communicate problems between the backup site and the user department
- Establish operations schedule for the backup site
- Resolve issues of priority
- Inform management of status and progress

Recommended Members

Coordinator:	Selected User Department Manager
Members:	User Department Managers
	Data Center Manager

Figure 3-17

ADMINISTRATIVE TEAM

Team Charter

The responsibility of the Administrative Team is to provide administrative support services to any team requiring such support. This includes the hiring of temporary help or the reassignment of other clerical personnel.

Typical Duties by Phase

Phase 2
- Notify all vendors and delivery services of change of address
- Contact all employees

Phase 3
- Process expense reports
- Account for the project cost
- Handle personnel problems

Phase 4
- Process expense reports
- Account for the project cost
- Handle personnel problems

Recommended Members

Coordinator: Personnel Department Manager
Members: Secretaries
Clerks
Personnel Department Staff

Figure 3-18

TRANSPORTATION TEAM

Team Charter

The Transportation Team will handle all transportation necessary during the disaster/recovery period. They will work with airlines, shippers, hotels, and so on to accomplish their task. Where travel is better handled by private car, they will arrange car pools (with reimbursement) and if necessary drive team members in their own cars.

Typical Duties by Phase

Phase 2
- Transport people to the backup site
- Transport off-site (and on-site) tapes to the backup site
- Transport disaster kits to the backup site

Phase 3
- Shuttle people, supplies, and material to and from the backup site

Phase 4
- Return all material to primary site

Recommended Members

Coordinator: Traffic Department
Members: Person(s) normally making travel plans
Drivers

Figure 3-20

SUPPLIES TEAM

Team Charter

The Supplies Team is responsible for purchasing all supplies during the disaster/recovery period. They are also responsible for the supplies in the disaster kit. The definition of supplies is extended here to include office furnishings for all departments dislocated by the disaster.

Typical Duties by Phase

Phase 2
- Distribute disaster kit supplies

Phase 3
- Order replacements
- Expedite shipments
- Ongoing distribution of supplies

Phase 4
- Restock supplies at restored site
- Restock disaster kit

Recommended Members

Coordinator: Purchasing Department Manager
Members: Data Center Clerk
Purchasing Department Staff

Figure 3-19

PUBLIC RELATIONS TEAM

Team Charter

The responsibility of the Public Relations Team is to pass appropriate information about the disaster/recovery process to the public, to customers, and to employees. Top management will want to give these groups reason to believe that the firm is viable and that everything possible is being done to mitigate losses and to ensure an early return to normalcy.

Typical Duties

All Phases
- Control information released to the public, to customers, and to employees
- Interface with journalists
- Publish employee letters
- Cajole customers, particularly the large accounts

Recommended Members

Coordinator: Public Relations Officer
Members: Personnel Staff
Union Officers
Public Relations Staff

Figure 3-21

AUDIT TEAM

Team Charter

The responsibility of the Audit Team is to verify that control mechanisms are providing data integrity even though there exists an emergency. Of particular concern is the condition of the databases that are recovered on the backup computer. All security measures will be constantly reviewed and criticized, with the expectation of immediate remedy.

Typical Duties by Phase

Phase 2
- Review procedures used to recover databases
- Audit databases and prove they are recovered

Phase 3
- Audit data security

Phase 4
- Prove databases are recovered properly at the primary site

Recommended Members

Coordinator: General Auditor
Members: Audit Staff
User Staff
Data Center Staff

Figure 3-22

In particular, the same level operating system and environment software will be loaded. All disk units can be configured to closely match the structure of the primary computer. Data communication, whether it be dial-up or lease line switch-over, will be brought up and tested.

The database will be recovered from the latest dump tapes. If the primary site is still somewhat functional, or at least intact, the current copy may be copied or disk media moved. If the database was open, it will require the normal restart procedure following any hard or soft fault. Database integrity checks should be made.

Applications software will be loaded from off-site tapes and tested via the recovered network. Further testing of data should be done by the user teams and audit team.

3.2.3 Interim Processing

After the applications have been restarted, quasinormal operations can resume. Operations will oftentimes be degraded, however, in the following ways:

1. Slower processor, which means slower response time
2. Slower processor, which means less capacity
3. Less disk space
4. Slower transfer rates
5. Less tape drives, which means single-threaded dumps
6. Slower line speeds, which means slower response times

Accordingly, the operations staff must be flexible and prepared to work long hours. Schedules may vary considerably from those at the primary site and may need to be redone from those envisioned in the contingency plan. Priorities can and should be preplanned, but will surely be changed as backlogs grow and the tensions of running degraded business operations mount. The User Liaison Team should balance the concerns of each user function.

If only a partial operations staff is provided at the backup site, management should consider augmenting the team to include the full staff or rotating staff to give a break to those who have worked long hours.

Tapes must be shuttled off-site from the backup site for the reasons already noted. If the backup site is remote, a new service must be chosen.

The contract for backup services generally has a fixed term or escalating fees to ensure that the client restores its primary site expeditiously so that other clients will have a backup site to use in case of their own disasters. It may be necessary to transfer to a second backup site, perhaps a cold site. If the necessity of a cold site can be foreseen, detailed planning must begin immediately after the backup site is stabilized.

3.2.4 Restoration

The Primary Site Recovery Teams (i.e., Salvage, New Facility, New Hardware) will begin evaluation within a few hours after the disaster. The Salvage Team will have primary responsibility for deciding whether the building and equipment can be repaired and, if so, how long that might take. Of course their decisions must be ratified by the DRMT, top management, and the insurance company. Regarding the insurance company, a public adjuster can assist in expediting those most critical negotiations and should be part of the salvage team (engaged in advance).

Decisions to be made include the following:

- Is the site of the data center to be repaired or replaced?
- Should any physical or environmental features be upgraded?
- What equipment is salvageable?
- Should the mainframe or peripherals be upgraded even if there is additional cost?

In general, you should count on only six weeks of operations at the backup site. But whatever the agreement calls for, it must be treated as the key parameter in the restoration process. If the constraint cannot be made, the new equipment can be moved to a cold site.

The restoration process should be carefully monitored. A PERT chart is a convenient tool to know whether you are on schedule.

After the physical restoration of the primary site is complete, the transfer of operations from the backup site to the restored site can begin. There are two main differences between this move and the earlier move to the backup site:

1. This move is not under emergency conditions.
2. All applications will be restored, not just the essential ones (although the essential ones will probably be moved first).

All the problems inherent in bringing up a site must be overcome. The environment must be checked and the new hardware must be assembled and tested. A formal acceptance procedure is recommended. By all means, get the hardware stable before proceeding.

A parallel environment can be set up at the restored site. When parallel runs indicate equivalency, the applications can then be transferred.

Applications that had been suspended must be restarted. They should be checked for integrity. Backlogged transactions will represent a substantial load on the restored site. Suspended reporting can be resumed as utilization levels off.

As the backup site is shut down, it is important to make copies of the production environment so that a quick return is feasible. The whole tape library will be transferred to the primary site and merged. It is important not to scratch tapes during this period. Off-site storage will resume at the primary site. A postmortem evaluation of the whole process can prove beneficial. Be sure to update the plan accordingly.

3.3 CRITICAL APPLICATIONS

3.3.1 Future Requirements

Critical applications and resource requirements necessary to support the continuity of business operations is the key ingredient for a successful disaster/recovery planning program. Backup strategies are specifically based on resources required to maintain normal business performance.

Management has the responsibility to determine those specific applications that are critical to the economic survival and business continuity of the organization.

Each organization will review key (critical) applications and determine what other support activities are required at the time of a disaster.

3.3.2 Applications Recovery Plans

Once the priorities (critical applications) have been established (including future needs), it is necessary to develop the applications recovery plans. These plans outline in detail what the specific requirements are to assure the maintenance and continuity of business performance.

The general methodology for the applications recovery plan is reviewed in detail, covering but not limited to:

- Application system
- Report or activity frequency
- On-line or batch
- Areas of exposure
- Unacceptable period of loss of availability
- Potential losses
- Prioritizing (including subsystem)
- System contact
- User contact
- Basic processing requirements
- Minimum hardware requirements
- Terminal requirements
- New developments

3.4 APPLICATIONS RECOVERY PLAN REQUIREMENTS FOR DATA CENTER PRODUCTION APPLICATIONS

An essential element of minimizing the business disruption that could result from a disaster in the data center is to be able to rapidly and effectively recover all production applications. The proper state of preparedness is to have an approved applications recovery plan for every application. An effective plan will require the coordinated efforts of the user department responsible for a system, the MIS (management information system) application team supporting the system, and the data center.

To reduce the level of business risk associated with the lack of a recovery plan, it is imperative that a completed and properly approved applications recovery plan for every production application be prepared and submitted to the data center (or disaster/recovery administrator) by _____, 19____. All systems without an approved plan by that date will have to be considered "development" work, with ultimate reassignment to the development priority queues. In accordance with the data center disaster/recovery plan, the data center will provide only standard file retention and backup for this development work, and recovery responsibility will lie totally with the MIS/user department(s) doing the work. At the same time, the data center will follow up with higher-level management regarding perceived production applications that are missing an application recovery plan in order to secure the completed plan.

The appropriate assumption to make in starting your application recovery plan is to assume a "worst case" situation—that the data center has been totally destroyed. You may assume that the off-site storage location is intact and that the data center will obtain the backup mainframe. For this scenario, the plan should specify a recovery process that coordinates source documentation media storage (the user department's responsibility) with the data center off-site file retention, and with consideration given to the processing cycle of the system. More details on factors to consider in your plan are outlined in Section 3.5, "Data Center File Retention Philosophy." This information provided is a summary of standard retentions.

Special consideration must be given to systems utilizing data from other facilities for information input. In your plan, you may assume that all other facilities are unaffected by the disaster. This allows the possibility that recovery would involve getting a resubmission of information. The cost benefit trade-offs need to be reviewed for each system by the appropriate MIS and user managers.

3.4.1 Applications Recovery Plan

Specific requirements for your application recovery plan are as follows:

1. The plan must be signed by the user department manager and the manager responsible for providing MIS application support.
2. Database applications must also have the concurrence of the appropriate database administrator or management.
3. The plan is to be submitted to the data center management and the disaster/recovery administrator for review and approval.
4. Distribution of the final, approved plan is the responsibility of the appropriate user manager. The data center will place a copy of the plan in the off-site storage facility.
5. The plan should be updated as often as required when the "system environment" changes. As a minimum, annual verification of the plan will be required.
6. The cover memo for this current submission must include the following information:
   a. Identification of any nonstandard file retention requirements.
   b. Identification of any specific telecom requirements if the system is to run at a backup site.
   c. A list of all user codes (by mainframe), databases, and non-DMS files associated with the application.
   d. The work and home telephone numbers for the MIS and user managers responsible for the application.

If you have any questions about the requirements, please do not hesitate to contact the data center management team or the disaster/recovery administrator. Please refer to Section 3.5, "Corporate Data Center File Retention Philosophy," for information about current retention practices.

3.5 CORPORATE DATA CENTER FILE RETENTION PHILOSOPHY

3.5.1 Overview

The generation and storage of backup tapes for the purpose of data recovery and retrieval is but one of the many important functions performed by the data center. Since this process is both expensive and time consuming, care must be taken to ensure that only pertinent (critical) data is selected, and that the storage of this data is done for a prudent duration of time. Retention may be dictated by company policy, government regulations, or industry standards. These should be verified before any arbitrary retention requirements are established (see Section 3.5.3).

Although it may be argued that no amount of backup can be considered too much, the storage of files which have exceeded their usefulness is unnecessary overhead. This overhead should be minimized, if not totally eliminated. Insufficient backup, on the other hand, may be much worse than too much backup if valuable files are subject to total loss owing to a pack malfunction and/or a tape failure. Good judgment must be exercised in the establishment of file retention policies in order to avoid the cost and wasted manpower associated with the unnecessary storage of information.

To fully address the issue of data file backup and storage (both on-site and off-site storage), the risk/loss factor versus the cost of data re-creation and the cost of storage must be taken into consideration. The probability of totally losing the on-site facility is relatively small when compared to the probability of losing a file on a given tape reel. If a tape is important enough to be created and duplicated, with the original tape stored at the on-site facility and a duplicate copy sent to the off-site facility, the retention of the duplicate reel should coincide with the retention of the original reel. This does not necessarily mean that the duplicate reel must reside at an off-site facility, but rather that the duplicate reel is not scratched until the original reel has lived out its usefulness.

3.5.2 File Management

The type of backup media, dump frequency, storage location, and storage duration must be based on several factors. Legal requirements, for example, may dictate how long a given file is to be saved, the method used in creating the backup media, the frequency of creation, as well as the location where the media is to be stored. File type and activity is another factor in determining creation frequency, form, location, and duration of storage. Other factors to be considered include file interdependencies and criticality, as well as the cost for backing up and storing the file(s).

3.5.3 Legal Requirements

All user departments should periodically check with the legal staff to ensure that they are in compliance with legislation concerning file retention and storage for their particular job applications. Compliance failure, in some instances, may result in costly litigation, potentially leading to fines and/or criminal prosecution.

3.5.4 Interdependencies

File interdependencies must be considered when backing up files. For example, it would not be sensible to save DMSII Audit tapes for a duration greater than that of their corresponding database dump tapes and vice versa. When retiring a system, care must be taken to ensure that all necessary object code, source code, data files, and all necessary operating system software are properly backed up so that the retired system can be reinstated if the need ever arises. If a necessary component, such as the operating system software, is missing, the reinstatement of that system would be made much more difficult, if not totally impossible.

3.5.5 File Activity and Dump Frequency

File activity, in most instances, will dictate the dump frequency required to ensure that a given system may be recovered in a timely fashion. A DMS-based system, for example, which has a low rate of update activity, may require that it be dumped once a month, whereas a high-activity system may require a weekly dump frequency. A month's worth of audit activity on a low-activity database may be very reasonable for recovery. A month's worth of audit activity for a high-activity system may result in a lengthy recovery period, placing costly demands on both machine and human resources. Therefore, the data center has a flexible schedule relative to the frequency of DMSII-database backups and methods of recovery. For more details on this subject, refer to the user guide and/or contact the user support section. Incorporating a tape library management system (TLMS) with the off-site storage program will automatically provide all the file activity, dump frequency, and retention schedules.

3.5.6 Criticality

The criticality of a given system is a major factor to be considered when establishing backup criteria. If a system is critical to the continued operation of the corporation, factors such as cost, activity, and dump frequency may, at first glance, be out of sync with similar considerations for a noncritical system. The criticality of a given system may necessitate that certain exceptions be made. For example, a critical system which has low-update activity may be dumped on a weekly, rather than a monthly, basis. Although this may generate additional cost, the added dimension of backup may be well worth the extra cost incurred.

3.5.7 Cost

Cost is yet another consideration in establishing frequency and duration of storage. It would not make sense to spend more time, effort, and money backing up files which could be readily re-created at a cost which involved less time and money than the backup and storage of files. It is the responsibility of users, application analysts, and the data center management team to optimize file backup and storage at a minimum cost to the corporation.

3.5.8 Summary of Standard Retentions

At the data center, retention of magnetic tape media may range from a few days to several years. Files of a temporary nature may only be needed for a day or two. Certain year-end master files, residing on tape media, may have retention requirements set in terms of years, owing to various legal requirements. The user has ultimate control of all its file retentions. By virtue of selecting the proper pack family and using the proper naming conventions and/or file-kind attributes, or by completing a magnetic tape action document which indicates the desired retention via the comment section or preestablished retention schedules set up on the basis of tape name, this may be the cost center and system ID number. Although there are exceptions to any established rules, the following briefly summarizes the standards for file retention for the data center. (This is a generic concept and may not be applicable to your file retention philosophy.)

1. Standard production archives:
   a. Pack residency is from 32 days to 45 days.
   b. Tape residency/retrieval is for 4 months.
   c. File danger reports are generated.
   d. Authorized file kinds include:
      (1) Flat data files (i.e., non-DMSII).
      (2) Card-image files/report specification files.
   e. Unauthorized files include:
      (1) Source and object files.
      (2) Production work files.
      (3) DMSII database files.
2. Long-term production archives:
   a. Pack residency is for 32 days.
   b. Tape residency/retrieval is for 15 months.
   c. Tape residency without archive retrieval is for 69 months.
   d. File danger reports are generated.
   e. Authorized/unauthorized files are included (same as previously stated).
3. Standard development archives:
   a. Pack residency is for 14 days.
   b. Tape residency/retrieval is for 4 months.
   c. File danger reports are generated.
   d. All file kinds, except for DMSII database files, are authorized.
4. Long-term development archives:
   a. Pack residency may range from indefinite period of time (SRCLIB pack on system Z) to a maximum of 7 days (SOURCELIB pack on system B).
   b. Tape residency/retrieval is for 15 months.
   c. Tape residency without archive retrieval is for 69 months.
   d. There are no restrictions or naming conventions on the object and source code.
   e. Various naming conventions are required for storing files other than object and source files.
   f. DMSII data and printer backup files are not allowed.
5. Production DMSII packs:
   a. There is indefinite pack residency.
   b. Residency on tape is based on an "as-needed" basis. This is usually four generations or 4 months.
   c. Only DMSII file kinds are authorized.
   d. Files to be saved must be registered and properly logged in the data center and on the TLMS.
6. Audit packs (production and development):
   a. Residency on pack is largely based on available pack resources. Maximum pack residency is usually 1 week.

b. Residency on tape with archive retrieval capabilities is based on the life cycle of the DMSII database utility tapes (usually 4 months).
c. Only DMSII audit files are authorized.
d. File danger reports are not generated.
7. Data processing operations library packs (DPOLIB):
   a. There is indefinite residency on pack.
   b. Residency on tape is 5 weeks.
   c. There are no user archive capabilities.
   d. Authorized files include:
      (1) Object.
      (2) Work files.
      (3) Small, static data files (screen generators).
   e. Unauthorized files include:
      (1) Source code.
      (2) Large data files.
      (3) Any file which has not been properly registered.
8. Nonarchived development packs:
   a. There is indefinite residency for registered files.
   b. Packs are dumped weekly via library maintenance.
   c. All registered file directories are allowed.
   d. Residency on tape is usually 5 weeks.
   e. Unauthorized files include all files which have not been registered for permanent residency.
9. Private packs:
   a. There is indefinite residency on pack.
   b. Residency on tape is normally the responsibility of the owner.
   c. The dump cycle is controlled by the owner.
   d. Only files authorized by the owner of the pack are allowed. All others are removed.
10. Disk and resource packs:
    a. There is indefinite residency of all operating system files.
    b. Residency for usercoded printed backup files is 3 days.
    c. The maximum residency for user-coded printer backup files on tape is 7 days.
    d. There are no archive capabilities.
    e. Operating system files are dumped monthly.
    f. Various billing and sumlog files are dumped daily.
11. Year-end master files:
    a. The retention cycle is determined by the user department.
    b. Specially named tapes which have long-term retentions established may be created.
    c. The user may specify special retentions for any tape currently in the tape library.
    d. The data center is responsible for making sure that the specified tapes are in fact retained for the period requested by the user department.
    e. Special retentions may be implemented for all data tapes including DMSII, unlabeled reels, and all library maintenance reels.

3.5.9 Conclusion

The data center has structured the operating environment of its computer mainframes to be conducive to the success of its user community. Users are provided with adequate computer resources, file security, media storage devices, usage guidelines for storage on pack/tape media, archival capabilities, as well as technical advice, support, and instruction. The operating environment is designed to meet all the users' day-to-day processing needs. There are ample facilities for the storage of necessary data and the ability to recover data which may be lost or corrupted as a result of hardware or software failures or accidental removal. However, storage of unnecessary files causes scarce pack resources to be unavailable for the use by other users and constitutes inefficient use of these resources. This inefficiency is compounded when these files are copied to tape for the purpose of file backup or when copied onto duplicate reels which are sent to the off-site storage facility.

Users should review their file residency requirements on a regular basis and remove unnecessary files. File retentions should likewise be reviewed and prudent retention schedules established and implemented. Since the data center cannot be expected to know all relevant file names and/or the content and/or the criticality of all files, users must work with data center personnel and the MIS application analysts to establish optimum retention schedules which meet their needs. Construction of application recovery plans and the requirements and facilities necessary to perform critical information processing operations can be developed using Figures 3-23, 3-24, and 3-25.

3.6 END USERS RECOVERY PROCEDURES

3.6.1 Outage Period Processing

Assuming that the data center is nonfunctional at the primary and backup sites, users will have to cope the best they can. This is very difficult for on-line users like the teller operations at the bank. With luck, the backup site will begin operations shortly.

It is important to get an estimate of downtime from operations so that an appropriate recovery strategy can be invoked. If the computer is used for making inquiries (as a bank does for account balances), you can reference the same information, although not as current, on a batch report. The last batch report produced will probably be useful only for a few days, depending on transaction volume. Some organizations go as far as printing the data on microfiche and providing quick access via various indexes.

If the application is an update process, transactions can be captured presently and updated later. This, of course, is true only if the transactions are basically independent of the database. "Capturing" can be manual or by use of an intelligent terminal with a memory buffer (i.e., diskette).

If there is a need to return to manual transactions, a

CORPORATE CONTROLLER'S ACTIVITY
Application Recovery Plan

Department: _____

Department Location: _____

System/Application: _____

Machine ID: _____

Section 1. General System/Application Information

1. Name and phone numbers (work and home) of contact personnel.

	Work	**Home**

   User manager: _____ _____

   MIS manager: _____ _____

2. Business purpose and criticality of application system.

   a. Business purpose:
   _____
   _____

   b. Criticality:
   _____
   _____

   c. Priority:
   _____
   _____

   d. Other:
   _____
   _____
   _____

3. General system application description and requirements:

   a. Frequency of access (e.g., daily, weekly): _____

      Processing window: From _____ To _____

   b. Volume of activity (average number of transactions/time period):
   _____
   _____

c. Type of access: On-line _____ %   Batch _____ %

d. Type of usage: Inquiry _____ %   Update _____ %

e. Required operating environment:

   (1) Mainframe (e.g., B1000, B6000): _____

   (2) User-required equipment (e.g., terminals, printers):
   _____
   _____

   (3) Other (e.g., LINC software):
   _____
   _____

f. Current release level: _____

g. Latest release number and date: _____

Section 2. Processing Procedures in the Event of a Disaster

1. Manual processing.

   a. In the event of a disaster, the Manual Processing Plan will be put into effect based on one of the following conditions:
   _____
   _____
   _____

   b. The notification process for initiating this plan is as follows:
   _____
   _____
   _____

   c. Indicate the required actions and responsibilities to initiate and maintain manual processing prior to the restoration of automated processing:

   (1) Individual(s) responsible for action(s):

	Work	**Home**
User:	_____	_____
MIS:	_____	_____

   (2) Title and location of input and output document(s):

Title	**Locations**
_____	_____
_____	_____
_____	_____

   (3) Specific communications warranted by the situation:
   _____
   _____
   _____

2. Application recovery.

   a. In the event that the data center is destroyed, the following actions are required to restore and maintain a minimal level of operations for the _____ system/application. This minimal level of operations is acceptable for _____ (time period):

   Actions   **Responsibility**
   _____   _____
   _____   _____
   _____   _____

   b. Each group will be requested to list its critical applications, and using the supplied application recovery plan, develop one or more plans during the exercise. This is performed to educate data processing and user management as to recovery requirements and to ensure that no critical application or any critical component of the application performance is omitted.

Figure 3-23

APPLICATION RECOVERY PLAN WORK SHEET
(****Name of Company Line1****)/(****Name of Division Line1****)
(Name of Company Line2)/(*Name of Division Line2)

I. General system/application information:

   A. Department name and location: _____

   B. System/application name: _____

   C. Machine ID: _____

   D. Name and phone numbers.

      1. User manager   Work: _____ Home: _____

      2. MIS analyst/manager Work: _____ Home: _____

   E. Business purpose:
      _____
      _____
      _____

   F. Criticality: high _____ medium _____ low _____

   G. Maximum tolerable downtime: _____ hours

   H. General application requirements:

      1. Frequency of access: daily _____ weekly _____ monthly _____

         Other (Explain) _____
         _____

      2. Critical processing window(s), days/times:
         _____
         _____
         _____

3. Volume of activity:

   a. Average on-line input transactions per day _____

   b. Average batch input per day _____

4. Type of access: Batch _____ %   Online _____ %

5. Type of usage: Update _____ %   Report _____ %   Inquiry _____%

6. Operating environment:

   a. Machine type (B5000, B6000, B7000, etc.): _____

   b. Minimum terminal requirements: _____

   c. MCP/MCSs (JCLs) used:

     (1) MCP Level _____

     (2) CANDE _____ Level _____

     (3) GEMCOS _____ Level _____

     (4) BNA _____ Level _____

     (5) Other _____ Level _____

II. Processing procedures during a disaster:

  A. Manual processing requirements:

    1. Title of individuals performing the process:

    _____

    2. Title and location of I/O documents:

      a. Input location/title of document(s):

        (1) Primary location/title: _____

        (2) Secondary location/title: _____

      b. Output type and distribution:

      _____
      _____
      _____

  B. Application recovery (attach additional sheets if needed):

    1. Work files/program execution syntax(es):

    _____
    _____
    _____

    2. Order of execution:

    _____
    _____
    _____

3. Control checks/interdependencies:

_____
_____
_____
_____
_____

III. Data center information:
   A. Nonstandard file retentions (if any):

_____
_____
_____

   B. Documentation/source file title/pack residency:
      1. Machine ID for source code: _____
      2. User Code(s) used: _____
      3. Directory name(s): _____
      4. Pack name(s): _____
      5. Tape name(s): _____

   C. Telecommunication requirements for development:
      1. MCS(s) required: _____
      2. Terminal(s): _____

   D. Queues used: _____ Normal priority assigned: _____

   E. Production file residency (object/data):
      1. Production machine ID _____
      2. Object code/WFL(s):

User Code(s)/Directory Name	On Pack Family
_____	_____
_____	_____
_____	_____

      3. Data files (Type = Data, DMSII, Guard, Seqdata, etc.):

Type	User Code(s)/Directory Name	On Pack Family
____	_____	_____
____	_____	_____
____	_____	_____
____	_____	_____

4. Tapes required:

Name	Input	Output	Quantity

F. Interface/dependency requirements (i.e., special file/tape):

G. Nonstandard software requirements (if any):

H. Special requirements: (forms/carriage tapes/etc.):

Figure 3-24

procedure must be prepared ahead of time and personnel trained in its use. One thing that is often overlooked is the need for special forms. Since the original manual system of processing may have been superseded many years ago, the appropriate forms were probably thrown away. Perhaps new forms will need to be designed. Samples of old manual forms and forms used for the current automated system can be used as models. Remember that an adequate supply of forms for manual processing is necessary until a new order can be processed. Whatever the final design of a form is, it must be used as a source document to update the database once computer operations have resumed. Therefore, it should be laid out in a similar fashion to the update screen, and it must capture all required information.

Some processes will not be able to be performed with any realistic efficiency if the computer is out, as in the case of inventory "picking" in a large factory warehouse. Here the shelves are often randomized. The computer generates the pick list in shelf order for efficient picking; it would be very inefficient to pick manually in a random order. Assembly explosions to generate the lists are also computerized; it would be nearly impossible to explode assemblies manually. In short, the computer just cannot go down. The assembly process will stop once the buffer of "picked" parts is exhausted, so the backup site must be in operation by that time.

3.6.2 Restart Processing

When the backup site has been activated (which includes hardware, system software, and environmental software), it will be possible to restart the applications. First of all it must be proven that the current database (taken from the primary site) is in a "quiet" state where all available transactions, in progress at the time of the disaster, have been applied. This is handled by the database software. If a problem is found, the database must be rolled back to a point of integrity. This will probably involve loading a database generator stored on tape (in off-site storage) and perhaps applying audited transactions from a separate file.

Once internal database integrity is confirmed by the database software, the user should run certain database totaling reports to confirm that the records have the values expected. For instance, if the database totaled to $1,000,000 two days before the disaster and $25,000 was added for each of the last two days, then the database should have $1,050,000. There is no substitute to which historical control totals can tie.

Auditors use integrity tests independently to assist in finding mistakes and also to spot any efforts to defraud. The information systems are quite vulnerable at this point. The only saving grace is that the disaster probably caught potential thieves unaware. This added scrutiny of the database may uncover frauds in progress as well.

The database specialist will note the time of the last successfully applied transaction and its transaction number. The user must relate these numbers to the source transactions in process within the department. Any transactions after the cutoff point must be reentered.

If transactions were done manually or in an off-line mode on a smart terminal during the outage period, these transactions must be submitted. A problem occurs if these transactions are rejected. A simple transcription error is easy to fix, but if the transaction is incompatible with the database (e.g., an employee number is not found), then the reconciliation is more difficult.

IS65-300-107
Disaster Avoidance

Developing a Business Resumption Plan

TABLE 1. THE APPLICATION ANALYSIS

```
APPLICATION: _____     ACRONYM: _____
SYSTEM: _____     CRITICAL RATING: _____
APPLICATION RUN FREQUENCY: _____     TELECOMMUNICATIONS? _____
APPLICATION REPRESENTATIVE: _____
USER REPRESENTATIVE: _____

Does the application control corporate assets/capital? _____
If yes, explain: _____
_____
_____
_____

If no, refer to SECTION 3.

Does it make good business sense to recover this application during the disaster recovery process? _____
If yes, explain: _____
_____
_____
_____

If no, refer to SECTION 3.

What is the bottom line impact to your company business if this application is not run either in whole or in part?
_____
_____
_____
_____

How soon must the process be recovered after processing has stopped at your data center to avoid a major business impact?
    Batch processing must be resumed within _____
                                                (Time Period)
    Telecommunications must be resumed within _____
                                                (Time Period)

Is the department or area that is requiring recovery willing to absorb the costs of production restoration within the time frame required?
_____
_____

Current cost to run this application?
    Daily? _____
    Monthly? _____

Is it necessary to run this entire application? _____
    If yes, refer to SECTION 1.
    If no, refer to SECTION 2.


SECTION 1. TOTAL APPLICATION RESTORATION

Determine current hardware/software configuration. Attach a hardware and software configuration diagram.

Can the application be run at a facility separate from the other applications that are being restored? _____
```

JANUARY 1986   © 1986 DATAPRO RESEARCH CORPORATION, DELRAN, NJ 08075 USA
REPRODUCTION PROHIBITED

Figure 3-25 The application analysis.

Since the application software used to process the outage period transactions is suspect (i.e., Is it the right generation?), each of the runs in the first cycle must be checked out with added emphasis. A sample of all transaction types should be traced through to the output reports. The auditor may be of assistance here as well.

3.6.3 Interim Processing

During the time that processing is carried out at the backup site, the user should be especially cognizant of error messages. All problems should be investigated fully. No new procedures or transactions should be implemented—keep to the status quo.

There will be a significant handicap to this philosophy if

IS65-300-108
Disaster Avoidance

Developing a Business Resumption Plan

TABLE 1. THE APPLICATION ANALYSIS (Continued)

Would additional hardware/software considerations need to be made if run separately?
If yes, what kind?

_____

Must the application be processed in the same time frame during the business day that it is normally processed? _____
If yes, explain:

_____

If no, what other time frame would be satisfactory?

_____

Must the application be processed in sequence with another application?
If yes, explain:

_____

What is the maximum tolerable telecommunications response time acceptable? _____

What is the current response time (average)? _____

Refer to SECTION 3 to determine which alternate method of processing needs to take place within the department until the application can be restored.

SECTION 2. PARTIAL RESTORATION OF THE APPLICATION

Define that portion of your application that must be recovered

_____

Determine hardware/software configuration for retained portion and attach a hardware/software configuration diagram.

Does it make better business sense to run the retained portion at your recovery site vs another site apart from the location where you will run all other critical applications? _____
Explain:

_____

© 1986 DATAPRO RESEARCH CORPORATION, DELRAN, NJ 08075 USA
REPRODUCTION PROHIBITED
JANUARY 1986

Figure 3-25 (Continued).

---

the capacity of the backup site is less than that of the primary site. For cost control reasons, this is often the case. Less capacity will imply less applications running, and within those that are, a reduced reporting schedule. The "bells and whistles" features will be suspended. Hours for on-line updating will be reduced and/or shifted. Batch updates will be consolidated. In summary, only a core system will be run on an abnormal schedule.

Since computer time will be restricted, priority issues will surface. The users should have a master schedule showing required dates for all essential (i.e., mandatory and necessary) applications. The application which is furthest behind should get automatic priority. Desirable jobs would not be rescheduled until all essential applications are processed. The User Liaison Team is the perfect vehicle to handle the prioritization function.

54 PART 1

IS65-300-109
Disaster Avoidance

Developing a Business Resumption Plan

TABLE 1. THE APPLICATION ANALYSIS (Continued)

Does the retained portion need to be processed at the same time of day as required during normal processing? _____
Explain:
_____
_____
_____
_____

What other applications interface with the retained portion?
_____
_____
_____
_____

What is the maximum tolerable telecommunications response time acceptable? _____

What is the current response time (average)? _____

Must additional processing take place due to not running the entire application? _____
Explain:
_____
_____
_____
_____

Are additional file/system retentions required?
Explain:
_____
_____
_____
_____

Refer to SECTION 3 for an analysis of that portion of the application that you will not process.

SECTION 3. NO RESTORATION OF THE APPLICATION

What procedures are in place within the department to capture and manually process the data that is currently run on the mainframe?
Explain:
_____
_____
_____

Does this application provide data to other applications? _____
If yes, list the applications: _____
_____
_____

If yes, how will the department ensure that data is provided to the application? . . . and in what form? . . . in the same day or a day delay? . . . what is the impact to the other application for a day delay?
Explain:
_____
_____
_____

JANUARY 1986 © 1986 DATAPRO RESEARCH CORPORATION, DELRAN, NJ 08075 USA
REPRODUCTION PROHIBITED

Figure 3-25 (Continued).

IS65-300-110
Disaster Avoidance

Section 3 Recovery Management Plan and Procedures Development **55**

Developing a Business Resumption Plan

TABLE 1. THE APPLICATION ANALYSIS (Continued)

Does the department affected provide other departments with data normally obtained from the computer? _____
Indicate the departments: _____
_____
_____
_____

How will the other department now obtain the required information?
Explain: _____
_____
_____

TABLE 2. THE ESTIMATED LOSS/DP INTERRUPTION SUMMARY

APPLICATION GROUP	1st DAY	3rd DAY	1st WEEK	2nd WEEK	3rd WEEK	4th WEEK	OVER
FINANCE/ GENERAL LEDGER							
PAYROLL							
SALES							
MARKETING							
MANUFACTURING							
MATERIALS INVENTORY							
PURCHASING							
BILLINGS/ COLLECTIONS							
TOTAL LOSSES ($,000)							

EDITOR'S NOTE: This is an example of a form to determine the financial impact and the length of time that the company can continue to operate without the services from the data processing community.

© 1986 DATAPRO RESEARCH CORPORATION, DELRAN, NJ 08075 USA
REPRODUCTION PROHIBITED   JANUARY 1986

Figure 3-25 (Continued).

Section 4
IMPLEMENTATION AND TESTING

4.1 CONTINGENCY PLAN IMPLEMENTATION

4.1.1 Preparedness Standards, Policies, and Procedures

4.1.1.1 Forms Book

It is critical that every form used by the company be cataloged. Although the use of forms is seldom thought of as part of an application, forms are often the heart of any procedure, particularly those that are still manual. The catalog can be put in the form of a looseleaf notebook. It should include a sample of each type of form and provide the name of the supplier, his or her name and address, the reorder quantity, the rate of usage, and lead time. A copy of this book must be stored off-site.

4.1.1.2 Vendor List

Various vendors furnish the supplies and services that keep the data center running. Therefore, it is important to have a list of all such vendors, their addresses and phone numbers, contact names, and products supplied. For each product the reorder quantity, rate of usage, and lead time should be noted. A copy of this list must be stored off-site.

4.1.1.3 Vendor Commitment

The firm experiencing a disaster should expect expedited service from each of its vendors. If a particular item cannot be supplied from inventory, it should be provided from the stream of work in progress. A preferred policy is to supply "the next off the line." It is proper to ask vendors, particularly hardware vendors, for a commitment to this policy in writing. A vendor's response will, in the planning stages, provide a constraint on which strategies are viable. In the disaster stage, it may mean the difference between success and failure of the plan.

4.1.1.4 Disaster Kit

Despite all the promises from the various vendors, it will still take some finite time for them to respond. Therefore, it makes sense to have a reserve stock of essential supplies stored off-site. Such supplies might include:

1. Tape (it takes several hundred tapes to handle the extra dumps done at the time of a disaster)
2. Diskettes
3. Pack media
4. Stock paper
5. Computer forms
6. Manual forms
7. Tape cutter
8. Tape path cleaner
9. Cotton swabs
10. Lint-free cloths
11. Ribbons
12. Office supplies (pens, paper, and so on)

There should be enough stock stored off-site to use for processing until a new order can be received. In order to keep these supplies fresh, they should be rotated into the normal supply.

4.1.1.5 Purchased Software and a Software Escrow Program

Purchase software should also be cataloged, complete with contacts. The system and environmental software will probably be easily replaced should any be lost, since most of the vendors in this area are solid, financially stable firms. However this may not be true of the specialty application software. Any software purchased from small firms that are subject to financial failure should be protected by a software escrow agreement. Normally source code is not released with such packages. If the vendor does not exist, there will be no updates to the application even though your firm may wish certain enhancements and is prepared to pay for them. Without the source code, enhancements are impossible.

Accordingly, the vendor and your firm should contract with an escrow agent (a third-party trustee) to hold the source code and to release that code to your firm in the event of vendor insolvency. The following should be included in the escrow contract:

1. Name of the vendor, the vendee, and the escrow agent.
2. The conditions under which the source code will be available.
3. The place where the documentation will be held.
4. Media on which code is available.
5. Contents of the escrow package (all contents related to the current release), including:
   a. Source code.
   b. Object code.
   c. Job control language.
   d. User documentation.
   e. Proprietary documentation.
6. Remuneration for the escrow agent.

4.1.1.6 Application Recovery Plans

In order to recover applications one at a time, application recovery plans should be prepared. They will state all resource requirements (see Figure 1-4 for a profile). In addition, the recovery plans should state what tape contains the database to be loaded and what tape contains the object code and job control language. It may be advantageous to create the job control language to orchestrate the reloading and recovery process.

4.1.1.7 Calling Tree

A complete address book listing the address and phone numbers (home and work) of each employee must be maintained. In the time of an emergency, this information is invaluable. If the number of employees is large, it will be impossible for any one person to contact the whole group. Therefore, a calling tree should be used. One employee would call all department heads, for example, and they would call all section managers, who in turn call all their employees. Alternates must be designated for all nonterminal nodes of the tree (i.e., management). This same calling tree scheme can be used with the recovery team structure for quick activation of the recovery plan.

4.1.1.8 Employee Commitment Letters

Employees should be polled for their availability to relocate during disaster/recovery operations. Management will also want to know about extended work hours or shift work. It should be made plain to employees that management is seeking candid answers and that whatever their answers are, they will not be used in any negative way during performance evaluations. A sample letter to employees is shown in Figure 4-1.

4.1.1.9 Security of the Plan

The contingency plan itself becomes a sensitive document. It is the key to all implemented security measures and contains private information, such as employees' home phone numbers. Accordingly, the plan (or parts of the plan) should only be given to those who have a need to know the information contained within. A recipient of the plan should sign for his or her copy, pledge not to divulge any sensitive information, and return his or her copy should he or she be terminated or transferred.

4.1.2 File Backup
4.1.2.1 Reasons for File Backup

There are three reasons to back up files. Backup files (audit trails), full-system dumps of source and object code, and data are all vital to normal as well as disaster conditions within an information processing environment. Each of the three reasons presented here serve specific purposes and should not be confused.

Immediate restart, for example, is required to bring back an operating system in the event of a system crash. The off-site storage backups are used in the event of a disaster, when on-site backup tapes are not available or have been destroyed.

1. *Immediate Restart*. Examples: (a) In batch mode the update task aborts. The previous generation of the database is used to reload and rerun. (b) A check-point dump can be used to restart a run if the machine fails. (c) A database audit trail can be used to roll back to a point of integrity. (d) A previous generation of an object file is reloaded when the new version fails. Backup (audit trail) tapes are stored on-site for quick restart. Some files may be stored on disk instead of tape to hasten this process.
2. *Disaster/Recovery*. This involves the periodic backup of all files which are then removed to a remote location, a place which is not subject to a common disaster. Should the data center be destroyed, the off-site files could be used to completely reconstruct the operations environment.
3. *Archive*. Key moments in history, such as month-end reports, are preserved on tape for future uses, such as audit. Transactions representing flows of data can also be preserved. Laws, regulations, and company policy affect how long this information will be saved. The backup tapes can be stored on- or off-site in a secure, environmentally proper place. Since these tapes will be in storage for up to seven years (or longer), the environment is critical. If the information is truly essential to the firm, duplicate copies should be stored in remote locations.

When developing a file backup scheme, it is possible to let the same physical dump tape serve all three purposes. The primary use changes from (1) to (2) to (3) as the generation becomes less current. However, some tapes used for immediate restart may never be used for any other use, and some disaster/recovery tapes are of little historical value. Also,

DISASTER/RECOVERY PLAN
Confidential
Employee Questionnaire

TO: _____

FROM: _____

The organization's office and/or the data processing facilities may be subject to an unexpected or undesirable event that may render one or more of our buildings or areas within the building unoccupiable. In the case of data processing, employees may be required to perform their jobs at a remote location for an extended period of time.

It is the objective of this questionnaire to determine your availability and/or restrictions that could prevent your participation in such company activities.

Your inability to temporarily relocate or remain at a remote location for any period of time will not jeopardize your position with the company during the declared emergency. The company reserves the right to reassign employees to other positions and/or locations that will not affect any personal or family responsibilities.

1. I ___ am/___ am not available for relocation in the event of a disaster.

2. I ___ can/___ cannot remain away from home for a period of time.

3. I can remain away from home for

    ___ one day   ___ two to four days   ___ one week

    ___ one month   ___ more than one month

4. I must return home

    ___ once a week   ___ more than once a week

5. I can remain away from home for the periods indicated in (3), but require assistance to care for ___ children and/or ___ aged parent(s).

6. I will require additional financial aid to provide for home care assistance.

    $ _____   ___ per week   ___ per month

7. I will require transportation to the site selected by the company. (Local relocation or reassignment.)

    _____

    _____

8. I ___ have/___ do not have any credit cards that will permit company charges in the event that I have to be relocated for any period of time.

In the event there has been any change of circumstances, please advise your superior to amend this questionnaire.

Name: _____

Department: _____

Title: _____

Date submitted: _____

Date amended: _____

Figure 4-1

immediate restart tapes are of most value if stored on-site; disaster/recovery tapes are of no value if left on-site.

4.1.2.2 Number of Generations Saved

In general, enough generations of a file should be on backup tapes to:

1. Satisfy reporting needs (including legal requirements).
2. Provide an audit trail deep enough to precede the introduction of an error.
3. Provide alternative copies in case of tape parities.
4. Ensure that a recent enough version is in existence to simplify recovery procedures.

This would, of course, imply that the more generations the better. However, the number of generations is limited by concern for cost. Costs of backup include:

1. Computer time to complete the dump
2. Operator time
3. Tape costs
4. Tape storage costs

As an absolute minimum, three generations of all files must be saved.

File interdependency in setting the generation depth should not be overlooked. For example, it makes no sense to save a database for 10 generations and to not save corresponding audit trails and access programs.

4.1.2.3 Backup Strategy

There are several types of files to be dumped:

1. Object code
2. Source code
3. Job control language
4. System files
5. Database files
6. Audit trail files

Small sites may choose to back up all files by making a complete copy of all disk units, perhaps once per week. Larger sites will back up disk units at different times. By convention, a particular disk unit may have only one type of file. Still larger sites will back up each type of file with a different strategy. Often a two-tiered backup strategy is used. Once a week a complete dump is taken and then each day a backup of only updated files is done. Sometimes dumps are not driven by time but by a major event, such as a large update run. In any case, recovery will only bring a file back to its status at the time of the last dump. For most businesses this must be no later than the previous night.

4.1.2.4 Scratch Tape Management

Scratch tapes that are old and worn out should be retired from service. If a tape fails to be readable, recovery is complicated, made less complete, or made impossible.

The simplest method of scratch tape management is to provide a pool of tapes that is used in a round-robin fashion for a particular use. For example, the nightly object code dump could be put in a pool of five tapes, one labeled Monday, another Tuesday, and so on. By noting the date of issue on the tape, the number of writes can be easily calculated (i.e., one per week since issue). Standards for the maximum number of reads and writes (for example, 100) can be put in place. Therefore, after two years the tapes should be discarded.

Another way to manage tapes is with a tape library management system (TLMS). With such a system, all scratch tapes are pooled together. Reads and writes can be counted by the system.

Standards must be made for the maximum age of a tape, the maximum number of reads and writes before cleaning and before discarding, and what to do with a tape having a hard parity error.

4.1.2.5 Microcomputers

Microcomputers have either hard disks or diskettes which must be backed up. Similar strategies used in backing up mainframes should be employed for file backup on micros or PCs. It is important to remember that many of the company's most important databases exist only on micros or PCs often without adequate backup.

4.1.2.6 Hard Copy

Paper copies of files often provide the ultimate backup. These may be saved input source documents or output reports. Microfiche is frequently an alternative media to paper. In any case, hard copy of documents should be safeguarded by proper storage, indexing, and retention policies.

4.1.2.7 Retention Policy

All files, machine readable or not, should have proper retentions. File management firms often advise their clients to throw away up to 75 percent of their files, because the files are too old to be of value or they can be stored in a much more economic fashion, such as on microfiche or magnetic tape. The savings in reclaimed floor space is impressive.

4.1.2.8 Desirable Features of an Off-site Storage Location

Choosing the proper off-site location for the storage of backup files is essential. The site can be another facility within the firm or a commercial facility, but in either case the same features are to be desired:

1. Convenient location
2. Open 24 hours per day, 365 days per year
3. Prompt response time to requests for emergency delivery

4. Pickup at the hours requested
5. Bonded delivery people
6. Liability insurance
7. Stability (longevity, financial solvency)
8. Safe location (all hazards)
9. Excellent environmentals (especially for long-duration archival storage)
10. Good containers for storage and shipment
11. Good security (low-profile building, guards, and so on)
12. Fire protection (Halon, sprinklers)
13. Micrographic services (optional)
14. Optional hard copy storage (optional)

4.1.2.9 Tape Rotation Strategy

Assuming the current on-line generation of a file is x, you would want the latest backup, $x-1$, on-site to aid in immediate restart. However, the $x-1$ backup must be off-site to ensure recovery from a disaster. This would suggest that optimally two copies of backup files must be made.

At least three copies or generations of a file should be stored off-site. One is not enough because a parity error would render the file useless. Storing extra generations of tape on-site in a vault offers some additional protection, but it is not a substitute since a buildingwide fire would bake the tape inside, making it unreadable.

A two- or four-hour fire rating on a safe will ensure that the tape stored inside does not burn, but it will not ensure that it will be readable. Since safes often fall to the basement in a major fire, the tapes would be subject to the greatest temperature for the longest time and additionally would suffer water damage since the water used to fight the fire floods the basement.

The optimum scheme is shown in Figure 4-2.

4.1.2.10 Recovery Up to the Minute

Off-site tape storage will probably ensure recovery only until the previous night. If a firm wanted to, it could request dumps of the audit trail every hour and send them off-site, but there is a limit to using this strategy.

The easiest way to get up-to-the-minute recovery is to use microcomputers for terminals and tank all transactions. (Of course, the terminals must be remote from the computer site.) Then, after recovery of the center (i.e., operations at the backup site), transactions could be replayed (downloaded) to the main computer.

Another way is to send each transaction to a second site (which becomes a real-time off-site storage facility) via data communications after it is received. These transactions could be replayed after recovery.

4.2 PERSONNEL TRAINING

4.2.1 Awareness Training

All employees have some involvement in disaster/recovery planning. Certainly every employee is responsible for his or her own safety in an emergency. All employees must know whom to contact at the onset of an emergency. Many employees are expected to take an active role, as in fire fighting. Most employees work in a department that is computer dependent. Accordingly, certain procedures must be followed when recovering from a service interruption. Everyone must be aware of privacy and confidentiality issues.

Therefore it is a good idea to give all employees some awareness training in emergency response and disaster/recovery principles. The security officer and members of the Disaster/Recovery Planning Team are excellent choices for instructors. Surely two hours of training would not be inappropriate.

Posters could remind employees to observe safety regulations. A letter of support from the CEO would be helpful. An Emergency Procedures Manual should be prepared and disseminated. A short overview of the disaster/recovery plan would also be desirable. A policy should exist for the handling of confidential information.

4.2.2 Technical Training

Standards should be set regarding who receives various technical training classes. The following standards are suggested:

- Fire fighting — All security personnel / All EDP personnel / All management
- First aid including CPR — 1 out of every 10 employees
- Bomb threats — All security personnel / All receptionists
- Recovery procedures — Relevant team members

The fire department will gladly demonstrate the use of fire extinguishers and explain what types (A, B, C) are to be used on which types of fire. The Red Cross offers courses in first aid. The police department and FBI are sources for information on how to handle bomb threats.

TAPE ROTATION TO OFF-SITE STORAGE	
Location	Generation
On-line	x
Vault on-site	$x-1$
Off-site	$x-1, x-2, x-3$
On-site	$x-4, \ldots$

Figure 4-2

4.3 VALIDATION TESTING

4.3.1 Scenarios Planning

The disaster/recovery plan is written for the worst case: total destruction of the EDP processing facilities (and perhaps user facilities if that is within the scope of the plan). It is assumed that all other less severe consequences can be handled as a subset of the worst case by only invoking those procedures (and teams) that are necessary.

All disasters fall into at least one of the following cases regardless of the immediate cause:

Case 1, damage to EDP facilities. If the damage is significant, then the decision will be made to go to the backup site immediately. Otherwise, the damage will be contained, and the center will operate at partial capacity. The viability of degraded configurations should be contemplated ahead of time.

Case 2, damage to terminal areas. Terminals can be set up in free office space, or users can find other terminals not in use.

Case 3, damage to the network. The network backup plan should be invoked for those facilities out of service. This will probably involve dial-up lines.

Case 4, damage to information services offices. These offices could be relocated elsewhere. However, day-to-day support will be hampered until terminals, listings, and other elements of the office environment are restored.

Case 5, damage to user offices. User offices can be relocated; phone number changes would have to be passed on to clients. The major loss would be any source documents unprocessed or any output documents for which there was no backup.

Case 6, areawide disaster. Business would be disrupted totally. No attempt to recover services could be made until the basics of commerce, such as transportation and utilities, are restored. If it would take some time to restore these prerequisites and the facility is a service center for other regions not affected, then some thought of moving the whole facility should be investigated. If this it not preplanned, then the business is in jeopardy.

A matrix can be set up as to which teams to invoke for each case. User teams have primary responsibility for disasters in user areas, but some assistance from EDP is advisable.

4.3.2 Preparedness Testing

Before doing a dynamic test of the plan, you should attempt to do a static evaluation, certifying the plan's completeness and correctness by logically analyzing its provisions and procedures. A checklist or questionnaire prepared by a third party will help in this process. A Preliminary Plan Evaluation Questionnaire is given in Figure 4-3.

4.3.3 EDP Restart/Recovery Testing

Dynamic testing involves actually doing part or all of the recovery procedures outlined in the contingency plan. This is crucial. *A plan that has not been tested is no plan at all.* This strong statement can be made because almost all plans fail their initial tests. There are just too many variables. In disaster/recovery operations, "a miss is as good as a mile." If you do not have the right generation off-site, the data is irrevocably lost.

A set of tests should be outlined from the basic to the full test. The complete series should be completed within one year of the plan's implementation. The following list will suffice as a guideline:

1. The on-site one application test
2. The on-site network backup test
3. The on-site all applications test
4. The surprise test
5. The off-site one application test
6. The off-site all applications and network test

4.3.3.1 The On-site One Application Test

This is the most basic test. The program, data, and job control language files are copied to a backup tape and set aside. All files are then removed. The files are reloaded from backup tapes normally stored off-site. The application is initiated. The status of the database is confirmed to be as recent as possible. A benchmark of transactions and reports is run. The system is shown to run normally. The versions of all files are compared to those live at the time as a double check.

4.3.3.2 The On-site Network Backup Test

The primary leased-line network is assumed down. This assumption is realized by turning off each incoming modem or by disconnecting its line. The alternate (probably dial-up) network is enabled. The terminals attached are enabled, and a normal transaction stream is initiated. The backup line should be able to handle a normal load. Although all lines may not be able to go to backup simultaneously, there should be provision for a significant number of lines to run concurrently. Each line must be tested. The user must know how to do his or her part at the remote end.

4.3.3.3 The On-site All Applications Test

All applications are dumped to a safety tape. All other files are also saved. Each disk spindle is scratched and the system is cold-started. The operating system and environmental software is recovered using off-site tapes. Each application is recovered from backup tapes. All applications are verified as before.

PRELIMINARY PLAN EVALUATION QUESTIONNAIRE

1. Has the contingency plan been approved by top management?
2. Does top management actively support contingency planning?
3. Is there a policy regarding the security of confidential documents?
4. Is there a permanent committee overseeing contingency planning?
5. Are all types of disasters planned for in an Emergency Response Manual?
6. Does the backup site plan provide for recovery within the maximum tolerated downtime?
7. Does the Disaster/Recovery Management Team have a recognized charter?
8. Is the data center protected against intrusion?
9. Can all employees be reached within a few hours after a disaster?
10. What provisions have been made for those working alone?
11. Does the company have an employee training program in CPR, first aid, and fire fighting?
12. Do employees know how to handle a bomb threat?
13. Is the data center protected with Halon and/or sprinklers?
14. Was the last fire drill within the last year?
15. Is food and drink prohibited from the data center?
16. Is smoking prohibited in the data center?
17. Is there more than one power outage each month?
18. Is there redundancy and extra capacity in the air-conditioning system?
19. Is all computer equipment under a maintenance agreement?
20. Are sensitive documents shredded when discarded?
21. Is a user code/password system universally used?
22. Are passwords changed frequently?
23. Are tapes with sensitive information degaussed when put into the scratch pool or discarded?
24. Are sensitive datacomm lines scrambled?
25. Do dial-in lines call the originator back?
26. Are there enforced standards for system development and support?
27. Does the user understand that he or she has primary responsibility for data integrity?
28. Are all permanent files backed up on tape?
29. Is the latest generation off-site?
30. Are there at least three generations or copies of critical data bases off-site?
31. Are source documents cataloged and stored safely?
32. Are microcomputers, files backed up?
33. Is there a retention schedule for all files and documents? Is it followed carefully?
34. Are well-used or older tapes systematically removed from the tape library?
35. Has a backup site strategy been implemented for each essential application?
36. Has spare office space been identified in case of a disaster?
37. Has the full network been tested at the backup site?
38. Is there a list of all employee addresses and telephone numbers?
39. Can all employees be notified within a few hours?
40. Is all vendor information listed off-site?
41. Is there a one month's supply of data center supplies stored off-site?
42. Is there a book of all forms in use and their suppliers?
43. Is all third-party software under escrow agreements?
44. Have all applications been prioritized?
45. Is there a probable backup site schedule?
46. Can money or credit easily be advanced to employees in case of disaster?
47. Is all computer equipment insured?
48. Are profits protected with business interruption insurance?
49. Does each recovery team have a charter and mini-plan?
50. Are users heavily involved in disaster/recovery planning?
51. If the entire computer equipment configuration were to be destroyed, with what models would they be replaced?
52. What is the maximum tolerated downtime?
53. Who must be notified within 24 hours of a disaster?
54. Have recovery teams been trained?
55. Do all data center employees know how to fight a fire?
56. How many employees know first aid and CPR?
57. Have all critical applications been tested at the backup site?
58. Has the user participated in testing?
59. Who approves plan updates?
60. Is the plan formally reviewed at least once a year?
61. Was a report filed at the conclusion of each test?
62. Are there multiple copies of the plan off-site?
63. Is confidential plan information restricted to those who have a need to know?
64. When employees leave the company, are their copies of the plan returned?
65. Have all holders of the plan signed a security agreement?

Figure 4-3

4.3.3.4 The Surprise Test

All previous tests were scheduled ahead of time so as not to interfere with production. It is now time to call a test unannounced. This confirms constant readiness. Files will not be at convenient cutoff points. Some personnel will be unavailable or on vacation.

4.3.3.5 The Off-site One Application Test

All aspects of the backup site test should be preplanned, since the cost of testing goes up considerably without such preparation. The backup site may be quite remote. Computer time is no longer "free." It is easy to forget to bring the correct files from the off-site storage vault. The network at the backup site must be made to work. One wrong generation and the test will fail.

4.3.3.6 The Off-site All Applications and Network Test

All backup files will be taken to the backup site. The operating system and environmental software will be brought up. Packs will be configured to the correct names. All applications will be recovered. The complete network will be brought up simultaneously. A full transaction load will be experienced. All results will be verified.

4.3.4 End-User Restart/Recovery Testing

The end user will participate in all the preceding tests. He or she will be responsible for doing anything that would normally happen at his or her location, including reconfiguration of datacomm for backup. He or she will be the party responsible for verifying the status of databases and the correctness of applications.

In addition, the user will be responsible for identifying to which point in the transaction stream the database has been recovered. This will involve procedures for batching and time stamping transactions and storing them for reprocessing if necessary. Each of the tests, particularly the surprise tests, will validate these procedures.

The end user should do one additional test. He or she should assume that his or her office has been destroyed. He or she should get copies of transactions stored off-site (perhaps at the branch of origin), identify the proper place to start, and then reenter the appropriate transactions. This implies that the user has a full recovery plan.

Section 5
ONGOING ADMINISTRATION AND PLAN MAINTENANCE

5.1 ONGOING TESTING

5.1.1 Test Plan

Ongoing testing of the plan will ensure that the organization is always ready for disaster. It is generally recommended that the plan be tested twice a year at the backup site. These tests can be one application, all applications, surprise, or not surprise. However, over the course of a year all essential applications must be tested.

It is best to put together a formal testing plan complete with dates and applications to be tested. It will be necessary to arrange time at the backup site well ahead of the test date. A weekend date will enable the testing not to interfere with normal operations, but it will not be as easy to get the full user community involved.

The following is a list of those items to be tested at least once a year:

1. Fire drill procedures
2. Use of fire extinguishers
3. Bomb threat procedures
4. Shutdown and evacuation procedures
5. Contents of the off-site tape library
6. Backup operations for all essential applications
7. Command center operations
8. Assembly of all disaster/recovery teams
9. Network Backup Procedures (e.g., dial backup)

It cannot be emphasized enough that the user community must participate in all testing procedures. It is their systems and their data.

For each test it should be decided *a priori* what the outcome is expected to be. This will make evaluation much more meaningful. A group should be appointed to evaluate all test results. This could be the Disaster/Recovery Planning Team or the Audit Team.

5.1.2 Results Analysis

In many ways the results of a backup site test are quite self-evident. Either the application recovered to the predetermined recovery point or it did not. The on-line transaction processors can be readily tested using live transactions. The report programs and database totaling programs can be run. All this should be done in parallel with the live database.

The most grievous fault is to be missing files in off-site storage. Data, programs, job control language, system software, and environmental software must match and must be the latest generation. Documentation should be complete, but the data center does run each day without reference to documentation. The most critical documentation is that used to recover the system, since recovery is not a procedure normally executed.

A log should be kept of any exceptions noted. A cause for each exception should be determined and accompanied by a recommended solution. Someone should be assigned to implement that solution. All implementations should be confirmed before the time of the next test.

5.2 ONGOING MAINTENANCE

5.2.1 Change Management

The contingency plan will require regular maintenance. First of all, many employees will change job assignments. These changes may or may not mean changes in recovery team assignments. It is recommended that all references to personnel include both name and title so that a replacement is immediately suggested. Applications will be added and removed. Procedures will be updated as testing identifies better ways to handle various issues. There are many security items to be checked. To serve as a reminder, a questionnaire is provided in Figure 5-1.

The Disaster/Recovery Planning Team is responsible for the maintenance of the plan as well as an ongoing facilities evaluation to determine any changes in the workplace or EDP environment. The interactive session (see Figure 5-4 on pp. 68–71) facilities evaluation assists in training personnel to identify threats and vulnerabilities. Any changes to the plan must be approved by the coordinator of that team using a sign-off log like the one in Figure 5-2.

The contingency planning ramifications of any implementation must be considered. It should be policy that any such

CONTINGENCY PLAN—UPDATE QUESTIONNAIRE

Date: _____

Quarterly = Jan/Apr/Jul/Oct   Semiannually = Jan/Jul   Annually = Jan

Administrative and Planning

Period	Question
Monthly	Terminations: 1. Are all terminations listed? 2. Have all keys and documentation been collected?
Quarterly	Has the disaster/recovery plan test schedule been met?
Quarterly	Have planned enhancements to the physical security system been made as scheduled?
Quarterly	Has the documentation off-site been checked?
Quarterly	Have the off-site requirements and contents been confirmed?
Semiannually	Has the disaster/recovery plan been tested?
Semiannually	Are the team assignments and telephone lists up to date?
Annually	Has the back-up site agreement been reviewed?
Annually	Has the off-site storage agreement been reviewed?
Annually	Are all purchased software products under a software escrow agreement?
Annually	Is the latest version of products in the software escrow?
Annually	Has the retention schedule been reviewed?
Annually	Has the backup site configuration been modified?
Annually	Has the computer insurance been reviewed?
Annually	Has the maintenance agreement been reviewed? (Are all units covered?)
Annually	Are employee commitment letters current?
Annually	Have new applications been added and are all disaster/recovery implications handled?
Annually	Is the disaster/recovery plan up to date?

Physical Security and Environment

Period	Question
Monthly	Has the training schedule been met?
Monthly	Are all alarms in working order?
Monthly	Have physical security breaches occurred? (If so, what preventative actions were taken?)
Monthly	Have new employees been instructed in fire safety?
Monthly	Are all emergency exits operable?
Quarterly	Are all locks in working order?
Quarterly	Has a fire drill been held?
Quarterly	Are the first aid kits full?
Semiannually	Are all fire extinguishers in working order?
Annually	Have all keys been changed?
Annually	Has the power quality been reevaluated?
Annually	Have air-conditioning requirements been reevaluated?

Software Security

Period	Question
Monthly	Have passwords been changed?
Monthly	Was the system log reviewed for security breaches?
Quarterly	Has old media been discarded?

Figure 5-1

UPDATE LOG			
Sections Affected	Description of Change	Date	Approved By

Figure 5-2

ramifications be addressed as part of the implementation process (i.e., required for acceptance) and not as an afterthought. This will make the job of maintaining the plan substantially easier.

5.2.2 Distribution

Copies of the plan must be in the hands of those who would use it. Since most disasters strike while the work force is not at work, each individual's copy would be best stored at home. Extra copies could be in a library at work and of course stored with other documentation off-site. All copies should be placed in a locked cabinet.

Since the plan contains much information of a confidential or private nature, those portions of the plan should be censored from some copies. Only those having a need to know should be privy to confidential information. The censoring function is the responsibility of the security officer. Of course the members of the Disaster/Recovery Management Team would require complete copies. The off-site copies should be complete as well. If adequate security of the plan is not provided, someone could use the plan to circumvent EDP security features.

The plan should be serialized and each holder should sign a confidentiality statement (see Figure 5-3). Upon termination or cessation of responsibility, the plan should be returned. No unauthorized copies should be made. Printing the book on colored (e.g., red) paper will discourage copies.

SECURITY AND CONFIDENTIALITY AGREEMENT

I _____

employed by _____

located at _____

have reviewed the confidentiality provisions of the organization's Prevention, Security, and Disaster/Recovery Plan and herewith agree to the following:

1. At no time will I divulge to any other individual the contents of the document(s) provided to me, unless such release of information is directly related to the performance of my responsibilities, and shall limit the dissemination of information contained in the referenced document to the specific authorization list.

2. At no time will I reproduce, transmit, or communicate any information covered by this confidentiality notice, or any other information that is considered confidential, secret, or proprietary to the organization and/or the plan herein referenced, to any other employee, contractor, or consultant (or any agency of the government) without the express written permission of an authorized officer of the organization. All authorizations *must* be in writing.

3. I understand that I am subject to disciplinary actions or discharge in the event of any breach of security or confidentiality herein described, or established in other provisions, policy, regulations, or communications of the company. The decision of the company to invoke disciplinary measures as herein described is final and cannot be waived by any member of the company.

Accepted by: _____

Date: _____

Authorized signature
Date: _____

Figure 5-3

INTERACTIVE SESSION:
Facilities Evaluation

(See Checklists Starting on Page 72)

I. Understanding the Threats, Hazards and Vulnerabilities, Risk, and Adverse Incidents That Contribute to Business Interruption

During this session, each organization will be provided with a methodology to view at their facilities in a manner not common to normal daily activity. Here you have the opportunity to look at the overall "security" of the business, and more specifically the information processing capabilities. The objective of this exercise is to create for management a greater sensitivity and understanding of the external and internal threats that could cause business discontinuance resulting from the loss of general facility availability or computer system failure.

Business interruptions and computer system failures may be disrupted by one of the following broad categories of threats:
1. Acts of God or nature
2. Errors and omissions—human imperfections
3. Deliberate acts of humans

II. Baseline Security

Further in this session, you will participate in an interactive session on baseline security. For the moment, it is important to observe the existing conditions and probable causes, and apply logical, common-sense strategies to ensure that adverse incidents, at least in categories 2 and 3, can be averted. Obviously, there are occurrences over which you have absolutely no control. This destruction is important in selecting the "prevention"/"protective" strategies that can be implemented to eliminate or mitigate an adverse incident.

III. Documenting Threats

Figure 2-4 shows a matrix of possible threats and their relative impact on the organization as well as computer operations. From this each organization must "individualize" its risk posture and their ultimate effects. Although we don't endorse the long-term risk analysis, each organization must consider the risks relative to its company and translate these risks into "maximum allowable downtime" and "cost of downtime." Both subjects are continuation sessions, based on the threats, hazards and vulnerabilities, risks, and adverse incidents that can create a condition of business discontinuity.

Exercise

There are three preliminary steps that have to be taken in documenting the potential threats, hazards and vulnerabilities, and the probability of an adverse incident that may affect one or more functional components of the organization.

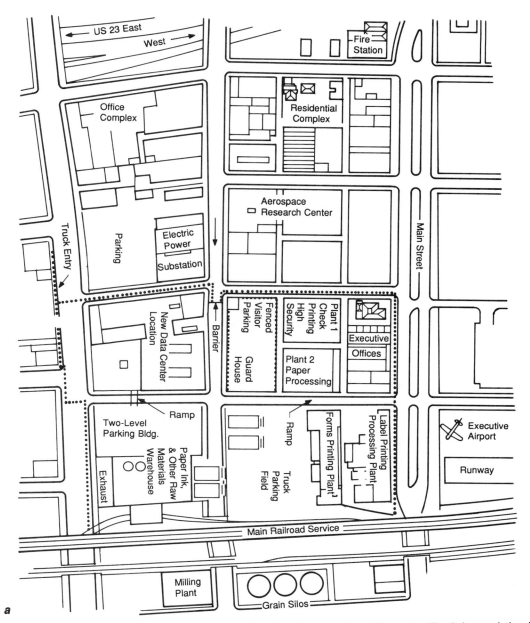

Step 1: Looking at the area. Figure 5-4a shows the method to view the area where the main office is located, the data center, main highways, traffic patterns, airports, utilities, dangerous industries, railroads, and so on.

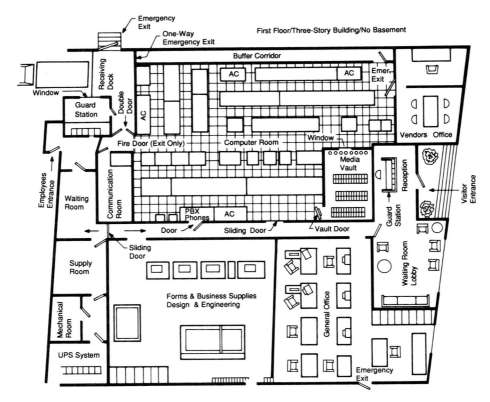

Step 2: Looking at the building. Figure 5-4*b* shows the building, the street(s) where it is located, the entries, loading docks, utility entries into the building, security features, parking lots or floors, power plants, HVAC, and communications.

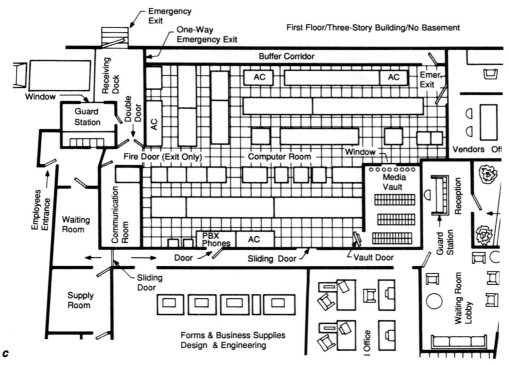

Step 3: Looking at the data center. Figure 5-4c shows the data center, the offices and activities surrounding the facility, and who is located above and below. Entries, security and detection devices, access control, and so on are also shown.

By reviewing each of the diagrams, it is relatively easy to understand those threats that can cause an interruption in the business activity or affect the timely performance of information processing.

It is also important to understand, as we proceed with the development of the plan, that the entire plan is based on a worst-case environment. Most planners realize that a disaster/recovery plan which becomes a single "monolithic" document will serve only the total catastrophic event. Understanding the individual threats and developing specific scenarios to respond to these threats help planners create a single plan that will cover numerous subplans.

In reviewing the diagrams you will come to understand how the various conditions surrounding or within the data center can create a catastrophic event; one that will assuredly interrupt the normal performance of the business.

1. *Identify.* Determine the potential threats.
2. *Estimate.* Use historical data and forecast how often the threat will occur.
3. *Calculate.* Determine losses by the multiplication of occurrences by the effective cost of the incident in business terms.
4. *Determine ALE.* That is, annual loss expectancy.

Figure 5-4

FACILITIES EVALUATION METHODOLOGY

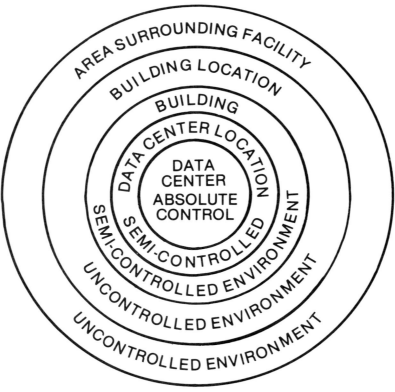

© Dia•log Management, 1987

Figure 5-5 Facilities evaluation methodology.

CHECKLIST 1			
Company and Data Processing Facilities Evaluation Checklists			
Several checklists are found on the following pages.			
Checklist	**Questions**	**Checklist**	**Questions**
A. Corporate Assets and Customer Services	12	L. Physical Security	3
		M. Security Planning	4
B. Vital Records	12	N. Software Considerations:	
C. Personnel and Facilities Protection	10	a. Controls	11
		b. Programming	14
D. Disaster/Recovery Plan	21	c. Documentation and Forms	10
E. Fire, Natural Disaster, and Safety Procedures	12	d. Data File Security	14
		e. Audit	9
F. Action Plans	10	O. Operations Considerations	18
G. Environmental Controls	10	P. Data Communications Considerations:	
H. Building Structure	10	a. General Security	14
		b. Terminal Security	16
I. Personnel Considerations	24	Q. Data Security Considerations	12
J. Facilities, Equipment, Programs, and Supplies	10	R. Communications Backup	16
K. Security	11		

A. CORPORATE ASSETS AND CUSTOMER SERVICES

Has senior management established a critical applications priority program to protect corporate assets and maintain the continuity of the business? If yes, then . . .

No.	Item	Yes	No	N/A
1	Does your plan provide for emergency recovery of critical applications?	☐	☐	☐
2	Have they in fact established a priority to implement the most vital programs?	☐	☐	☐
3	Does the data center have an approved list of priority applications?	☐	☐	☐
4	Do you have a facilities backup capability that can institute a high-priority operations within 24 hours?	☐	☐	☐
5	Are all user groups aware of the priority list of critical applications?	☐	☐	☐
6	Do the user groups agree with the priority ratings?	☐	☐	☐
7	Has provision been made to bring up on-line customer services within a reasonable period of time?	☐	☐	☐
8	Are cash deposits protected by making direct deposits?	☐	☐	☐
9	Is there a method to inform customers to make direct cash deposits?	☐	☐	☐
10	Is there a plan to control press releases in the event of a disaster?	☐	☐	☐
11	Is there a plan to protect the corporate stature and credibility in the event of a disaster?	☐	☐	☐
12	Is there a security plan to protect classified documents or vital records during an emergency when the organization is most vulnerable?	☐	☐	☐

B. VITAL RECORDS

Has the company implemented a vital records management program? If yes, then . . .

No.	Item	Yes	No	N/A
1	Is there a written vital records management program?	☐	☐	☐
2	Does the corporate policy conform to the requirements of federal, state, and local government regulations and laws?	☐	☐	☐
3	Is the corporate policy based on a careful review of its specific industrial requirements?	☐	☐	☐
4	Are vital records protected on-site? How? _____	☐	☐	☐
5	Is there an active program to protect and secure vital records and computer-generated data off-site?	☐	☐	☐
6	Is there an internal security control program for:			
	a. Hard copy?	☐	☐	☐
	b. Microfilm (microfiche)?	☐	☐	☐
	c. Software?	☐	☐	☐
	d. Critical records?	☐	☐	☐
	e. Computer data?	☐	☐	☐
	f. Historical records?	☐	☐	☐
	g. Classified information?	☐	☐	☐
	h. Department of Defense or government contractual requirements?	☐	☐	☐
7	Has the auditing department implemented controls to maintain an audit control on all computer-generated or stored information?	☐	☐	☐
8	Is documentation, run instructions, and so on, protected, duplicated, and secured off-site?	☐	☐	☐
9	Has the legal department approved the vital records and retention programs?	☐	☐	☐
10	Are computer data cycled frequently enough to ensure proper recovery in the event of an emergency?	☐	☐	☐
11	Is there a written prevention and protection plan in effect relating to:			
	a. Assets of the corporation?	☐	☐	☐

      b. Stockholders?
      c. Employee records?
      d. Tax records?
      e. Other vital records?
12 Is there a specific corporate policy relating to:
      a. Records retention?
      b. Records protection?
      c. Classification of records?
      d. Data processing documentation?
      e. What records are to be computer generated?
      f. What data processing records are contained in this information?
      g. How are they created?
      h. Who controls input?
      i. Who is responsible for data (information) dissemination?

C. PERSONNEL AND FACILITIES PROTECTION

Are there specific procedures for the protection of personnel and the facility in the event of an emergency? If yes, then . . .

No.	Item	Yes	No.	N/A
1	Has your entire staff been trained to properly respond to:			
	a. Environmental failure?	☐	☐	☐
	b. Weather conditions?	☐	☐	☐
	c. Fire?	☐	☐	☐
	d. Bomb threats?	☐	☐	☐
	e. Equipment failure?	☐	☐	☐
	f. Communications failure?	☐	☐	☐
	g. Power failure?	☐	☐	☐
	h. Other abnormal event?	☐	☐	☐
2	Has the senior data processing staff been properly trained to react to an emergency?	☐	☐	☐
3	Do personnel understand the procedures to institute in the event an alarm is sounded or an emergency is announced?			
	a. Do they immediately vacate the building?	☐	☐	☐
	b. Are they allowed to retrieve personal possessions?	☐	☐	☐
	c. Are they required to report to specific supervisory personnel?	☐	☐	☐
	d. Are supervisors assigned specific designated locations and actions?	☐	☐	☐
	e. Is management apprised of the situation (condition)?	☐	☐	☐
	f. Is there a second/third shift (or weekend) action plan?	☐	☐	☐
4	Do all staff members have specific calling instructions in the event of an emergency?	☐	☐	☐
5	Are there emergency phones outside the data center?			
	a. Are they toll-free phones?	☐	☐	☐
	b. Do they reach emergency services automatically?	☐	☐	☐
	c. Are all emergency alarm systems monitored by a central station system?	☐	☐	☐
	d. Has the response time in daytime hours, evening hours, and weekends been checked?	☐	☐	☐
6	Do emergency plans and procedures make protection of personnel their prime objective, including disaster/recovery team members?	☐	☐	☐
7	Has management approved all emergency plans?	☐	☐	☐
8	Is management notified in the event of any emergency? How? _____	☐	☐	☐
9	Are there provisions for securing the data center in the event of an emergency?			
	a. Internally?	☐	☐	☐
	b. Externally?	☐	☐	☐
10	Is there one person responsible and in authority on each shift to declare an emergency?	☐	☐	☐

D. DISASTER/RECOVERY PLAN

Do you have an effective disaster/recovery plan? If yes, then . . .

No.	Item	Yes	No	Don't Know
1	Does your plan cover the entire organization or only the data processing activity?	☐	☐	☐
2	Has the plan been approved by senior management?	☐	☐	☐
3	Has senior management provided the resources (funding and manpower) to implement the plan and run periodic tests?	☐	☐	☐
4	Is there a well-defined responsibility assigned to the disaster/recovery team?	☐	☐	☐
5	Are the procedures clear and simple enough for use during an emergency situation?	☐	☐	☐
6	Have all operating personnel been properly trained pertaining to their specific activities in the event of an emergency?	☐	☐	☐
7	Does the plan include disaster actions and recovery plans and backup for:			
	a. Data files?	☐	☐	☐
	b. Data backup?	☐	☐	☐
	c. Computer hardware?	☐	☐	☐
	d. Operating system software?	☐	☐	☐
	e. Application programs?	☐	☐	☐
	f. User requirements?	☐	☐	☐
	g. Personnel?	☐	☐	☐
	h. Other resources?	☐	☐	☐
8	Is the plan in an organized manual?	☐	☐	☐
9	Is the plan up to date?	☐	☐	☐
10	Is the plan revised when new resources are added to your system?	☐	☐	☐
11	Has the plan been tested? When? _____	☐	☐	☐
12	Are tests performed on at least the critical applications?	☐	☐	☐
13	Are the applications user and internal auditor involved in the test?	☐	☐	☐
14	Are results logged and reviewed?	☐	☐	☐
15	Do the people on the disaster/recovery team have assigned specific responsibilities? Are these responsibilities clearly defined by individual and position?	☐	☐	☐
16	Are all user groups briefed on emergency procedures?	☐	☐	☐
17	Has the plan been retested at least once?	☐	☐	☐
18	Do you consider the disaster/recovery plan appropriate for your organization?	☐	☐	☐
19	Is the plan, based on tests, adequate and workable?	☐	☐	☐
20	Do you have the continued support of senior management?	☐	☐	☐
21	Do you think the plan should be redone based on its inability to provide the recovery necessary in the event of an emergency?	☐	☐	☐

E. FIRE, NATURAL DISASTER, AND SAFETY PROCEDURES

Have the proper steps and personnel training been initiated to cover these risks?

No.	Item	Yes	No	N/A
1	Has the facility (plant, building, data processing center, etc.) been recently reviewed by OSHA, the local fire department, the loss-prevention engineers of the insurance companies, outside consultants?	☐	☐	☐
2	Is there a written report on their findings?	☐	☐	☐
3	Is there a person, in authority, responsible for the administrative and corrective measures necessary to correct findings?	☐	☐	☐

No.	Item	Yes	No	N/A
4	Is there a written emergency procedure for the following:			
	a. Equipment and environmental controls (air-conditioning, etc.) power cutoff?	☐	☐	☐
	b. Fire fighting?	☐	☐	☐
	c. Alerting police, emergency medical services (EMS), fire assistance, power company, communications, etc.?	☐	☐	☐
	d. Protecting and securing data files and programs?	☐	☐	☐
	e. Facilities evacuation and routes?	☐	☐	☐
5	Other than fire, are there procedures for action in the event of a natural disaster:			
	a. Hurricane?	☐	☐	☐
	b. Tornado?	☐	☐	☐
	c. Flood?	☐	☐	☐
	d. Earthquake?	☐	☐	☐
6	Has anyone contacted the various federal agencies to identify the threats and probabilities of:			
	a. Hurricanes?	☐	☐	☐
	b. Tornadoes?	☐	☐	☐
	c. Floods?	☐	☐	☐
	d. Earthquakes?	☐	☐	☐
	e. Rain?	☐	☐	☐
	f. Snow?	☐	☐	☐
	g. Other?	☐	☐	☐
7	Is there a specific individual who has been assigned the responsibility for supervising and training others in the performance of emergency procedures?	☐	☐	☐
8	Are there posted emergency procedures in the building and data center?	☐	☐	☐
9	Have the emergency procedures (all areas) been tested at least once this year?	☐	☐	☐
10	Has a test (or inspection) been held for the fire detection systems and fire extinguishing systems in the past six months?	☐	☐	☐
11	Have all below-floor drains (or pumps) been checked within the past six months?	☐	☐	☐
12	Is there a controlled-access system for admittance to the data center requiring positive identification or special access "key"?	☐	☐	☐

F. ACTION PLANS

Has the organization initiated a test program schedule to see if the "actions" designed for specific conditions work? If yes, then . . .

No.	Item	Yes	No	N/A
1	Are all on-site personnel trained and equipped to take immediate action against incipient or relatively small fires?	☐	☐	☐
2	Have personnel been trained as to the procedures to report fire, water damage, building damage, environmental failure, bomb threats, malicious or erratic employee actions, etc.?	☐	☐	☐
3	Are all operating personnel, on all shifts, assigned individual responsibilities in the case of any emergency?	☐	☐	☐
4	Are there fire detectors, intrusion detectors, water detectors located in the:			
	a. Area surrounding the data center?	☐	☐	☐
	b. Computer room?	☐	☐	☐
	c. Tape library?	☐	☐	☐
	d. Input control area?	☐	☐	☐
	e. Communications room?	☐	☐	☐
	f. Power input room?	☐	☐	☐
5	Are portable fire extinguishers placed in strategic locations in the:			
	a. Area surrounding the data center?	☐	☐	☐
	b. Computer room?	☐	☐	☐
	c. Tape library?	☐	☐	☐
	d. Input control area?	☐	☐	☐
	e. Communications room?	☐	☐	☐
	f. Power input room?	☐	☐	☐
6	Is there a central fire extinguishing system in any of the areas in (5a–f)?	☐	☐	☐

7. When was the last time these systems were tested under simulated circumstances?
   _____

8. If smoke detectors are used [in lieu of or in addition to fire (heat) detectors], have they been installed:
   a. In the ceiling? ☐ ☐ ☐
   b. Under the raised floor? ☐ ☐ ☐
   c. In air return ducts? ☐ ☐ ☐
   d. In the computer room? ☐ ☐ ☐
   e. In tape storage areas? ☐ ☐ ☐
9. Are all emergency detector and alarm equipment on a separate (uninterruptable) power supply? ☐ ☐ ☐
10. Do the annunciators (alarm bells) operate both at on- and off-premises locations (central station)? ☐ ☐ ☐

G. ENVIRONMENTAL CONTROLS

Do building maintenance personnel or vendors check the condition or serviceability of the environmental controls? If yes, then . . .

No.	Item	Yes	No	N/A
1	Does the data center have a separate air-conditioning (and humidity) control system, or is it part of the building's air-conditioning with separate air handlers?	☐	☐	☐
2	Is there an alarm system for the: a. Building system? b. Data center?	 ☐ ☐	 ☐ ☐	 ☐ ☐
3	Are the controls for the air-conditioning system secured, preferably located within the center?	☐	☐	☐
4	Are any of the duct linings of the air systems combustible?	☐	☐	☐
5	Are there permanent recording devices (disc or chart) that continuously measure: a. Temperature? b. Humidity?	 ☐ ☐	 ☐ ☐	 ☐ ☐
6	Has the environmental system been provided with an audible alarm system to indicate temperature and humidity limits, and if those limits have been exceeded?	☐	☐	☐
7	Has emergency lighting that operates in the event of a power failure been provided for the computer center?	☐	☐	☐
8	What systems does the emergency power support: a. Environmental? b. Hardware? c. Alarm systems? For how long? _____	 ☐ ☐ ☐	 ☐ ☐ ☐	 ☐ ☐ ☐
9	Can the air-conditioning system for the data center be quickly and easily shut down from within the computer center (whether separate from or part of the central system)?	☐	☐	☐
10	Are all safes and constructed vaults rated by either Safe Manufactures National Association (SMNA) or Underwriters Laboratories (UL)?	☐	☐	☐

H. BUILDING STRUCTURE

Do building maintenance personnel and administrator review the changing conditions of the facility and building periodically? If yes, then . . .

No.	Item	Yes	No	N/A
1	Has a review of the data center building been made by a professional to determine: a. Fireproofing? b. Waterproofing? c. Noncollapsible support? d. Vapor barriers?	 ☐ ☐ ☐ ☐	 ☐ ☐ ☐ ☐	 ☐ ☐ ☐ ☐

No.	Item	Yes	No	N/A
2	Is the data center located in a multiple tenant building? What are the threats and hazards from any of these areas? _____	☐	☐	☐
3	Are all fire exits equipped with exit alarms?	☐	☐	☐
4	Are all windows (if any) to the outside protected from breakage or intrusion?	☐	☐	☐
5	Are all exterior doors of sufficient strength to deter impulse intrusion (i.e., pushing)?	☐	☐	☐
6	Are all the entry doors to the data center fire doors?	☐	☐	☐
7	Is the data center protected by a solid wall on all sides?	☐	☐	☐
8	Is there a problem of water seepage: a. Under the floor? b. From sewer drains? c. From condensation or leaks from overhead or vertical pipes?	☐ ☐ ☐	☐ ☐ ☐	☐ ☐ ☐
9	Are there high-pressure steam lines inside of, or adjacent to, the data center?	☐	☐	☐
10	Are there any other water mains or pipes within the data center?	☐	☐	☐

I. PERSONNEL CONSIDERATIONS

No.	Item	Yes	No	N/A
1	Does the company maintain good employee relations to promote motivation and loyalty?	☐	☐	☐
2	Is there a career path plan for employees? Are they aware of it?	☐	☐	☐
3	Has management provided training sessions for employees? Has there been interest created in prevention of fire and theft?	☐	☐	☐
4	Have employees readily accepted their responsibilities and assignments pertaining to an emergency situation?	☐	☐	☐
5	Have data processing supervisors appointed alternates for all assignments in case of absence of the primary assignee?	☐	☐	☐
6	Are all assignment records kept up to date with original and copies?	☐	☐	☐
7	Are programmers, operators, and other data center employees rotated on jobs and programs and on different shifts?	☐	☐	☐
8	Are data center employees trained to undertake more than one responsibility and function?	☐	☐	☐
9	Are there preemployment investigations performed for all levels of employees?	☐	☐	☐
10	Are preemployment polygraphs given?	☐	☐	☐
11	Are periodic work habits and stability polygraphs given to employees?	☐	☐	☐
12	In the event of an incident, are specific polygraphs given?	☐	☐	☐
13	Are personnel debriefed prior to discharge or resignation?	☐	☐	☐
14	Are persons discharged (terminated) with cause required to leave the premises immediately?	☐	☐	☐
15	Do all employees, required to have ID badges, wear them at all times? Is this strictly enforced?	☐	☐	☐
16	Is access denied to a known employee in the event he or she does not have his or her ID badge?	☐	☐	☐
17	Is the personnel department immediately notified in the event of termination?	☐	☐	☐
18	Is security notified in the event of a termination?	☐	☐	☐
19	Is data processing management notified of a termination?	☐	☐	☐
20	Are ID badges collected from terminated employees before they leave the premises?	☐	☐	☐
21	Do you provide lectures and bulletins on disaster recovery, emergencies, and criminal acts?	☐	☐	☐
22	Is this security education continuing?	☐	☐	☐
23	Are all supervisory, programming, and operations personnel trained in backup and recovery procedures?	☐	☐	☐
24	Can operations personnel perform normal work flow in the event of an emergency?	☐	☐	☐

Section 5 Ongoing Administration and Plan Maintenance   79

J. FACILITIES, EQUIPMENT, PROGRAMS, AND SUPPLIES

Has every precaution been taken to protect the facility, equipment, programs, data, and supplies in the event of an emergency? If yes, then . . .

No.	Item	Yes	No	N/A
1	Has the organization an OSHA administrator?	☐	☐	☐
2	Has a fire (or other disaster) program been implemented under OSHA?	☐	☐	☐
3	Have you used the services that are provided by fire departments or safety loss prevention engineers of insurance companies?	☐	☐	☐
4	Have their recommendations been acted upon?	☐	☐	☐
5	Have the fire and safety systems in the facility (building and data processing department) been reviewed by an independent professional?	☐	☐	☐
6	Have the security systems been evaluated by an independent professional?	☐	☐	☐
7	Have programs been established with equipment and supply vendors as to their support and response time in the event of an emergency?	☐	☐	☐
8	Has there been a recent review of the documentation level of programs and the existence of updated backup copies of the programs and the documentation?	☐	☐	☐
9	Is there a complete listing of all supplies and an inventory of forms and supplies off-site?	☐	☐	☐
10	Can these be retrieved quickly, no matter what time or day?	☐	☐	☐

K. SECURITY

Is there a written security plan for the whole company and the data center? If yes, then . . .

No.	Item	Yes	No	N/A
1	Does the security plan cover external and internal security of the entire facility, including the data processing center?	☐	☐	☐
2	Is there a formal security manual, approved by management, which outlines the company's security procedures, defines responsibilities, and provides emergency actions?	☐	☐	☐
3	Is there a continuing problem of pilferage, shortages of materials, etc.?	☐	☐	☐
4	Has in-house security, if any, been involved in such instances?	☐	☐	☐
5	Have you, or the risk management department, performed a risk assessment program for the company; for the data center?	☐	☐	☐
6	Is there a formal group involved in prevention planning for the company and for the data center to reduce the possibility of:			
	a. Fire?	☐	☐	☐
	b. Theft?	☐	☐	☐
	c. Power failure?	☐	☐	☐
	d. Communications failure?	☐	☐	☐
	e. Other hazards and vulnerabilities?	☐	☐	☐
7	Is there a formal procedure which outlines procedures and assigns responsibilities for			
	a. Education?	☐	☐	☐
	b. Supervision?	☐	☐	☐
	c. Housekeeping?	☐	☐	☐
	d. Maintenance?	☐	☐	☐
8	Is there a separation-of-duties policy and specific responsibility with respect to internal controls involving corporate assets?	☐	☐	☐
9	Is the company located in a:			
	a. High-risk area (physical, environmental, personnel safety, etc.)?	☐	☐	☐
	b. Moderate-risk area?	☐	☐	☐
	c. Low-risk area?	☐	☐	☐

10	Is there a complete access-control system and procedure for the:			
	a. Grounds?	☐	☐	☐
	b. Building?	☐	☐	☐
	c. Data center?	☐	☐	☐
11	Does the company have a security force that obtains or provides professional training?	☐	☐	☐

L. PHYSICAL SECURITY

Is there a documented security plan for vital records? If yes, then . . .

No.	Item	Yes	No	N/A
1	Are all working areas and storage areas of the company protected by some form of fire extinguishing equipment (sprinklers, Halon, hand-held extinguishers, fire blankets)?	☐	☐	☐
2	Is there a safe or specially constructed vault to protect vital documents and/or magnetic media?	☐	☐	☐
3	Is access to these facilities limited to employees?	☐	☐	☐

M. SECURITY PLANNING

Does the company have a security policy in force? If yes, then . . .

No.	Item	Yes	No	N/A
1	Is there a person specifically designated as the security administrator for the:			
	a. Company?	☐	☐	☐
	b. Data processing?	☐	☐	☐
2	Is there a formal security staff?	☐	☐	☐
3	Is there a written, comprehensive plan in the event of a disaster?	☐	☐	☐
4	Does this plan provide for:			
	a. Quick decision and action plans?	☐	☐	☐
	b. Complete delegation of authority?	☐	☐	☐
	c. Vital records protection?	☐	☐	☐
	d. Recovery procedures?	☐	☐	☐

N. SOFTWARE CONSIDERATIONS
a. Controls

Is there a management approved procedure for internal controls? If yes, then . . .

No.	Item	Yes	No	N/A
1	Are there written guidelines for standards, specifications, system analysis, and programming functions?	☐	☐	☐
2	Have these guidelines been verified and approved by the audit department (internal and external)?	☐	☐	☐
3	Are the development of procedures and programmed controls sufficient and appropriate for the detection and correction of errors?	☐	☐	☐
4	Is there a procedure for verification of control totals with users? Is provision made to initiate corrective measures?	☐	☐	☐
5	Has the system designer identified the most critical files and established priorities in the event of an emergency?	☐	☐	☐

No.	Item	Yes	No	N/A
6	Are all the critical files backed up?	☐	☐	☐
7	Are all source documents secured after entry? Are there written procedures for their storage?	☐	☐	☐
8	Are critical files protected by written and auditing standards against:			
	a. Programming errors?	☐	☐	☐
	b. Operator errors?	☐	☐	☐
	c. System errors?	☐	☐	☐
	d. User errors?	☐	☐	☐
9	Are there established schedules for production runs?	☐	☐	☐
10	Is the schedule written out for each operator on a shift and day-to-day basis?	☐	☐	☐
11	Do all confidential or vital records covering cash receipt and disbursement applications (and other critical applications) have the following controls:			
	a. Total dollar controls?	☐	☐	☐
	b. Counts of documents?	☐	☐	☐
	c. Matching and sequence checks?	☐	☐	☐
	d. Batch controls?	☐	☐	☐
	e. Total run controls?	☐	☐	☐
	f. Hash totals?	☐	☐	☐
	g. On-line protection controls?	☐	☐	☐

N. SOFTWARE CONSIDERATIONS
b. Programming

Is there an organizational policy covering the application of new software? If yes, then . . .

No.	Item	Yes	No	N/A
1	Is a complete system flowchart prepared as part of the documentation before programming is initiated?	☐	☐	☐
2	Is all preliminary program documentation approved before programming is started?	☐	☐	☐
3	Has there been a system and specific rules for file handling?	☐	☐	☐
4	Is a final program listing:			
	a. Created at the time of program acceptance?	☐	☐	☐
	b. Held in secure storage?	☐	☐	☐
	c. Periodically checked against the operating program?	☐	☐	☐
5	If a change is required, are all program changes made at the source level, then recompiled and tested?	☐	☐	☐
6	Do audit, user, and data processing managerial personnel approve changes and check programs before entry into the program library?	☐	☐	☐
7	Do user personnel verify, test, and approve changes prior to placing programs into production?	☐	☐	☐
8	Is there an audit control trail printed out in all financial, scientific, or mathematically oriented programs?	☐	☐	☐
9	Are adjustments printed out in detail with the audit trail?	☐	☐	☐
10	Are all accounts payables printed out with the audit trail?	☐	☐	☐
11	Is there a special exception program control for unusual (excessive amount) payments in the audit trail?	☐	☐	☐
12	Have all programs placed into production tested with data with known output results?	☐	☐	☐
13	Is there a person (or group) responsible for maintaining the currency of all documentation?	☐	☐	☐
14	Does all documentation reflect the latest changes in systems and programs?	☐	☐	☐

N. SOFTWARE CONSIDERATIONS
c. Documentation and Forms

No.	Item	Yes	No	N/A
1	Are all system, program, and procedure documentation routinely and formally approved?	☐	☐	☐
2	Are there run manuals for all operating programs containing:			
	a. Problem definition?	☐	☐	☐
	b. System description?	☐	☐	☐
	c. Program description?	☐	☐	☐
	d. Operating instructions?	☐	☐	☐
	e. Listing of controls?	☐	☐	☐
	f. Acceptance record?	☐	☐	☐
3	Are all duplicate master run manuals kept in a secure location as well as in an off-site secure storage facility?	☐	☐	☐
4	Is there a systematic program change control?	☐	☐	☐
5	Is there a specific procedure for the user to sign off on all revised documentation and programs with details kept in a secure file?	☐	☐	☐
6	Is the documentation for programs kept on tape?	☐	☐	☐
7	In the event of an emergency, is there a procedure for manual entry and keypunching included in the documentation?	☐	☐	☐
8	Are all critical forms and blank checks secured in a locked storeroom or vault and accessed only by authorized personnel?	☐	☐	☐
9	Are all these forms checked properly, inventoried, and controlled?	☐	☐	☐
10	Is there a forms control program with management support?	☐	☐	☐

N. SOFTWARE CONSIDERATIONS
d. Data File Security

Has the data management department considered all the requirements necessary to secure the database? If yes, then . . .

No.	Item	Yes	No	N/A
1	Does the organization have large, critical, highly sensitive databases?	☐	☐	☐
2	Are the most important records maintained on magnetic files?	☐	☐	☐
3	Is there a hard-copy backup maintenance for all critical magnetic files?	☐	☐	☐
4	Has management established a database classification schedule?	☐	☐	☐
5	Does the data processing department classify files in accordance with management policy, such as critical, sensitive, important, useful, or nonessential?	☐	☐	☐
6	Does the data center have a database management system? Are all backup files secured and stored off-site? Are they:	☐	☐	☐
	a. Outside the data center building?	☐	☐	☐
	b. Accessible within reasonable (less than 2 hours) travel time?	☐	☐	☐
	c. Covered under another insurance risk?	☐	☐	☐
7	Is the data center:			
	a. Served by another power grid?	☐	☐	☐
	b. Served by a different municipal fire department?	☐	☐	☐
	c. Not subject to the normal municipal services deterioration (water main breaks, power failures, etc.)?	☐	☐	☐
	d. A facility where only authorized personnel are allowed to enter the media vault where backup files are stored?	☐	☐	☐
	e. A facility where company employees do not have access to backup data or know where it is stored?	☐	☐	☐
	f. A facility where your total anonymity is maintained?	☐	☐	☐
	g. A facility with 24 hour, 365 day emergency access and retrieval?	☐	☐	☐

No.	Item	Yes	No	N/A
8	Are security control entry and disaster/recovery procedures rehearsed at least twice a year?	☐	☐	☐
9	Are all off-site librarian and courier personnel tested at least once a year for fidelity, etc.?	☐	☐	☐
10	Are all delivery vehicles owned and environmentally controlled and secured?	☐	☐	☐
11	Are all vaults environmentally controlled?	☐	☐	☐
12	Does the data center maintain at least three copies of the most current generations (grandparent, parent, child) of all data files?	☐	☐	☐
13	Is the library (i.e., the data files and program storage facilities) segregated from the remainder of the data center in a backup or secondary storage site, as well as a third-party security vault?	☐	☐	☐
14	Is access to the library restricted to designated librarians?	☐	☐	☐

N. SOFTWARE CONSIDERATIONS
e. Audit

Has the data processing/audit department established adequate internal controls in data processing to ensure the integrity, security, and audit trail of data? If yes, then . . .

No.	Item	Yes	No	N/A
1	Is it the responsibility of, and has the data processing/audit group (internal and external) formally approved each new cash receipts of disbursements system? Has it established sufficient controls, division of responsibility, control totals, and other internal control features?	☐	☐	☐
2	Has the audit group (internal and external) reviewed and formally approved all other applications where potential for theft, embezzlement, or data manipulation exists? Has it established sufficient controls, division of responsibility, control totals, and other internal control features?	☐	☐	☐
3	Does the auditing department verify, through special audit programs, periodically, the results of critical applications (e.g., process payroll data through audit programs to check out the totals produced by the standard payroll program)?	☐	☐	☐
4	In regards to audit trails:			
	a. Are the programs designed with detailed audit trails that record each action?	☐	☐	☐
	b. Does the data processing/audit group specify and review all applications of cryptographic software protection and, if used, hardware?	☐	☐	☐
	c. Do all such applications automatically produce access journal (AJ) records?	☐	☐	☐
	d. Are the AJ records generated independently of system accounting data?	☐	☐	☐
5	Have the internal or external auditors recently reviewed the procedures of users and the control group to determine their adequacy?	☐	☐	☐
6	Has the auditing group reviewed all processing procedures to ensure that they are processed in a safe and sound manner?	☐	☐	☐
7	Has the auditing group had recent exceptions concerning operating procedures?	☐	☐	☐
8	Have revisions been made and are they being found acceptable? Has written approval for their use been granted?	☐	☐	☐
9	Has the auditing group verified that data processing conforms and adheres to management standards, policies, applicable law, and regulations?	☐	☐	☐

O. OPERATIONS CONSIDERATIONS

No.	Item	Yes	No	N/A
1	Are operator instructions clear and complete?	☐	☐	☐
2	Do data processing management periodically review and update operations instructions?	☐	☐	☐
3	Are complete run manuals readily accessible to operators?	☐	☐	☐

4. Can changes be made, or are they allowed to be made, or can operators examine production state data or programs from system consoles? If yes, who?
   a. Operators?
   b. System programmers?
   c. Other personnel?
5. Is there a training program for computer center personnel concerning the security and confidentiality of data?
6. Are several of the summary breakdowns of computer use compared to detect deviations from normal patterns?
7. Is there a regular production job schedule?
8. Is the operation properly supervised and monitored for compliance with the established schedule?
9. Are all periods of downtime verified?
10. Is a detailed schedule of a shift prepared by authorized personnel?
11. Is shift performance checked against the schedule to detect deviations?
12. Is security a part of the employee review and performance evaluation?
13. Is emergency "patching" permitted? If yes, who can authorize changes? _____
14. Is there a formalized procedure to assure that periodic checks are made to determine whether:
    a. Specific jobs were actually run by checking outputs?
    b. Individual run times conform to schedule?
    c. Frequency of runs of specific programs conform to schedule?
    d. Total time charged to a specific program over a given time period is reasonable?
    e. An unusual number of runs or amount of time is charged to reruns?
15. Has a determination been made as to the causes of reruns:
    a. Program errors?
    b. Input errors?
    c. Hardware errors?
    d. Operator errors?
    e. Unclear instructions?
    f. Power conditions?
    g. Other?
16. If there has been a large amount of rerun time, has the problem been identified and has corrective action been taken?
17. Are checks, purchase orders, invoices, and other sensitive forms and documents under tight physical and auditable controls?
18. Is there an updated distribution list with control of all reports sent out?

P. DATA COMMUNICATIONS CONSIDERATIONS
a. General Security

No.	Item	Yes	No	N/A
1	Is there a configuration chart showing: a. Location of terminals? b. Number of lines? c. Types of lines? d. Location of modems? e. Types of modems (by manufacturer/model)? f. Branches off the main? g. Cryptographic transformation facilities, if available? h. Locally developed software? i. Software that has been crypto-protected?			
2	Have you verified that reserved, unused lines are still active?			

No.	Item	Yes	No	N/A
3	Is there a written procedure for actions to be taken in the event of an emergency for the:			
	a. Lines?	☐	☐	☐
	b. Modems?	☐	☐	☐
	c. Terminals?	☐	☐	☐
4	Are users briefed as to emergency procedures?	☐	☐	☐
5	Are personnel trained in emergency procedures in a centralized location?	☐	☐	☐
6	Has the facility been checked to verify whether there are bugs or taps on any communication lines?	☐	☐	☐
7	Are incoming lines secured within the data center?	☐	☐	☐
8	Are incoming lines secured outside the building?	☐	☐	☐
9	Are lines switchable in case one or more of them are down?	☐	☐	☐
10	Are lines switchable:			
	a. Automatically?	☐	☐	☐
	b. Manually?	☐	☐	☐
11	Can provision be made for lines to be provided from two or more central stations?	☐	☐	☐
12	Did an auditor participate in the systems designs?	☐	☐	☐
13	Has the scattering or loss of source documents been discussed with the auditors? Has programming been supplied to give an audit trail?	☐	☐	☐
14	Is control and audit information supplied from both input transaction data and subsequent adjustment?	☐	☐	☐

P. DATA COMMUNICATIONS CONSIDERATIONS
b. Terminal Security

No.	Item	Yes	No	N/A
1	Are specific terminals, designated as security terminals, allowed to operate in the unattended mode?	☐	☐	☐
2	Are leased or dedicated lines used to transmit secure teleprocessed information?	☐	☐	☐
3	Are checks made to verify the security of teleprocessed information on dial-up lines?	☐	☐	☐
4	Is there a procedure for using terminal ID and operator ID as well as identification by password?	☐	☐	☐
5	Are codes changed frequently?	☐	☐	☐
6	Is there documented control of these codes or other security features in the teleprocessing system?	☐	☐	☐
7	Are the documented controls separate from the programming and operations group?	☐	☐	☐
8	Is there restricted access to specific areas of the system depending on a need-to-know basis?	☐	☐	☐
9	Are terminals locked when not in use?	☐	☐	☐
10	If an operator continues to cause a series of errors, irrespective of the type, what action is taken? _____			
11	Are the following hardware controls part of the security of the teleprocessing procedures:			
	a. Key locks?	☐	☐	☐
	b. System-controlled keyboards?	☐	☐	☐
	c. Operator ID (readers, i.e., card, voice, fingerprint, code)?	☐	☐	☐
	d. Alternate path and line configurations?	☐	☐	☐
	e. Multiplexing of transmission lines?	☐	☐	☐
	f. Synchronous transmission of data?	☐	☐	☐
	g. Parity checks?	☐	☐	☐
	h. Encryption devices?	☐	☐	☐
	i. Time-of-day control locks?	☐	☐	☐
	j. Identification and authentication checks?	☐	☐	☐

      k. Automatic callback procedures?
      l. Authorized activity checks?
      m. Security logs?
      n. Encryption algorithms?
      o. Automatic timed log off?

12. Is a security log kept to record unauthorized attempts to gain access to the system?
13. Is a log maintained to record users' attempts to gain information above their level of accessibility?
14. Is there a detailed transaction file kept for each application?
15. Is there a history of line penetration/interception by:
      a. Masquerading (unauthorized user attempting to gain access)?
      b. Eavesdropping (tapping or cutting into transmission lines)?
      c. Piggybacking (monitoring transmitted messages to intercept, modify, or replace)?
      d. Between lines (inserting a compatible terminal in the telecommunications lines)?
      e. Line grabbing (inserting a compatible terminal and waiting until signoff)?
16. Who is responsible for implementing security programs? Are the security programs accomplished upon notification of security breach?

___

Q. DATA SECURITY CONSIDERATIONS

No.	Item	Yes	No	N/A
1	Can master files only be updated from selected (restricted or security) terminals?	☐	☐	☐
2	Are terminal operators checked?	☐	☐	☐
3	Are computer room operators continuously checked?	☐	☐	☐
4	Is sufficient audit trail material retained?	☐	☐	☐
5	During outages, is there sufficient material retained to ensure that nothing is lost?	☐	☐	☐
6	If individual records are lost, is there a procedure for their reconstruction?	☐	☐	☐
7	Is vital master information duplicated?	☐	☐	☐
8	When all received messages are placed on file (tape, disk), does the received data contain:			
	a. The identification of the transmitting station (terminal)?	☐	☐	☐
	b. A message number or other means of identifying a message or the sender?	☐	☐	☐
9	Are all data communications cables and transmission wires in weather-tight conduits?	☐	☐	☐
10	Are all security lines inside (or outside) the buildings secured in conduits?	☐	☐	☐
11	Are all lines tested periodically to ascertain whether or not they are live and in good operating order?	☐	☐	☐
12	Does the center continuously switch (on a schedule) to reserve lines?	☐	☐	☐

R. COMMUNICATIONS BACKUP

No.	Item	Yes	No	N/A
1	Are written backup procedures at all terminal sites?	☐	☐	☐
2	Are the backup procedures periodically tested?	☐	☐	☐
3	Has the company established a priority listing on the data communications work?	☐	☐	☐
4	In the event of a disaster, are there alternate communications (incoming and outgoing) routes provided for, and can they be readily accessed?	☐	☐	☐
5	Have alternate routes been tested?	☐	☐	☐

6	Is there another method of moving documents, data, and so on (vehicle, air express, mail, etc.) in the case of an emergency?	☐	☐	☐
7	Are the alternate methods affected by weather, strikes, etc.?	☐	☐	☐
8	Is there a datacomm backup facility from which to operate in the event of an emergency or disaster?	☐	☐	☐
9	Is there a data communications plan and procedure?	☐	☐	☐
10	Is there a backup off-site that can deliver data communications programs and data to any alternate site within a reasonable (under 2 hours) time?	☐	☐	☐
11	Are there programs available that will handle data communications jobs in the batch mode on-site or at an alternate site?	☐	☐	☐
12	Have these programs been tested on both the primary and backup computer?	☐	☐	☐
13	Have backup processing methods at the terminal ends been considered?	☐	☐	☐
14	In an emergency mode, are there routines that provide for logging of all transactions in a form that will create an audit trail and that can be rapidly traced?	☐	☐	☐
15	Are there checkpoints in large programs where information is recorded on tape?	☐	☐	☐
16	Are all communications security provisions in place during an emergency operating mode?	☐	☐	☐

PART 2

EFFECTIVE DISASTER/RECOVERY PLAN: MODEL PLAN—TABLE OF CONTENTS REVIEW

OVERVIEW

The Table of Contents Review was written in lieu of providing the reader/planner with a "cookbook" plan. The cookbook approach has been specifically avoided for this handbook since the approach fosters a tendency to fill in the blank spaces and avoid the work required to "fingerprint" a plan for the organization. The plan you develop *must* be *your* plan, a plan designed for *your* organization; a plan that embodies the culture, idiosyncrasies, uniqueness, and operating style of *your* company. It must be based on *your* computer dependencies and respond to your time requirements to recover critical applications. Therefore, the plan is provided in concept form to assure that the interpretation of its components, paragraphs, decisions, and so forth, are created by the planning group.

Note: The numbering system incorporated in the Table of Contents Review is identical to the sections and paragraphs used for the Effective Disaster/Recovery Planning Workshop—Master Plan. Every main heading of the plan is shown in this review. Subparagraphs under the main headings may or may not be described in the copy in every instance. Where this occurs, the explanation under the main heading covers the details necessary to complete the subunits. Every aspect of the plan has been included, and intent and concept have been noted regardless of whether subparagraphs appear.

BUSINESS INTERRUPTION PREPAREDNESS PLANNING

Purpose of the Table of Contents Review

This document has been prepared to assist the disaster/recovery coordinator or workshop presenter and participants in understanding the relationship of the material requested to be assembled during the data collection phase and the actual plan to be written at the Plan Preparation Workshop.

The disaster/recovery coordinator at your preplanning seminar will assist you in relating the data collection requirements to the master plan components. The information in this document follows the format and table of contents of the master plan, and together with the data collection manual, the table of contents, and the supplementary data collection list, this phase of the disaster/recovery planning process should proceed quickly and easily.

DISASTER/RECOVERY PLAN

Table of Contents

[*Note:* The numbering system for this table of contents is independent of the numbering system used elsewhere in this handbook.]

1.	**Introduction**
1.1	*Purpose of the Plan*
1.2	*Plan Postscript*
1.3	*Management's Letter of Commitment*
1.4	*Information Security Policy*
1.5	*Personal Liability of Officers*
2.	**Preplanning and Assumptions**
2.1	*The Disaster/Recovery Planning and Data Collection Team*
2.1.1	*Charter*
2.1.2	*Members*
2.2	*Scope*
2.3	*Risk Assessment*
2.3.1	*Probabilities of a Disaster*
2.3.2	*Disaster Scenarios*
2.3.2.1	*Sample Scenarios*
	[Other scenarios will be added by the participants]
2.3.3	*Costs of Downtime*
2.3.4	*Assumed Downtime Maximum*
2.3.5	*Worst-Case Assumption*
2.4	*Other Manuals*
2.5	*Role of the Disaster/Recovery Management Team*
2.6	*Plan Initialization*
3.	**Prevention/Security**
3.1	*Physical Security*
3.1.1	*Layouts*
3.1.2	*Locks*
3.1.3	*Guards*
3.1.4	*Cameras*
3.1.5	*Telephone Emergency Numbers*
3.1.6	*Intrusion Alarms*
3.1.7	*Medical Alert*
3.1.8	*Access Control to the Data Center*
3.2	*Medical Security*
3.2.1	*First Aid Kit*
3.2.2	*Nurse*
3.2.3	*General Employee First Aid and CPR Training*

3.2.4	*EMS (or Fire Rescue Squad)*
3.3	*Fire Protection*
3.3.1	*Fire Alarms*
3.3.2	*Hand-held Fire Extinguishers and Fire Hoses*
3.3.3	*General Fire Safety*
3.3.4	*Passive Fire-Fighting Measures*
3.3.5	*Fire Drills*
3.4	*Emergency Response Plan*
3.4.1	*Contents of Emergency Response Plan*
3.4.2	*Emergency Response/Action Coordinator (Building or Plantwide)*
3.4.3	*Data Center Emergency Response—Miniplan*
3.5	*Housekeeping*
3.5.1	*Data Center Prohibitions*
3.5.2	*General Cleanliness*
3.6	*Power*
3.6.1	*Isolation Transformer*
3.6.2	*Motor Generator*
3.6.3	*Uninterruptible Power Supply (UPS)*
3.6.4	*Engine Generator*
3.6.5	*Reference Grid*
3.7	*Air-Conditioning*
3.8	*Maintenance Agreements*
3.9	*Document Security*
3.9.1	*Classification*
3.9.2	*Storing: Security and Retention*
3.9.3	*Shredding: Security*
3.9.4	*Forwarding: Security Methods for Distribution*
3.10	*Software Security*
3.10.1	*User Code/Password Protection*
3.10.1.1	*System Level*
3.10.1.2	*Application Level*
3.10.2	*Required Password Change*
3.10.3	*Other Application Security Measures*
3.10.4	*Pack Lockout*
3.10.5	*Program Library Control System (PLCS)*
3.10.6	*Erasure of On-line Copies of Sensitive Data*
3.11	*Datacomm Security*
3.11.1	*Physical Security of Terminals*
3.11.2	*Data Encryption Standard (DES) Devices*
3.11.3	*Leased Lines*
3.11.4	*Dial-back Feature for Dial-up Lines*
3.12	*Application Development Standards*
3.13	*Application Support Standards*
3.14	*Application Purchase*
3.15	*Role of the Auditing Staff*
3.16	*Role of the User Department(s)*
4.	**Disaster Preparedness**
4.1	*File Backup*
4.1.1	*Decision Criteria*
4.1.2	*Pack Family Usage Rules*
4.1.3	*Backup Strategies*
4.1.3.1	*Operating System Code*
4.1.3.2	*Application Code*
4.1.3.3	*Databases*
4.1.4	*Dump Schedule*
4.1.5	*Source Documents*

4.1.6	*Terminal Floppy Dumps*	
4.1.7	*Microfiche Reports*	
4.1.8	*Period-End File (Tape) Retentions*	
4.2	*Off-site Storage*	
4.2.1	*Off-site Management*	
4.2.2	*Off-site Strategy*	
4.2.3	*Off-site Rotation Schedule of Current Tapes*	
4.2.4	*Off-site Archive of Period-End Tapes*	
4.2.5	*Documents Off-site*	
4.2.6	*Tape Library Management System*	
4.2.7	*Scratch Tape Management*	
4.2.8	*Scratch Floppy Disk Management*	
4.3	*Backup Site Strategy*	
4.3.1	*General Alternatives*	
4.3.2	*Specific Alternatives*	
4.3.3	*Chosen Alternative (Period of Time)*	
4.4	*Mainframe Configurations*	
4.4.1	*Primary Site*	
4.4.2	*Backup Site*	
4.5	*Datacomm Configuration*	
4.5.1	*Primary Site Configuration*	
4.5.2	*Backup Site Configuration*	
4.5.3	*Additional Equipment Required*	
4.6	*Emergency Phone Numbers of Employees*	
4.6.1	*Alphabetical List of Phone Numbers*	
4.6.2	*Calling Tree for Disaster/Recovery Teams*	
4.6.3	*Calling Tree for All Employees*	
4.7	*Vendor Information*	
4.7.1	*Vendor Addresses, Phone Numbers, and Contacts*	
4.7.2	*Vendor Product List*	
4.7.3	*Vendor Support Letters*	
4.8	*Disaster Kit*	
4.9	*One-Month Supply List*	
4.10	*Forms Book*	
4.10.1	*Facility and Office Space Requirements*	
4.11	*Vital Record Retentions*	
4.12	*Software Escrow*	
4.13	*Employee Commitment Letters*	
4.14	*Priorities for a Limited Processing Environment*	
4.14.1	*Current Site Operations Schedule*	
4.14.2	*General Priority Scheme*	
4.14.3	*Probable Backup Site Operations Schedule*	
4.14.4	*Application Recovery Plans*	
4.15	*Logistical Considerations*	
4.15.1	*Credit Cards, Cash, Etc.*	
4.15.2	*Housing*	
4.15.3	*Local Transportation*	
4.15.4	*Office Space*	
4.15.5	*Control Center*	
4.15.6	*Employee Notification by Broadcast (and Other) Media*	
4.16	*Insurance*	
4.16.1	*Business Property Insurance*	
4.16.2	*Extra Expense Insurance*	
4.16.3	*Business-Interruption Insurance*	
4.17	*Disaster/Recovery Action Teams—Charters*	
	[See Figures 3-6 to 3-23 (Part 1/Section 3) for sample charters.]	
4.17.1	*Disaster/Recovery Management Team*	

4.17.1.1	*Team Charter*
4.17.1.2	*Typical Duties of the Disaster/Recovery Management Team*
4.17.1.3	*Role of the Coordinator*
4.17.1.4	*Role of the Administrator*
4.17.1.5	*Membership*
4.17.2	*Transportation Team*
4.17.2.1	*Team Charter*
4.17.2.2	*Typical Duties by Phase*
4.17.2.3	*Membership*
4.17.3	*Off-site Storage Team*
4.17.3.1	*Team Charter*
4.17.3.2	*Typical Duties by Phase*
4.17.3.3	*Membership*
4.17.4	*Supplies Team*
4.17.4.1	*Team Charter*
4.17.4.2	*Typical Duties by Phase*
4.17.4.3	*Membership*
4.17.5	*Administrative Team*
4.17.5.1	*Team Charter*
4.17.5.2	*Typical Duties by Phase*
4.17.5.3	*Membership*
4.17.6	*Software Team*
4.17.6.1	*Team Charter*
4.17.6.2	*Typical Duties by Phase*
4.17.6.3	*Membership*
4.17.7	*Applications Team*
4.17.7.1	*Team Charter*
4.17.7.2	*Typical Duties by Phase*
4.17.7.3	*Membership*
4.17.8	*Communications Team*
4.17.8.1	*Team Charter*
4.17.8.2	*Typical Duties by Phase*
4.17.8.3	*Membership*
4.17.9	*Operations Team*
4.17.9.1	*Team Charter*
4.17.9.2	*Typical Duties by Phase*
4.17.9.3	*Membership*
4.17.10	*Data Preparation Team*
4.17.10.1	*Team Charter*
4.17.10.2	*Typical Duties by Phase*
4.17.10.3	*Membership*
4.17.11	*Salvage Team*
4.17.11.1	*Team Charter*
4.17.11.2	*Typical Duties by Phase*
4.17.11.3	*Membership*
4.17.12	*New Hardware Team*
4.17.12.1	*Team Charter*
4.17.12.2	*Typical Duties by Phase*
4.17.12.3	*Membership*
4.17.13	*New Facilities Team*
4.17.13.1	*Team Charter*
4.17.13.2	*Typical Duties by Phase*
4.17.13.3	*Membership*
4.18	*Other Teams*
4.18.1	*User Team*
4.18.1.1	*Team Charter*
4.18.1.2	*Typical Duties by Phase*

4.18.1.3	*Membership*
4.18.2	*Audit Team*
4.18.2.1	*Team Charter*
4.18.2.2	*Typical Duties by Phase*
4.18.2.3	*Membership*
4.18.3	*Hardware Team*
4.18.3.1	*Team Charter*
4.18.3.2	*Typical Duties by Phase*
4.18.3.3	*Membership*
4.18.4	*User Liaison Team*
4.18.4.1	*Team Charter*
4.18.4.2	*Typical Duties by Phase*
4.18.4.3	*Membership*
4.18.5	*Security Team*
4.18.5.1	*Team Charter*
4.18.5.2	*Typical Duties by Phase*
4.18.5.3	*Membership*
4.19.1	*Third-Party Liaison Team*
4.19.1.1	*Team Charter*
4.19.1.2	*Typical Duties by Phase*
4.19.1.3	*Membership*
4.19.2	*Public Relations Team*
4.19.2.1	*Team Charter*
4.19.2.2	*Typical Duties by Phase*
4.19.2.3	*Membership*
4.19.3	*Alternate Strategies Coordinating Team*
4.19.3.1	*Team Charter*
4.19.3.2	*Typical Duties by Phase*
4.19.3.3	*Membership*
5.	**Disaster/Recovery Action Plans**
5.1	*Action Plan Overview*
5.2	*Definition of the Disaster/Recovery Phases*
5.2.1	*Phase 1. Emergency Response*
5.2.2	*Phase 2. Backup Site Activation*
5.2.3	*Phase 3. Recovery of Home Site*
5.2.4	*Phase 4. Return*
5.3	*The Master PERT Chart for Disaster/Recovery Operations*
5.4	*Elaboration of the PERT Chart—Phase 1*
5.4.1	*Disaster Strikes (Box 1)*
5.4.2	*Emergency Response Procedures Overview (Box 2)*
5.4.2.1	*Incident Reporting*
5.4.2.2	*Fire*
5.4.2.3	*Water*
5.4.2.4	*Power Failure*
5.4.2.5	*Environmental System Failure*
5.4.2.6	*Weather and Natural Phenomena*
5.4.2.7	*Sabotage (Causing Denial of Service)*
5.4.2.8	*Bomb Threat*
5.4.2.9	*Other Special Conditions*
5.4.3	*Notification of the Disaster/Recovery Management Team (Box 3)*
5.4.4	*Disaster/Recovery Management Team Meeting (Box 4)*
5.4.5	*Secure the Area (Box 5)*
5.4.6	*Assess Damage (Box 6)*
5.4.7	*Determine Downtime Estimate (Box 7)*
5.4.8	*Decision to Declare the Disaster (Box 8)*
5.5	*Elaboration of PERT Chart—Phase 2*

5.5.1	*Assemble Teams (Box 9)*	
5.5.2	*Activate Control Center (Box 10)*	
5.5.3	*Notify Other Parties (Box 11)*	
5.6	*Elaboration of PERT Chart—Phases 2, 3, and 4*	
5.6.1	*Transportation Team*	
5.6.2	*Off-site Storage Team*	
5.6.3	*Software Team*	
5.6.4	*Applications Team*	
5.6.5	*Communications Team*	
5.6.6	*Operations Team*	
5.6.7	*Supplies Team*	
5.6.8	*Salvage Team*	
5.6.9	*Administrative Team*	
5.6.10	*Data Preparation and Unit Record Team*	
5.6.11	*New Hardware Team*	
5.6.12	*New Facilities Team*	
5.6.13	*Disaster/Recovery Management Team*	
5.7	*Phase 4—Cancellation of Disaster/Recovery Mode*	
5.8	*Phase 4—Disaster/Recovery Postmortem*	
5.9	*Detailed Team Plans*	
6.	**Training for Disaster Recovery**	
6.1	*Recovery Team Training*	
6.2	*All Employee Training*	
6.3	*CPR and First Aid Training*	
7.	**Validation and Testing**	
7.1	*Validation*	
7.2	*The On-Site Application Test*	
7.3	*The On-Site Complete Test*	
7.4	*The Backup Site Test*	
7.5	*The Surprise Test*	
7.6	*Test Schedule*	
7.7	*Problem Log*	
8.	**Plan Update/Control**	
8.1	*Update Log*	
8.2	*Approval*	
8.3	*Formal Review—Once a Year*	
8.4	*Checklists*	
9.	**Plan Distribution and Security**	
9.1	*List of Recipients*	
9.2	*Screening of Information*	
9.3	*Serialization*	
9.4	*Termination or Transfer of Copy Holder*	
9.5	*Copies*	
9.6	*Physical Security of the Plan*	
9.7	*Security and Confidentiality Agreement*	
10.	**Planned Enhancements**	

Addenda

C-1	*Data Center Layout*
C-2	*Building Layout*
C-3	*Area Layout*
C-4	*Emergency Call List*
C-5	*Card Key Holders*
C-6	*Authorized Entrants*
C-7	*Authorizing Management*

C-8	*Data Center Sign-in Sheet*
C-9	*Evacuation Route*
C-10	*Copies of Maintenance Agreements*
D-1a	*Off-site Agreement*
D-1b	*Vendor Information*
D-2	*Tape Library Report*
D-3	*Backup Site Alternatives*
D-4	*Backup Site Agreement (Initial)*
D-5	*Backup Site Agreement (Secondary)*
D-6	*Primary Site Mainframe Configuration*
D-7	*Backup Site Mainframe Configuration*
D-8	*Primary Site Datacomm Configuration*
D-9	*Backup Site Datacomm Configuration*
D-10	*Required Ancillary Equipment*
D-11	*Home Phone Numbers*
D-12	*Calling Tree for Disaster/Recovery Teams*
D-13	*Calling Tree for Other Employees*
D-14	*Vendor List*
D-15	*Vendor Product List*
D-16	*Vendor Support List*
D-17	*Disaster Kit*
D-18	*One-Month Supply List*
D-19	*Vital Record Retention*
D-20	*Escrow Agreement*
D-21	*Purchased Software*
D-22	*Disaster/Recovery Plan Confidential Employee Questionnaire*
D-23	*Current Site Operations Schedule*
D-24	*Probable Backup Site Operations Schedule*
D-25	*Application Recovery Plan Worksheet*
D-26	*Copies of Policies and Annual Premiums*
E-1	*Inventory of Damaged Assets*
E-2	*Time to Replace Hardware Assuming Emergency Expedite Order*
E-3	*Team Plans*
G-1	*Certification Checklists*
G-2	*Treat Evaluation Form*
G-3	*Testing Schedule*
G-4	*Problem Log*
G-5	*Characterization of Test*
H-1	*Update Log*
H-2	*Questionnaire/Disaster/Recovery Plan—Periodic Update*
I-1	*Security and Confidentiality Agreement*
J-1	*Planned Enhancements*
K-1–12	*Disaster/Recovery Plan Monthly Checklists*
L-1	*Backup Facility Requirements and General Information*
L-2	*Backup Facility Control Manual*
L-3	*Disaster/Recovery Authorization*
L-4	*Backup Site Operating Personnel*
L-5	*Special Authorization*
L-6	*Backup Facility Personnel*
M-1	*Location and General Information about the Area of the Backup Facility*
M-2	*Directions from the Airport*
M-3	*Directions from the Railroad*
M-4	*Direction to the Backup Facility by Car*
N	*Tapes/Supplies/Equipment/Documentation*
O	*Local Transportation*
P-1	*Local Supplies*

P-2	*Transportation*
P-3	*Office Services*
P-4	*Off-site Storage Services*
P-5	*Service Bureaus*
P-6	*Temporary Personnel*
Q	*Lodgings*
R	*Banking Services*
S	*Emergency Services at the Backup Facility*
T	*Shopping Facilities*
U	*Procedures for Reporting and Obtaining Personal Property and Clothing Damaged or Lost*
V	*Substance Abuse*

Preface

Corporate and data processing executives recognize that the data processing operation is an integral operation of the business, and the loss of this *support* service will dramatically affect the organization's capability to survive in a highly competitive business environment. Business interruption preparedness planning and crisis management planning (contingency-disaster/recovery planning) can significantly reduce the susceptibility of an installation to damage, improve the ability of any organization to survive outages, and reduce the duration of the outage and the cost of recovery.

This business interruption preparedness plan is divided into primary units and an addendum as follows:

- Preplanning and Assumptions
- Prevention/Security
- Disaster Preparedness
- Disaster/Recovery Action Plans
- Training for Disaster Recovery
- Validation and Testing
- Plan Update
- Plan Distribution and Security
- Planned Enhancements (addenda)

The purpose of this plan content review is to assist workshop participants during the data collection phase of the program to correlate the material requested with the plan components. The review is also provided to assist the participants, in advance of the plan preparation workshop, to understand the overall plan concepts and the depth of the actual plan in relation to other organizational components, functions, and users—computer and noncomputer operations.

At the plan preparation workshop, each participating organization will receive further documentation concerning the development of special miniplans designed to interface with the organization's master plan. These other plans include:

- Data Center User Guide
- Branch and User Disaster/Recovery Plan Guidelines and Outline
- Users Guide

Three generic reports are provided to assist workshop participants in developing:

1. A management overview report on prevention, security, emergency response, and disaster/recovery planning (distributed after the preplanning seminar).
2. Twenty questions (and answers) about disaster/recovery planning to assist participants in supplementing or reinforcing their management report (distributed after the preplanning seminar).
3. A business interruption preparedness management report describing the

functions of the disaster/recovery plan (distributed after the plan preparation workshop).
(Both documents are available from DIA*log Management, Inc.)

Program Enhancement

DIA*log Management has made every effort to ensure that the material presented in this handbook and at seminars and workshops meets or surpasses the current technology of business interruption preparedness planning (contingency or disaster/recovery planning). The concepts, strategies, and action plans provide organizations with a viable, implementable, testable, and maintainable plan, all of which are represented in the new master plan used at the DIA*log four-day plan preparation workshop.

The Master Plan Table of Contents Review found in this document provides planners with a method to correlate the data collection phase with the actual plan requirements. This is covered in as much detail as possible to ensure that the data collection is performed in the most expeditious and cost-effective process. The master plan, plus the additional miniplans previously described, will ensure that at the time of an unexpected or undesirable event, the organization can perform in a highly structured, disciplined, rational manner in order to reduce the interruption to a minimal time frame.

As part of the master plan program, DIA*log has prepared an innovative, productive, and cost-effective program for the implementation of a prevention/security program within the organization. *Baseline Security Concepts and Strategies* introduces to those responsible for the protection, preservation, security, and integrity of software and data a new approach to evaluating security needs and implementing a security program without long-term risk analysis justification normally used in this process. This entire document is available from DIA*log Management, Inc. (See Part 10.)

In addition to the problems inherent in operating a central (mainframe/large-scale) computer operation, attention must be called to the problem of the protection, security, and preservation of personnel computer operations. The *Personal Computer Security Techniques* manual (also available from DIA*log Management, Inc.) addresses yet another project that will have to be dealt with not only by data processing professionals and corporate management, but more specifically by the user departments (either on-line or individual PCs) within the operating environments of the organization.

Data Collection Phase

In order to develop an effective business interruption preparedness plan, one of the first responsibilities of the Disaster/Recovery Planning Team, upon completion of the Preplanning Seminar, is to gather as much existing documentation as possible concerning the functions, critical activities, resources, and operation requirements of the organization.

The Disaster/Recovery Planning Team will have in their possession the Disaster/Recovery—Data Collection/Decision List distributed at the Preplanning Seminar. This list will have been reviewed in detail during the session concerning the type and extent of information needed to prepare the actual plan. If properly administered, the Disaster/Recovery Planning Team should find that it has very little raw data (information, documentation) to collect. The data collection requirements are based on how the company presently operates, where it is located, how it provides its services, how these services are performed, and the resources necessary to support the organization in the event a disaster strikes. Most of the information that is required probably exists somewhere within the organization. Information, existing standards, procedures, and policies can usually be found in the user departments and within the environs of data processing. Using the data collection and various other checklists, the Disaster/Recovery Planning Team should, as early as possible after completing the Preplanning Seminar, establish the various data collection assignments and responsibilities and send out requests-for-information lists to the appropriate groups throughout the organization.

The extent of the information to be gathered to meet the plan requirements will vary greatly among organizations. Although each organization is required to fulfill the entire list, the volume and complexity will be considerably different. The complexity of material will depend on the number of users, the size and configuration of data processing operations, the communications (on-line) support requirements, noncomputer operations, and so on. The information is simply basic data on what is occurring within the organization. The material is required so that the Disaster/Recovery Planning Team can grasp the size of the problem, the relationships involved, their effect on the performance of the organization under normal conditions, and the implications resulting from a short- or long-term outage. Not unlike the process performed in developing a strategic plan for data processing and the organization, the data processing department cannot work in a vacuum. The development of an effective disaster/recovery plan *must* have the input from the entire organization.

DATA COLLECTION AND DECISION REQUIREMENTS
BUSINESS INTERRUPTION PREPAREDNESS PLANNING FOR THE DATA CENTER
USER AND MANAGEMENT FUNCTIONS
Data Collection Information Requirement

---

1. **Checklists**
    1.1 Company and data processing facilities evaluation (applicable sections)
    1.2 Management considerations checklist(s) (applicable sections)

2. **Manuals/Handbooks (Already in Existence)**
    2.1 Organization policy statements relating to emergencies
    2.2 Security standards, procedures, and policies
    2.3 Managerial, supervisor, and general employee job descriptions
    2.4 General employee personnel manual (personnel policies)
    2.5 Security, standards, and procedures for PCs (mini- and microcomputers)
    2.6 Operations manual (documentation) for stand-alone PCs/microcomputers/minicomputers
    2.7 Software and security policies (PCs/microcomputers/minicomputers)
    2.8 General operations manual (covering all noncomputerized operations)
    2.9 Application run books for special department PCs/mini-operations/micro-operations
    2.10 General user-specific function policy and procedure manual(s)
    2.11 Other special departmental operations or functional performance documentation
    2.12 Data center users guide (see Item 16)
        (1) _____
        (2) _____
        (3) _____
        (4) _____
        (5) _____

3. **Emergency Response Plan (Department-Specific)**
    3.1 Personnel performance and responsibility
    3.2 Accountability procedures (assembly locations)
    3.3 Evacuation routes

3.4 Fire-fighting guidelines (warden assignments, and so on)

3.5 Other emergency conditions

3.6 Emergency phone numbers

3.7 Building emergency response plan

3.8 Handicapped employee procedures and plan

4. Organization Chart (Senior Management: Department or Function: Alternative Line of Authority)

5. Vital Records Management

5.1 Vital records

5.2 Daily input source documentation

5.3 Daily input log

5.4 Off-site storage program (federal regulatory requirements)

5.5 Records retention schedule

5.6 Off-site storage agreement

6. Forms Book (Department-Specific)

7. Department (Functional Unit) Layout

7.1 General floor layout

7.2 Emergency evacuation routes

7.3 Security features [card key entry, cameras, alarms (motion detectors, electric eyes, and so on)]

7.4 Number of desks, work pods, private offices

7.5 Number of people in each department (total per floor)

7.6 Number of emergency exits

8. Ancillary Equipment (With Vendor List)

8.1 PC terminals (dumb)—departmental configuration

8.2 PC terminals (intelligent)—departmental configuration

8.3 PC clusters (nonmainframe), and associated modems, printers, and so on. Also, special security features of specific terminals, access capabilities, read and write capabilities, and so on.

8.4 Copy equipment (style, brand)

8.5 Facsimile(s) [show fax number(s)]

8.6 Teletype(s)

8.7 Telephone systems (telephone company, purchased, leased) (installation diagram showing extensions and numbers)

8.8 Special monitors, displays, or other unique devices

8.9 File cabinets

8.10 Security storage

8.11 Typewriters

8.12 Other
  (1) _____
  (2) _____
  (3) _____
  (4) _____
  (5) _____
  (*Note:* List all equipment associated with computers, modems, printers, special furniture, and so forth.)

9. Supply List

List one month's supply of forms, materials, papers, diskettes, hard disks and/or any consumables that would be required to sustain operations at an alternate location for a period of one month. Identify supply area(s) on the floor layout.

10. Power Requirements

List all power requirements that would be unique to your specific department or function.

11. Air-Conditioning Requirements

List all power requirements that would be unique to your specific department or function.

12. Vendors List

13. Employee Telephone and Address List

List office extensions, special outside dial numbers, and home numbers.

14. All Maintenance Contracts

List equipment, cleaning, and so forth.

15. Insurance Policies

    15.1 Business interruption

    15.2 Personal liability

    15.3 Equipment, fixtures, and so on

    15.4 Extra expense

    15.5 Data reconstruction

    15.6 Software reconstruction

16. Mainframe Interface and Dependencies

List in order of priority the specific schedules and applications that must be maintained in order to sustain your department's operations. This list should include any time-phased secondary reports or functions that would indirectly impact on the performance of your operations or function.

17. Purchased or Leased Software

List all purchased or leased software by name and vendor; also list any special enhancements that have been provided by the vendor, or that the company has made, that will not permit recovery using the standard package.

18. Hard and/or Floppy Disk Backup Procedures

Each of the individual workstations should provide a report on their backup procedures and off-site storage procedures.

19. Terminal Security Provisions

Describe present terminal security provisions—physical and logical.

20. User-Specific Applications

Mainframe-based

    20.1 _____

    20.2 _____

    20.3 _____

    20.4 _____

    20.5 _____

    20.6 _____

PC-based
20.7 _____
20.8 _____
20.9 _____
20.10 _____
20.11 _____
20.12 _____

21. **Safes and Fire-Resistant File Cabinets**
Supply location, designation of type, and items stored in safes and/or fire-resistant file cabinets.

Decisions To Be Made (items to be considered)

---

22. **Disaster/Recovery Administration**
    - 22.1 Designation of the Disaster/Recovery Management Team and noncomputer operations
    - 22.2 Designation of the disaster/recovery *administrator* and *coordinator* for each of the organization's functional departments
    - 22.3 Designation and/or determination of the specific teams for each of the functional departments
    - 22.4 Designation of the team members (by functional/recovery unit)
    - 22.5 Specific computer center liaison assignment

23. **Corporate Administration and Preplanning**
    - 23.1 Designation of a senior corporate executive as the corporate disaster/recovery administrator
    - 23.2 Designation of an internal auditor as the disaster/recovery auditor
    - 23.3 Designation of the line of authority at the time of a disaster (working under the assumption that senior management and/or the board of directors are either unavailable, unable to reach the company, or seriously injured by the catastrophe)
    - 23.4 Designation of proper legal authority (internal or corporate lawyer)
    - 23.5 Designation of personnel (or adjuster) to perform negotiations with the insurance company adjusters
    - 23.6 Modification or addition to the corporate minutes to provide for:
        - 23.6.1 Line of authority
        - 23.6.2 Ability to perform company business away from or at another location
        - 23.6.3 Ability of the board of directors to vote and pass rulings without a quorum
        - 23.6.4 Specific designation of functional authority and authorizations to fulfill or honor contracts
        - 23.6.5 Engagement of outside "agencies" or "agents" to assist in the performance of certain tasks as designated by the management committee and/or the board of directors
        - 23.6.6 Establishment of corporate policy on disaster/recovery administrative actions (certification of the disaster/recovery plan)
        - 23.6.7 Establishment of provisions to override procedures on monetary limits, credit limits, extension of deadlines for payments, and so on

27. Designation of Authority to Document the Entire Activity during the Emergency Declaration

27.1 Chronological events

27.2 Key action steps

27.3 Authorizations at the time of an emergency (any changes in plan)

27.4 Role of outside agencies

27.5 The obtaining of all evidential materials, photographs, drawings, special agency reports, and VCR recordings

27.6 Collection of all other reports

27.7 Declaration of the end-of-emergency period

27.8 Summary report

28. Assembly Location(s)/Control Center(s)/Employee Reporting Center(s)

28.1 Determine all assembly locations for employees to ensure full accountability

28.2 Select and establish a disaster/recovery management team control center

28.3 Select an employee reporting center for those employees not specifically assigned to a disaster/recovery team

29. Scope of Plan

30. Risk Assessment/Business Impact Analysis

30.1 List the top 10 risks (vulnerabilities, hazards, and so on)

30.2 Evaluate the impact by functional unit as to degradation of business in the event of a disaster

30.3 Identify maximum allowable downtime per functional unit

30.4 Estimate the cost of downtime

31. Business Stand-by Provisions and Allowable Function Deterioration

31.1 Identify all user functions that must be performed

    31.1.1 Determine the maximum (or minimum) personnel required to maintain a particular service level

    31.1.2 Determine the maximum (or minimum) space required to maintain the service level and support critical user functions

    31.1.3 Determine the resources necessary to ensure proper support of critical user functions

31.2 Identify all other stand-by user provisions to ensure the continuity of maintenance and business

31.3 Define those functions that will be abandoned during the time of the catastrophic event

31.4 Define methodology for recovery of abandoned functions

31.5 Identify (if possible) all alternate capabilities (stand-by provisions) that can be mobilized to maintain the continuity of *all* business functions

31.6 Select and document user stand-by provisions working from the highest priority down to those functions that can be abandoned

32. Preplanning Stand-by Practices and Tasks Review

32.1 Establish policy for all user fallback procedures

32.2 Establish criteria to determine when stand-by procedures should be put into action

32.3 Establish full authorizations and directions to initiate stand-by operational mode
    32.3.1 For computer functions
    32.3.2 For noncomputer functions
    32.3.3 For all other activities: administration, personnel, and so on
32.4 List user-manager responsibilities
32.5 Reassign (or realign) user personnel to concentrate on the vital or prioritized tasks not specifically designated in the disaster/recovery assignments
32.6 Develop methodologies for nonrecovered computer applications
32.7 Establish reconciliation procedures to enable users to reinitialize the system, update files, and restore normal operations, and provide a method for avoiding double transaction entry or transaction "gaps"
32.8 Assure continued security policy
32.9 Establish training procedures for all user personnel
32.10 Provide for authorizations for overtime
32.11 Obtain (user) personnel commitments for activity at the backup facilities

33. General Review—User Data Collection

Each of the following require predecision documentation, notification lists, resources, special instructions, network design and alternate topology, alternate data entry procedures, backup diskettes, hardware, manuals, vendor support, and other appropriate facilities in order to maintain the continuity of operations and the integrity of data. This is the total responsibility of the user planning process. The "user" has the responsibility to provide to the planning group information concerning how each of the following are to be handled at the time of a disaster.

33.1 Public information
33.2 Administrative and personnel services
33.3 Insurance adjusting (damage assessment)
33.4 Scheduling
33.5 Computer liaison and computer operations
33.6 System support coordination
33.7 Communications
33.8 Data entry
33.9 Off-site storage (logistics of support supplies)
33.10 Supplies
33.11 Salvage (insurance)
33.12 Transportation
33.13 Corporate, divisional, branch, liaison
33.14 Accountability
33.15 Alternate facilities
33.15.1 Control center
    33.15.2 Employee reporting center
    33.15.3 Alternate operating locations
33.16 Special critical period identification
33.17 Log-on records
33.18 Evaluation of new applications and new products

EFFECTIVE DISASTER/RECOVERY PLAN PREPARATION WORKSHOP
Data Collection Information Status Report

Date _____

Company/Division _____

Completed by _____

Part 1: Documents To Be Collected or Completed

	Yes	No
1. Checklists, particularly:		
1.1 Company and data processing facilities evaluation	☐	☐
1.2 Management checklists	☐	☐
2. Manuals, handbooks (already in existence):		
2.1 Personnel policies	☐	☐
2.2 Data center and user job descriptions	☐	☐
2.3 Operators manual	☐	☐
2.4 Data centers users guide	☐	☐
2.5 Security policy	☐	☐
2.6 Application development standards	☐	☐
2.7 User department policy and procedures	☐	☐
2.8 _____	☐	☐
2.9 _____	☐	☐
2.10 _____	☐	☐
2.11 _____	☐	☐
2.12 _____	☐	☐
3. Emergency response plan		
3.1 Evacuation procedures	☐	☐
3.2 Evacuation routes	☐	☐
3.3 Fire-fighting guidelines	☐	☐
3.4 Procedures for each type of emergency	☐	☐
3.5 Emergency phone numbers	☐	☐
4. Organization charts (top, data center, user)	☐	☐
5. Records management		
5.1 Vital record retentions and storage policy (a copy of the records management policy stored off-site)	☐	☐
5.2 Forms book (a copy of every preprinted and computer-generated form stored off-site)	☐	☐
6. Layouts (data center, user, building, area) showing all utility lines and security features	☐	☐

7. Power distribution system (diagram or description) ☐ ☐
8. Air-conditioning system (diagram or description) ☐ ☐
9. Employee telephone number and address (home and office) ☐ ☐
10. Vendor list including contacts and products ☐ ☐
11. Supply list (one month's usage) ☐ ☐
12. Mainframe configuration ☐ ☐
13. Mainframe data flow diagram ☐ ☐
14. Pack family layout and usage ☐ ☐
15. Datacomm configuration ☐ ☐
16. Maintenance contracts (list all equipment, including micros) ☐ ☐
17. Insurance policies ☐ ☐
18. Software escrow agreements ☐ ☐
19. Off-site storage agreements ☐ ☐
20. Backup site agreement (including configuration) ☐ ☐
21. Tape library system sample reports ☐ ☐

Part 2: Decisions To Be Made (Or at Least Considered)

22. Members of the Disaster/Recovery Planning and Data Collection Team (coordinator or administrator) ☐ ☐
23. Members of the Disaster/Recovery Management Team (coordinator, administrator, alternates) ☐ ☐
24. Other teams and their membership (one summary page per team) ☐ ☐
25. Scope of plan ☐ ☐
26. Risk assessment (top 10 risks) ☐ ☐
27. Maximum downtime ☐ ☐
28. Cost of downtime ☐ ☐
29. File backup strategy (source, object, job, database, transactions, and so on) ☐ ☐
30. Off-site storage strategy (How many generations are off-site?) ☐ ☐
31. Mandatory, necessary, and desirable applications and reports ☐ ☐
32. Vital records strategy ☐ ☐
33. Develop versus buy strategies ☐ ☐
34. Available backup site options ☐ ☐
35. Available office space ☐ ☐
36. Employee commitment in the time of a disaster ☐ ☐
37. User recovery plans ☐ ☐
38. Custodian versus ownership of information assets
    As the owner of the applications, the data user departments must accept their role in the development of an effective disaster/recovery plan. They must provide the necessary input as part of the data

collection and in the development of the application recovery plans, as well as the master disaster/recovery plan. In addition, they must develop their own user plans. This establishes the true nature of the relationship between the data processing center (The Custodian) as a support organization and the user as the *Owner* of the application software and the data.

As part of the user responsibility in the development of an effective disaster/recovery plan, the user must contribute to the decisions on the critical applications that must be performed, and those applications that are deferable (noting for how long they can be deferred). This activity must be integrated with the development of the master plan. Management must provide the final input for the selection of those applications that it deems critical to the ongoing performance of the organization.

The Model Plan Table of Contents Review will assist the Disaster/Recovery Planning Team to understand the application of the "data requested" and the actual plan components. It is suggested that this be carefully reviewed with all the users and other teams to ensure that all aspects of the data collection requirements are fulfilled.

EFFECTIVE DISASTER/RECOVERY PLAN
Model Plan
Actual Table of Contents and Review

[*Notes:* (1)The numbering system in this section is for the plan only. It has no relation to the numbering system used elsewhere in this handbook. (2) When unit titles are followed by "(DC)," it means that the unit is part of the data collection process.]

1.	**Introduction**	
1.1	*Purpose of the Plan*	
1.2	*Plan Postscript*	
1.3	*Management's Letter of Commitment*	
1.4	*Information Security Policy*	
1.5	*Personal Liability of Officers*	
2.	**Preplanning and Assumptions** (DC)	
2.1	*The Disaster/Recovery Planning and Data Collection Team (DC)*	
2.1.1	*Charter*	
2.1.2	*Members*	
2.2	*Scope*	
2.3	*Risk Assessment*	
2.3.1	*Probabilities of a Disaster (DC)*	
2.3.2	*Disaster Scenarios (DC)*	
2.3.2.1	*Sample Scenarios* [Other scenarios will be added by the participants]	
2.3.3	*Costs of Downtime (Maximum Allowable Downtime)*	
2.3.4	*Assumed Maximum Downtime (DC)*	
2.3.5	*Worst-Case Assumption (DC)*	
2.4	*Other Manuals (DC)*	
2.5	*Role of the Disaster/Recovery Management Team*	
2.6	*Plan Initialization*	
3.	**Prevention/Security**	
3.1	*Physical Security (DC)*	
3.1.1	*Layouts (DC)*	
3.1.2	*Locks (DC)*	
3.1.3	*Guards (DC)*	
3.1.4	*Cameras (DC)*	
3.1.5	*Emergency Telephone Numbers (DC)*	
3.1.6	*Intrusion Alarms*	
3.1.7	*Medical Alert*	
3.1.8	*Access to the Data Center (DC)*	
3.2	*Medical Security*	
3.2.1	*First Aid Kit*	
3.2.2	*Nurse*	
3.2.3	*General Employee First Aid and CPR Training*	

3.2.4	*EMS (or Fire Rescue Squad)*
3.3	*Fire Protection (DC)*
3.3.1	*Fire Alarms*
3.3.2	*Hand-held Fire Extinguishers and Fire Hoses (DC)*
3.3.3	*General Fire Safety (DC)*
3.3.4	*Passive Fire-Fighting Measures (DC)*
3.3.5	*Fire Drills (DC)*
3.4	*Emergency Response Plan*
3.4.1	*Contents*
3.4.2	*Emergency Response/Action Coordinator (Building and/or Plantwide)*
3.4.3	*Data Center Emergency Response Miniplan*
3.5	*Housekeeping*
3.5.1	*Data Center Prohibitions*
3.5.2	*General Cleanliness*
3.6	*Power (DC)*
3.6.1	*Isolation Transformer (Required)*
3.6.2	*Motor Generator (Optional)*
3.6.3	*Uninterruptible Power Supply (UPS) (Optional)*
3.6.4	*Engine Generator (Optional)*
3.6.5	*Reference Grid (Optional)*
3.7	*Air-conditioning (DC)*
3.8	*Maintenance Agreements (DC)*
3.9	*Document Security*
3.9.1	*Classification*
3.9.2	*Storing: Security and Retention*
3.9.3	*Shredding: Security*
3.9.4	*Forwarding: Security Methods for Distribution*
3.10	*Software Security (DC)*
3.10.1	*User Code/Password Practices*
3.10.1.1	*System Release Level*
3.10.1.2	*Application Level*
3.10.2	*Required Password Change*
3.10.3	*Other Applications Security Measures*
3.10.4	*Pack Lockout*
3.10.5	*Program (Tape) Library Control System (PLCS)*
3.10.5.1	*Tape Library Management System*
3.10.6	*Erasure of On-line Copies of Sensitive Data*
3.11	*Datacomm Security*
3.11.1	*Physical Security of Terminals*
3.11.2	*Data Encryption Standard (DES) Devices*
3.11.3	*Leased Lines*
3.11.4	*Dial-back Feature for Dial-up Lines (DC)*
3.12	*Application Development Standards (DC)*
3.13	*Application Support Standards*
3.14	*Application Purchase*
3.15	*Role of the Auditing Staff*
3.16	*Role of the User Department(s)*
4.	**Disaster Preparedness**
4.1	*File Backup (DC)*
4.1.1	*Decision Criteria (DC)*
4.1.2	*Pack Family Usage Rules (DC)*
4.1.3	*Backup Strategies (DC)*
4.1.3.1	*Operating System Code*
4.1.3.2	*Application Code*

4.1.3.3	*Databases*	
4.1.4	*Dump Schedule (DC)*	
4.1.5	*Source Documents (DC)*	
4.1.6	*Terminal Floppy Dumps (DC)*	
4.1.7	*Microfiche Reports (DC)*	
4.1.8	*Period-end File (Tape) Retentions*	
4.2	*Off-site Storage (DC)*	
4.2.1	*Off-site Management (DC)*	
4.2.2	*Off-site Strategy (DC)*	
4.2.3	*Off-site Rotation Schedule of Current Tapes*	
4.2.4	*Off-site Archive of Period-End Tapes*	
4.2.5	*Documents Off-site*	
4.2.6	*Tape Library Management System (DC)*	
4.2.7	*Scratch Tape Management*	
4.2.8	*Scratch Floppy Disk Management*	
4.3	*Backup Site Strategy (DC)*	
4.3.1	*General Alternatives*	
4.3.2	*Specific Alternatives*	
4.3.3	*Chosen Alternatives (Period of Time)*	
4.4	*Mainframe Configurations*	
4.4.1	*Primary Site*	
4.4.2	*Backup Site*	
4.5	*Datacomm Configuration*	
4.5.1	*Primary Site Communications*	
4.5.2	*Backup Site Configurations*	
4.5.3	*Additional Equipment Requirement*	
4.6	*Emergency Phone Numbers of Employees*	
4.6.1	*Alphabetical List of Phone Numbers*	
4.6.2	*Calling Tree for Disaster/Recovery Teams*	
4.6.3	*Calling Tree for All Employees*	
4.7	*Vendor Information*	
4.7.1	*Vendor Addresses, Phone Numbers, and Contacts*	
4.7.2	*Vendor Product List*	
4.7.3	*Vendor Support Letters*	
4.8	*Disaster Kit*	
4.9	*One-Month Supply List*	
4.10	*Forms Books*	
4.10.1	*Facility and Office Space Requirements*	
4.11	*Vital Records Retention*	
4.12	*Software Escrow*	
4.13	*Employee Commitment Letters*	
4.14	*Priorities for a Limited Processing Environment*	
4.14.1	*Current Site Operations Schedule*	
4.14.2	*General Priority Scheme*	
4.14.3	*Probable Backup Site Operations Schedule*	
4.14.4	*Application Recovery Plans*	
4.15	*Logistical Considerations*	
4.15.1	*Credit Cards, Cash, Etc.*	
4.15.2	*Housing*	
4.15.3	*Local Transportation*	
4.15.4	*Office Space*	
4.15.5	*Control Center*	
4.15.6	*Employee Notification by Broadcast (and Other) Media*	
4.16	*Insurance*	
4.16.1	*Business Property Insurance*	
4.16.2	*Extra Expense Insurance*	

4.16.3 *Business Interruption Insurance*
4.17 *Disaster/Recovery Action Teams*
4.17.1 *Disaster/Recovery Management Team (DRMT)*
4.17.1.1 *Team Charter*
4.17.1.2 *Typical Duties of the Disaster/Recovery Management Team*
4.17.1.3 *Role of the Coordinator*
4.17.1.4 *Role of the Administrator*
4.17.1.5 *Membership*
4.17.2 *Transportation Team*
4.17.2.1 *Team Charter*
4.17.2.2 *Typical Duties by Phase*
4.17.2.3 *Membership*
4.17.3 *Off-site Storage Team*
4.17.3.1 *Team Charter*
4.17.3.2 *Typical Duties by Phase*
4.17.3.3 *Membership*
4.17.4 *Supplies Team*
4.17.4.1 *Team Charter*
4.17.4.2 *Typical Duties by Phase*
4.17.4.3 *Membership*
4.17.5 *Administrative Team*
4.17.5.1 *Team Charter*
4.17.5.2 *Typical Duties by Phase*
4.17.5.3 *Membership*
4.17.6 *Software Team*
4.17.6.1 *Team Charter*
4.17.6.2 *Typical Duties by Phase*
4.17.6.3 *Membership*
4.17.7 *Applications Team*
4.17.7.1 *Team Charter*
4.17.7.2 *Typical Duties by Phase*
4.17.7.3 *Membership*
4.17.8 *Communications Team*
4.17.8.1 *Team Charter*
4.17.8.2 *Typical Duties by Phase*
4.17.8.3 *Membership*
4.17.9 *Operations Team*
4.17.9.1 *Team Charter*
4.17.9.2 *Typical Duties by Phase*
4.17.9.3 *Membership*
4.17.10 *Data Preparation Team*
4.17.10.1 *Team Charter*
4.17.10.2 *Typical Duties by Phase*
4.17.10.3 *Membership*
4.17.11 *Salvage Team*
4.17.11.1 *Team Charter*
4.17.11.2 *Typical Duties by Phase*
4.17.11.3 *Membership*
4.17.12 *New Hardware Team*
4.17.12.1 *Team Charter*
4.17.12.2 *Typical Duties by Phase*
4.17.12.3 *Membership*
4.17.13 *New Facilities Team*
4.17.13.1 *Team Charter*
4.17.13.2 *Typical Duties by Phase*
4.17.13.3 *Membership*

4.18		Other Teams
4.18.1		User Team
4.18.1.1		Team Charter
4.18.1.2		Typical Duties by Phase
4.18.1.3		Membership
4.18.2		Audit Team
4.18.2.1		Team Charter
4.18.2.2		Typical Duties by Phase
4.18.2.3		Membership
4.18.3		Hardware Team
4.18.3.1		Team Charter
4.18.3.2		Typical Duties by Phase
4.18.3.3		Membership
4.18.4		User Liaison Team
4.18.4.1		Team Charter
4.18.4.2		Typical Duties by Phase
4.18.4.3		Membership
4.18.5		Security Team
4.18.5.1		Team Charter
4.18.5.2		Typical Duties by Phase
4.18.5.3		Membership
4.19		Additional Teams to Be Considered
4.19.1		Third-Party Liaison Team
4.19.1.1		Team Charter
4.19.1.2		Typical Duties by Phase
4.19.1.3		Membership
4.19.2		Public Relations Team
4.19.2.1		Team Charter
4.19.2.2		Typical Duties by Phase
4.19.2.3		Membership
4.19.3		Alternate Strategies Coordinating Team
4.19.3.1		Team Charter
4.19.3.2		Typical Duties by Phase
4.19.3.3		Membership
4.19.4		Legal Team
4.19.5		Crisis Management Team (Management Level)
5.		**Disaster/Recovery Action Plans**
5.6		Elaboration of PERT Chart—Phases 2, 3, and 4
5.6.1		Transportation Team
5.6.2		Off-site Storage Team
6.		**Training for Disaster Recovery**
6.1		Recovery Team Training
6.2		All Employee Training
6.3		CPR and First-Aid Training
7.		**Validation and Testing**
7.1		Validation
7.2		The On-site One-Application Test
7.3		The On-site Complete Test
7.4		The Backup Site Test
7.5		The Surprise Test
7.6		Test Schedule
7.7		Problem Log
8.		**Plan Update/Control**
8.1		Update Log

8.2	*Approval*
8.3	*Formal Review (Once a Year)*
8.4	*Checklists*
9.	**Plan Distribution and Security**
9.1	*List of Recipients*
9.2	*Screening Information*
9.3	*Serialization*
9.4	*Termination or Transfer of Copy Holder*
9.5	*Copies*
9.6	*Physical Security of the Plan*
9.7	*Security and Confidentiality Agreement*
10.	**Planned Enhancements**

Addendum Review

C-1	*Data Center Layout (DC)*
C-2	*Building Layout (DC)*
C-3	*Area Layout (DC)*
C-4	*Emergency Call List (DC)*
C-5	*Card Key Holders (DC)*
C-6	*Authorized Entrants (DC)*
C-7	*Authorizing Management (DC)*
C-8	*Data Center Sign-in Sheet (DC)*
C-9	*Evacuation Route (DC)*
C-10	*Copies of Maintenance Agreements (DC)*
D-1a	*Off-site Agreement (DC)*
D-1b	*Vendor Information (DC)*
D-2	*Tape Library Report (DC)*
D-3	*Backup Site Alternatives (DC)*
D-4	*Backup Site Agreement (Initial) (DC)*
D-5	*Backup Site Agreement (Secondary)*
D-6	*Primary Site Mainframe Configuration (DC)*
D-7	*Backup Site Mainframe Configuration (DC)*
D-8	*Primary Site Datacomm Configuration (DC)*
D-9	*Backup Site Datacomm Configuration (DC)*
D-10	*Required Ancillary Equipment (DC)*
D-11	*Home Phone Numbers (DC)*
D-12	*Calling Tree for Disaster/Recovery Teams (DC)*
D-13	*Calling Tree for Other Employees (DC)*
D-14	*Vendor List (DC)*
D-15	*Vendor Product List (DC)*
D-16	*Vendor Support Letters (DC)*
D-17	*Disaster Kit*
D-18	*One-Month Supply List (DC)*
D-19	*Vital Records Retention (DC)*
D-20	*Escrow Agreement*
D-21	*Purchased Software*
D-22	*Disaster/Recovery Plan: Confidential Employee Questionnaire*
D-23	*Current Site Operations Schedule*
D-24	*Probable Backup Site Operations Schedule*
D-25	*Application Recovery Plan Worksheet*
D-26	*Copies of Policies and Annual Premiums*
E-1	*Inventory of Damaged Items*
E-2	*Time to Replace Hardware—Assuming Emergency Expedite of Order*
E-3	*Team Plans*

G-1	Certification Checklists
G-2	Threat Evaluation Form
G-3	Testing Schedule
G-4	Problem Log
G-5	Characterization of Test
H-1	Update Log
H-2	Disaster/Recovery Plan—Periodic Update Questionnaire
I-1	Security and Confidentiality Agreement
J-1	Planned Enhancements
K-1–K-12	Disaster/Recovery Plan Monthly Checklists
L-1	Backup Facility Requirements and General Information
L-2	Backup Facility Control Manual
L-3	Disaster/Recovery Authorization
L-4	Backup Site Operating Personnel
L-5	Special Authorization
L-6	Backup Facility Personnel
M-1	Location and General Information about the Area of the Backup Facility
M-2	Directions from the Airport
M-3	Directions from the Railroad
M-4	Direction to the Backup Facility by Car
N	Tapes/Supplies/Equipment/Documentation
O	Local Transportation
P-1	Local Supplies
P-2	Transportation
P-3	Office Services
P-4	Off-site Storage Services
P-5	Service Bureaus
P-6	Temporary Personnel
Q	Lodgings
R	Banking Services
S	Emergency Services at the Backup Facility
T	Shopping Facilities
U	Procedures for Reporting and Obtaining Personal Property and Clothing Damaged or Lost
U-2	Replacement of Damaged Clothing
U-3	Funds and Credit Cards
U-4	Payroll
U-5	Family Care and Assistance
V	Substance Abuse

EFFECTIVE DISASTER/RECOVERY PLAN MODEL

Paragraph-by-Paragraph Review; Section and Paragraph Analysis and Overview

1. Introduction

1.1 Purpose of the Plan

This paragraph should reflect the organization's specific requirements for a business interruption preparedness plan. In rewriting this paragraph, the Disaster/Recovery Planning Team should determine and reflect management's purpose for the plan.

Sample Paragraph

"The purpose of this Business Interruption Preparedness Plan is to reduce the number of decisions which must be made when, and if, an incident occurs. It is this plan that will provide the action plans that will guide the various teams and executives of the _____ Company in responding to an unexpected or undesirable event and the recovery of company operations following the incident."

The Plan Preparation Team can further enhance the purpose by addressing any other area of activity that management, users, the data processing department, and noncomputer operations have indicated.

1.2 Plan Postcript

This section indicates management's commitment, based on organizational policy, and its intention to provide the necessary resources to support the program.

1.3 Management's Letter of Commitment

If the plan is to be successful, it requires the full support of management. Every functional unit of the organization (computer and noncomputer activities) will accept the necessity and requirements of the plan, cooperate in its development and implementation, and provide the necessary input and maintenance only if management shows its willingness to support the activity, provide the necessary resources (personnel and funds), and demonstrate that it is a vital program for the "health and survival" of the business. A letter indicating management's support and mandating all personnel to cooperate in the activities required should be placed here. It should also be distributed to all the groups that are needed to develop and perform under the plan. The purpose of the plan should be carefully stated, even if it is repetitious of the purpose outlined in 1.1.

Sample Purpose Statement

"In the event that critical support capabilities, primarily data processing operations, are affected by an unexpected or undesirable event, the management of the _____ has required that all functional units of this organization have a written (well-documented) Business Interruption Preparedness Plan, commonly known as a Contingency Plan, or Disaster/Recovery Plan. It is the purpose of this plan to provide our organization with highly structured, well-documented (predecision) procedures to ensure the continuity of the business and the integrity of data if critical support services are lost."

Throughout the workshop, the instructor "consultants" will emphasize the fact that the Business Interruption Preparedness Plan (disaster/recovery planning) is a management/business plan rather than a technical plan. The management letter supports this claim.

1.4 Information Security Policy

Each organization, during the preplanning and data collection phases of the workshop, will have the opportunity to review the various concepts concerning the protection, preservation, and security of information assets for their operation. Here, each organization should state simply its affirmation, confirmation, and implementation of a physical and logical security program in order to *prevent* any disaster that could result in a short- or long-term interruption. This should be a broad-based policy statement, since the details concerning security are addressed further in the plan.

1.5 Personal Liability of Officers

Contingency disaster/recovery planning has become a business issue, reflected in recent and proposed public law and the dissemination of public policy that "information" is an *asset* and must be protected in the same manner as other corporate assets.

The Foreign Corrupt Practices Act of 1977; Office of the Comptroller of the Currency circulars (BCC 177, 226); Federal Home Loan Bank Board, Memorandum 67; as well as privacy and computer crime laws establish a mandate for management to take prudent steps to protect the viability and security of their organizations. Legal requirements by the EPA, Department of Commerce, EEOC, OSHA, SEC, IRS, and so forth add even more credibility to security procedures and prevention mechanisms. Add the state and local laws and management has a major task at hand.

Under these federal (and other state and local laws), directors and officers of organizations can be held liable for violations of the law; other abuses may lead to civil suits, fines, and/or imprisonment.

The organization's legal advisors should provide the necessary research concerning the laws and regulatory requirements to which the company is subject, and these should be described here.

2. Preplanning and Assumptions (DC)

The first task of the disaster/recovery administrator and coordinator once they have been appointed, is to determine the threats, hazards, and vulnerabilities to which the organization, the data center(s), and/or the various users are exposed. This information provides the basis for the design of the plan. These threats, hazards, and vulnerabilities are defined in terms of "effect" (damage/downtime) to the organization, and compared with the maximum allowable downtime. Assumptions have to be made based on the results of the facilities evaluation and other methodologies used by the organization to determine risk. These establish the methods, strategies, and action plans that will deal with short- or long-term outages. It is important to note that the master plan will be generated using the concept of a "worst-case" condition. Recovery from a lesser incident(s) may only require use of subsets of the plan and the mobilization of only a few of the action teams.

Basic Assumptions

1. What are the threats, hazards, and vulnerabilities in the facility (building) which affect data processing operations (machine room), the information stored on magnetic media, and the software and data stored in the computer and at the user on-line or PC location?
2. What are the critical business functions and/or computer-dependent applications that will keep the organization running? (The critical computer applications may be on PCs as well as on mainframe and large-scale computers. Also, critical functions may be identified as computer or noncomputer operations.)
3. What are the critical computer resources necessary to support the critical business functions and/or critical computer applications?
4. What are the potential losses that will result from the failure to perform these business functions or computer operations? (Losses can be money, customer satisfaction, loss of opportunity, exposure or disclosure of critical—classified proprietary or trade secret—information, and so on.)
5. What are the strategies and action plans necessary to ensure prompt, accurate, and timely backup and recovery?

In most instances, data processing personnel are not in the position to respond to any question other than question 3—*the resources required to support the organization's critical computer operations.* Systems development groups assigned to specific users are in an excellent position to assist in developing this material. The user(s)/(owner) department performing the input of data and users of the specific applications and other related computer and noncomputer operations will have to provide the answers. It is management's responsibility to make the final determination as to what critical functions and computer applications are necessary to support the organization. The first and most appropriate steps during the data collection phase is to prepare statements of the objectives, definitions, and assumptions.

- *Objectives* are relatively easy to develop based on the organization's requirements. They may be stated in a manner suitable to company policy and operation methodology.
- *Definitions* relate to the terms to be used; the defining of a "disaster" as it relates to the individual organization, functional unit, user community, and so on; and all the areas that are assumed to be within the purview (scope) of the disaster plan, both computer and noncomputer operations. These definitions should avoid any technical terminology.
- *Assumptions* can be more difficult to achieve. When dealing with assumptions relating to the probable threats, hazards, and vulnerabilities, it is important that each assumption have a viable basis, and not a supposition. Selecting the probable risks without a firm basis, just to create management paranoia, may invalidate all efforts to develop a plan. On the other hand, if assumptions are based on good judgment and/or historical data, the ease of obtaining management's continued support can be relatively easier.

Dealing with assumptions that include guidelines established by management as mandatory actions or organizational policy provides the team with an excellent starting point. These assumptions could be, but are not limited to, the critical guidelines:

- The scope of the plan, organizational functions, specific operations, departments, research and development (product or computer systems), distribution, on-line users (defined by criticality), and so on.
- The time frame required (mandated) to restore specific or all affected activities: definitions of criticality (applications) determined by the time of year, time of month, and so on, as well as the resources required to support those critical applications on a time-phase situation.
- The specific threats, hazards, and vulnerabilities to be addressed, the level of criticality of the exposure, or, on the other hand, those risks that do not have to be addressed.
- The commitment of resources (funds and personnel) to the project. The budget for the plan and other resources will not be part of the normal department or operations budgets (data processing or users).
- The steps necessary (in broad terms, since the details are covered in other sections of the plan), for fire protection, fire fighting, and emergency response procedures.

The Data Collection Manual contains detailed lists and study methodology to assist the Disaster/Recovery Management Team in preparing definitions or identifying the threats, hazards, vulnerabilities, risks, and exposures. Definitions and assumptions should also relate to all critical computer and noncomputer functions of the organization regardless of their locations.

2.1 The Disaster/Recovery Planning and Data Collection Team (DC)

Before any team(s) can be formed and the project gotten underway, it is necessary to appoint a disaster/recovery administrator (coordinator, team leaders, and so on).

The disaster/recovery administrator is the senior executive responsible for the total administration of the plan. The disaster/recovery coordinator is the individual responsible for the development, implementation, maintenance, and operation of the plan.

The specific title(s) selected will depend on organizational structure and policy, the level within the organization established by management, and the size of the group(s) to be formed. In a small organization, the position of disaster/recovery administrator may be assigned to someone as a responsibility second to his or her normal position and responsibilities.

Regardless of the size of the organization and the title selected, the position assignment is important for a number of reasons:

- To give high profile to the project
- To ensure continuity of activities
- To ensure continued support
- To ensure continuous funding

The disaster/recovery administrator and coordinator should be appointed, if possible, prior to attending the three-day preplanning seminar. It is important that when seminar participants return to their home environment to perform the data collection phase that they be placed in charge. This appointment establishes the leadership well in advance of any activities that have to be assigned, directed, or coordinated. (The Disaster/Recovery Preplanning Questionnaire outlines these appointments and workshop team attendance.) The disaster/recovery coordinator should have complete authority to proceed with any activity within the scope of the plan without further approval. One person, the disaster/recovery administrator (and his/her alternate) has to be totally familiar with every phase of the plan.

Second to this appointment should be the formation of two other teams: the Disaster/Recovery Management Team and the Disaster/Recovery Data Collection Team.

It is important to note, that at the time of a disaster, those individuals who have been given the authority (disaster/recovery administrator, specific team members) will supercede the higher executive authority. Therefore, selection of individuals to perform these tasks is extremely important.

Disaster/Recovery Administrator: Duties and Responsibilities (DC)

Before other teams can be developed, charters written, assignments made, and action plans created, it is important that a single person (with an officially appointed alternate) have the overall scope of his or her activities defined. Here, the areas of responsibility should be briefly outlined. They may be greater or lesser in degree, depending on the size of the organization, complexity of the data processing activity, the level of computer dependency, and company policy.

The disaster/recovery administrator and the disaster/recovery coordinator:

1. Assume the responsibility for all follow-up on the preplanning necessary to develop the business interruption preparedness plan.
2. Assign teams to perform the data collection required to prepare the plan and maintain an active system to keep all lists of personnel, vendors, emergency services, and so on up to date.
3. Establish an interactive program with the internal auditing department to certify and verify all activities involved in the disaster/recovery plan and verify all physical and logical security measures.
4. Establish test procedures, verify actions with the internal auditing department, establish procedures, and have Internal Auditing monitor all tests—static or dynamic.
5. Have periodic meetings with other teams, verify team miniplans, and update team personnel changes.
6. Verify all contractual arrangements that have their foundation within the scope of the disaster/recovery plan.
7. Hold meetings with users and operations groups concerning security and prevention implementation or violations.
8. Disseminate plan components to appropriate members of teams, users, and management.

The disaster/recovery administrator's position is a staff job which is management related. In general, the disaster/recovery administrator should have a broad knowledge of all company operations and business activities. He or she should also be somewhat technically proficient in computers and understand systems, programming, operations, communications, and so on. Some background in security, both physical and logical, is also helpful.

In addition, the disaster/recovery administrator (coordinator), during the preplanning, data collection, plan development, and plan implementation phases will be responsible for:

1. Convening and chairing meetings with other teams, management, users, security personnel, auditors, and so on.
2. Assigning all team plan development activities that interface with the master plan. He or she will also be verifying plans of various teams to ensure that they agree with the intent of the master plan and that they will perform.
3. Distributing all data collection assignments and integrating information into the plan in appropriate sections.
4. Working with organization departments (users) to establish test programs, administering tests, and reporting on the results of all test activities.
5. Confirming all action plans, critical activities, and so on with operations, data processing, and users to ensure performance, and modifying activities where appropriate.
6. Preparing all security lists for the distribution of the disaster/recovery plan. He or she will also be responsible for maintaining lists of all personnel with plans in their possession. Other personnel who have responsibilities within the scope of the plan are to receive sections deemed necessary for them on a need-to-know basis.
7. Coordinating all team assignments and action plans.
8. Entering the plan in some form of word processing in order to make updates easier, as provided by DIA*log/UNISYS.

Disaster/Recovery Plan Preparation Team
Each organizational group participating in the seminar/workshop will have selected the members of the company that will be responsible for the actual writing of the plan. The make-up of this group has been discussed earlier in this document.

2.1.1 Charter

This section covers the charter of the team designated as the Disaster/Recovery Plan Preparation Team. This charter should be carefully reviewed and modified to meet the specific policy and requirements of the organization. The charter should detail management's authorization, a description of the activities of the team, authorities, authorization capabilities, and so on. (See Figure 3-7.)

2.1.2 Members

In addition to the charter, a list of the team members, addresses and phone numbers (work and home) should be part of this section.

2.2 Scope

The Disaster/Recovery Management Team, the Disaster/Recovery Plan Preparation Team, and the disaster/recovery administrator have to detail the scope of the plan. The plan's scope defines all the organizational units that are covered by the plan. The scope should be developed and explained in detail to ensure that every functional unit of the organization is covered and that management has defined which operations are critical to the organization. The basis for determining the scope of the plan is to *ensure that every unit of the organization (computer or noncomputer operations) responsible for the continuity of the business activity in the event of a disaster is included into the strategies and action plans of the master plan.* It is conceivable that separate plans will have to be developed for many operations outside the scope of the central data processing activities. Each of these functional units will have to develop mini–disaster/recovery plans for their sites (or activities) in the event an adverse incident affects their performance—their ability to continue activities. The scope of the plan also includes a description of each of the risks and exposures to which the "organization" facility, user, and data center(s) are vulnerable.

2.3 Risk Assessment

In previous sections of this document, methodologies are discussed relating to risk assessment. The new approach to risk assessment and security enhancement, the *baseline security concept,* is designed to meet the needs of those organizations which can introduce computer security on the basis of good judgement, rather than lengthy, costly risk analysis/risk assessment programs.

Detailing the requirements of a risk assessment/risk analysis program will be reviewed in other sections of the manual and discussed by the instructor(s). It is not appropriate nor is there sufficient space here to discuss all the alternatives or methodology.

As a brief review, in order to establish some form of methodology in the approach to understanding the threats, hazards, vulnerabilities—risks and exposures—the following is provided for your edification as general guidelines. These can be documented, using the Threat Evaluation Form (see Addendum G-2).

Purpose of Risk Assessment

In order to obtain a quantitative statement of the potential problems to which the organization, facility, user, and/or data center is exposed, as well as to establish an appropriate, cost-effective safeguard to be selected, a risk assessment can be performed. The risk assessment process classifies data into a measurable manner, thereby allowing risks (exposures ranked by their potential effects on the organization). Risks can be classified into three levels:

1. *Mandatory.* Those risks or exposures that must be addressed since they have the highest rating or potential to affect the performance of the organization, or cause an interruption (short- or long-term) to one or more operations of the organization, user, or data center. Mandatory controls are those security requirements that must be implemented to ensure continuity.
2. *Necessary.* Those security exposures (physical or logical) and control requirements that will affect the performance of the organization, or delay the mission of the organization, but which will not create a substantial impact on the performance of the organization. Necessary controls may be parallel with mandatory requirements, yet they can be deferred for a reasonable period of time.
3. *Desirable.* Those security exposures and control requirements that would enhance the total operation of the organization, but which, even without their implementation, would not affect, to any great degree, the operation or performance of the organization. The items identified in the *desirable* category can be reasonably deferred for a selected period of time, but they should be addressed periodically to determine if any changes in operation have reclassified items into the mandatory or desirable categories.

The risk assessment program has a number of specific steps during the preplanning, study, and determination of security prevention controls that have to be implemented as well as the determination of what will be initiated to secure the work and data processing environment.

1. *Management.* Participation and commitment by management is mandatory. This is an absolute requirement; without the support of management and the commitment to resources, the risk assessment cannot be initiated or completed, and the recommendations cannot be implemented within the organization.
2. *Risk assessment task force(s).* The risk assessment task force(s) must be formed from every functional (operating) unit of the organization and computer

(data center, users, PC, and minicomputer) and noncomputer operations. With the ultimate objective of the plan being the development, implementation, training, and maintenance of effective disaster recovery for the entire organization, total participation is absolute, especially from those groups or departments that are critical to the continued (daily) performance of the business. Not all the groups will be required to perform tasks simultaneously, but substantial time will be necessary to complete the study. It is appropriate that the disaster/recovery administrator be assigned the task of risk assessment.

3. *Defining and assessing.* The key to a successful and rapid completion of the project is to define each of the functions that will be required to participate in the study. Then each of these groups will define those critical activities, operations, and applications required to sustain the mission of their activity which ultimately results in the continued performance of the organization. Here we are looking for the criticality and sensitivity of the specific operation.

   Then it is necessary to identify those threats, hazards, and vulnerabilities that could affect performance and cause short- or long-term interruptions.

   Finally, it is important to determine the prevention, security, mechanisms, and controls that have to be imposed to prevent the incident from occurring.

4. *Potential loss.* Determination of loss potential can be illustrated in a number of ways. Regardless of the specific effects of an interruption, or whom they harm, the usual result is, or can be expressed in, money (profits) lost. Data losses are usually considered intangible asset losses, yet they have a monetary effect owing to the cost of reconstruction, the effects upon the owner of the data, the changed performance of the business, and the hard asset losses that can be incurred. Delayed performance in any area can be ultimately expressed in one or more forms of asset losses.

5. *Control functions.* Most systems, operations, and organizational functions have some control functions involved in their methodology. Each task force member of each functional group must determine whether these controls are sufficient to provide the protection and security necessary to prevent an adverse incident from affecting continued performance. Here *effectiveness* is the measurement tool. Good judgment and common sense in many cases supercede technological methodologies.

6. *Facility evaluation checklists (see Checklist 1 and Figures A–R, pp. 72–87).* The checklists covering facilities evaluation; threat, hazard, and vulnerability probability; and others relating to facility performance will guide the disaster/recovery administrator/coordinator and the task forces in determining the most critical exposure to their operations.

7. *Standard risk assessment.* The next step in a standard risk assessment (risk analysis as discussed in Section 2.3) is the development of an annual loss expectancy (ALE) for each function, operation, application, user, and so on involved in the study.

   Here, we impose a modification to the standard procedures by viewing the potential interruption caused by an adverse incident in two distinct manners: (a) How long can the specific operation, function, user, and/or application be interrupted before it will seriously (and *seriously* has to be qualified) affected? (b) Do the specific threats, hazards, and vulnerabilities identified fit into any of the categories of the baseline security concepts? If yes, then consideration should be given to the specific recommendations presented. If there is no baseline security reference to the specific threat or vulnerability, then the task force member who has identified the exposure (vulnerability) should develop a specific baseline security worksheet using the format presented. It is important to note that almost all the potential threats, hazards, vulnerabilities, risks, and exposures—adverse, unexpected, or undesirable incident—that can be identified are covered in the Disaster/Recovery Data Collection Manual either in single work identification or in full disruption definitions.

There are potentially other conditions that can exist (i.e., one facility evaluated located a data center less than 200 yards across a river from a military munitions storage facility. This same facility is located above all the incoming natural gas lines to a food processing plant. In addition, a gasoline fuel storage tank is located within 50 yards from the building, and the fuel pumps are located just under the outer walls, on both sides of the building housing the data center). Obviously, no one can account for the potential threat this kind of environment creates not only for the safety of individuals, but also for the numerous functions (computer and noncomputer) that could be totally destroyed in the event one or more of the hazardous elements ignited or blew up.

This type of condition doesn't require much risk assessment or calculation of ALE. It requires immediate action to remove the operation to a safer environment—whether or not the removal is cost-effective.

8. *Cost versus benefit.* The next phase (or stage) of the risk assessment process requires the calculations of cost versus benefit in order to justify the implementation of the required security/prevention mechanisms to protect the facility, operation, and so on.
9. *Final report.* The results of the entire study should be collated into a final report addressed to management. This report has to include the findings as well as the specific recommendations of the task force. The report establishes the basis on which the future activities (action plans) are developed and implemented—for a prevention, security, and controls program, as well as the development of the business interruption preparedness plan (disaster/recovery plan or contingency plan).

2.3.1 Probabilities of a Disaster (DC)

During the data collection phase and using the material presented in the risk assessment section and the activity just described, the Disaster/Recovery Planning Team will be required to identify the disaster (adverse incident) probabilities. In Section 2.3.2 following, they will be asked to develop a series of disaster scenarios. Each organization should select one or more of the methodologies presented in the data collection section.

2.3.2 Disaster Scenarios (DC)

The use of disaster scenarios, rather than a list of potential disaster items, provides those groups who will be participating in the plan advance knowledge of potential incidents, described in detail, thus allowing a better understanding of the problem and providing a basis for the development and implementation of emergency response plans and action plans based on some form of reality, not guesswork.

2.3.2.1 Sample Scenarios

"A power failure in the area in which the main headquarters and central data processing operation are located will render all activities null and void. Historically, this can be expected to happen at least once every three years, and the extent of the power failure has been 8 to 24 hours. Most of our activities are generated from regional and divisional offices; therefore, a power failure would render these offices totally inoperative. Operations could reasonably cease for 8 hours, but beyond that, remedies would have to be taken to rectify the condition.

"The Missouri River has a history of overflowing its banks at least once each year. This has created sufficient power outages and in some cases employee inability to get to work. Sufficient prevention mechanisms have been taken to protect the property of the company. At the time of a flood period, it is expected

that employees will remain with their families; those who have no family responsibilities may be unable to reach the building. Therefore, all operations will cease until the flood waters recede. The maintenance staff will be available within a reasonable period of time to clean up the area surrounding the building and any water or debris in the building. Employees will be notified as to when to report to work.

"Regional and district offices and other manufacturing facilities that are outside the flood area will be required to develop a disaster plan to interface with the backup site that will support all computer operations."

2.3.3 Costs of Downtime

There is a direct relationship between the loss of data processing operations and a "cost" or "loss of profits" to the company. This can be related to direct financial loss. In simpler terms, a travel agent dependent on booking clients through major airlines systems (e.g., Sabre) would be hard pressed if through a communication failure they lost access to the system. Their first step toward recovery would mean a reversion to manual operations (direct voice access to an airline) without benefit of alternatives, extensive delays in determining the most appropriate flight, the inability to provide the most economical flight(s) for most of the clients, hand-written tickets which are time consuming, and so on.

This degradation of service would result in a series of cost events:

1. A reverting to manual operation (potentially increasing the need for additional personnel).
2. The cost of time for each transaction times the number of transactions.
3. The inability to service customers means loss of revenue.
4. The loss of revenue reduces profits (and cash flow).
5. The inability to obtain authorization for credit card transactions; potential return of charges from credit card companies.
6. Time to obtain payments.

Each organization has to view its specific activity in this light and determine the cost-effectiveness (financial loss potential) relating to any degradation of activities based on computer failure or loss, or the loss of any other service unit within the organization as a result of a disaster. The scope of this review covers every activity, from the origination of a revenue activity to the ultimate collection of funds in payment for services or products. One of the more intangible figures in a loss calculation is the specific loss of a customer base, the cost associated originally in obtaining the customer, and the cost to get the customer back.

2.3.4 Assumed Maximum Downtime (DC)

Each organization, each functional unit within an organization, divisions, subsidiaries, and manufacturing, research and engineering activities all operate differently. Therefore, each of these units, as well as the organization as a whole, must determine at what period in time after an adverse incident the critical support services and the performance of the company reach a point where it is no longer cost-effective to perform the services, deliver the products, or achieve the organizational mission.

Again, each group within the organization, together with a management committee, must determine what the critical (computer and noncomputer) support services are. Without these services, how long can the organization maintain a reasonable business activity?

Here, it is necessary to define those critical operations and determine at what point the performance of the organization will be degraded to a degree where potential losses will render the organization inoperative.

2.3.5 Worst-Case Assumption (DC)

The master plan that will be developed during the plan preparation stage is based on a *worst-case condition*. To define a worst-case condition (assumption), state that an adverse incident has totally destroyed or damaged (1) the organization's central operating location (computer and noncomputer operations), (2) a major user's location, (3) the central and/or satellite data processing center, (4) the database, (5) the software (which may have been illegally altered or drastically disrupted, and no current version is available), or (6) power or communications [which may have been severed to the degree that, for all practical purposes, the disaster will result in an interruption of considerable time before the initial site(s) (functions) can perform in an accustomed, normal manner].

In designing the plan based on a worst-case condition, every backup, recovery, team assignment, and function is required to be detailed in a highly definitive manner. Less than a worst-case environment disaster requires a determination by management of the level of disaster—the period of time the disaster will last—and implementation of various subsets of the master plan that meet the specific incident results directed towards the recovery of operations.

2.4 Other Manuals (DC)

During the data collection phase, the disaster/recovery coordinator should make every effort to obtain those manuals that represent every aspect of organizational policy, emergency response procedures, organizational charts (executive level, line, and staff), divisional and subsidiary descriptions, and internal organizational functions. Other documentation includes those publications that deal with users, applications, data centers activities, and so on. The titles may be (but are not limited to):

1. Corporate Organization Manual
2. Government Handbook (agency, ministry, commission)
3. Personnel Manual [organization and data processing (critical)]
4. Emergency Procedures Manual (covering specific user and data center sites) (critical)
5. Data Center Emergency Response Manual (critical)
6. Corporate Information Management Policy (critical)
7. Records Management Policy and Program (including vital records and retention policy)
8. User Application Documentation (critical)
9. System Development Standards and Policies
10. Data Center Standards and Procedures Manual (critical)
11. Data Center Users Guide
12. Data Center Operations Manual (critical)
13. System Software Documentation (critical)
14. Computer Hardware Documentation (critical)
15. Internal Audit Standards and Procedures Manual (critical)
16. Annual Report (one or two representative years)
17. Forms Book

We recognize that these and other manuals represent substantial documents (if available), and therefore are impossible to transport. The disaster/recovery coordinator should list those manuals that are available, and provide an abstract of each publication. If there is a reference number, this should be included. Those publications that are critical to a rapid and timely backup and recovery and which should be part of the disaster/recovery coordinator's library should be so noted; a copy of these should be stored off-site to ensure that the policies, standards, and procedures are readily available in the event all other documentation is destroyed. *The objective of the overall business interruption preparedness plan is to ensure a timely, rapid, disciplined, structured, management-directed, orderly recovery of any or all the business functions of the organization in the event of a catastrophic event.* Procedural and policy documentation provides this organized methodology and leaves little to decisions at a crucial time during an organization's struggle to survive.

2.5 Role of the Disaster/Recovery Management Team

Each organization will structure its Disaster/Recovery Management Team differently, but in essence, the charter of the DRMT is established for two primary purposes:

1. During the preplanning and plan development stage, it is the DRMT's authority and responsibility to approve all activities, strategies, action plans, and general activities dedicated to the business interruption preparedness plan.
2. At the time of disaster, the DRMT has the authority of top management to proceed in initiating any activity and utilize any resource of the organization to recover the business operation and affected site or function—to return to a normal operating environment as quickly, efficiently, and cost-effectively as possible.

It is the further responsibility of the DRMT, based on observation and best information available (computer operations, vendors, emergency services, building engineers, and so forth), to declare a disaster (initiate the disaster/recovery plan based on the assessment of the damage), expend the funds necessary to back up and/or recover the affected site in order to maintain the continuity of operation, achieve the corporate mission, and ensure the integrity and accuracy of the organizational and information assets of the company.

Those selected for a position on the DRMT have the ultimate directive and mandated responsibility of reconstructing the company (damaged or destroyed facility) at the time of any adverse incident, regardless of the activity affected. The DRMT will remain mobilized and perform their function until the disaster has been neutralized and operations have been recovered to the level that is acceptable to senior management.

2.6 Plan Initialization

Only the Disaster/Recovery Management Team, and in their absence the highest echelon of an organization, can declare a disaster and mobilize the people and resources necessary to maintain the operational structure and viability of the organization.

3. Prevention/Security

The primary obligation of management is to provide a sufficient level of security within the organization that (1) will prevent a potential disaster, and (2) in the event a security breach occurs, will establish the necessary response plans to mitigate the damage to and interruption of business activities. This is not a one-time project. Sections 3.1 to 3.16 cover the minimum subjects that should be addressed in order to implement a prevention/security program. They cover both physical and logical security—directing such programs to every functional unit of the organization. The best plan for disaster/recovery is to prevent the disaster in the first place.

In this section it is important that details of the prevention/security strategies and mechanisms be outlined. Where security provisions are nonexistant or where they might be considered "soft," the enhancement of those aspects of prevention and security that must be provided should be carefully listed (obtain management approval to implement them, assign the responsibilities, and decide when they will be implemented).

The disaster/recovery coordinator or the presenter should use the facilities evaluation checklists, starting on page 72, to determine the status of prevention/security within the organization as well as in the data processing department. Once the "soft" areas have been identified and the security mechanisms are put in place, security reviews using the facilities evaluation checklist must be performed periodically.

There is no substitute for personal experience and knowledge about the organization. All that the checklists can provide are guidelines. It is up to the data collection task force to assign determination responsibilities; evaluate the responses, checklists, and questionnaires; and determine those potential threats, hazards, vulnerabilities, risks, and exposures to the organization, data center, user, noncomputer support operations, and so on.

3.1 Physical Security (DC)

The cornerstone of the entire business interruption preparedness plan is *prevention*. Physical security provides the bulwark to assure that people, facilities, user centers, data centers, data, documentation, and source documents are safe. The determination as to what they have to be safe from has to be made by individuals on-site who can recognize risk and exposures. The following sections address the basic items only; further enhancements based on specific problems must be addressed by individuals within the organizations, and enhancements to existing security procedures and mechanisms must be reinforced.

3.1.1 Layouts (DC)

Layouts of all operating facilities provide security management with the opportunity to evaluate prevention and security, primarily on a physical basis. In addition to the evaluation process, the preparation and display of such layouts and diagrams assist operating personnel to quickly identify those areas where a signal (annunciation) has been actuated, and take remedial action.

These layouts are part of the disaster/recovery plan and appear in Addenda C1, C2, and C3. The number of layouts (drawings) are based on the scope of the plan. If the plan encompasses areas outside the data processing area, then appropriate layouts should be prepared similar to the ones requested in C1, C2, and C3.

Layouts provide additional reminders as to what hardware, communications, equipment, office machines, chairs, file cabinets, desks, and so on will be required in the event the facility will have to be reconstructed.

3.1.2 Locks (DC)

This unit should describe in detail the physical application of restricting access control to the building, to the working areas, to the data center, to user areas, and so on. Again, the content of this section should restrict itself to the general scope of the plan.

3.1.3 Guards (DC)

If there were a specific procedure relating to the activity of premises guards, then the entire document rather than any simplified description would be included here. Many organizations overlook the value of guards on a permanent (off-hours) basis, or at minimum, the hiring of guards to monitor the facility during shift closing times, holidays, long vacation periods, and so on.

3.1.4 Cameras (DC)

Surveillance cameras offer a level of protection at a minimum of cost while maximizing coverage. Monitors located at critical points of the building during the daytime working hours provide a substantial level of security, especially if the activity picked up by the cameras is viewed by more than one person. A single observation post for off-hours creates maximum coverage but limited response capability. Moreover, observers tend to get vision paralysis when viewing screens for any length of time.

In general, do not depend on monitors alone. Other alarm systems that bring attention to the area are also important.

Surveillance cameras should be placed at strategic locations inside and outside the building to achieve maximum coverage. Additional lighting may be required to obtain the most effective observation condition.

3.1.5 Emergency Telephone Numbers (DC)

In this section you should list all emergency telephone numbers:

1. Air-conditioning service (heating)
2. Bomb squad
3. Building superintendant and alternate
4. Central alarm system headquarters
5. Emergency medical services
6. Environmental Protection Agency (EPA)
7. Federal Bureau of Investigation (FBI)
8. Field engineer(s)
9. Fire department
10. Fire extinguisher service company
11. Hospital
12. Insurance company (day and emergency number)
13. National guard
14. Police department
15. Power company (day and emergency number)
16. Telephone company (day and emergency number)

3.1.6 Intrusion Alarms

To support the surveillance cameras and monitors, intrusion alarms are extremely important. There are few companies who feel comfortable leaving their premises unoccupied for a period of time (weekends, extended holiday periods) without having the additional support of intrusion alarms. Consideration should be given to having the intrusion alarm (fire, water, and so on) annunciate at a central station (e.g., the police department or fire department). Delayed response to any alarm could, in fact, be advantageous to the illegal intruder.

This paragraph should describe the type, distribution, coverage, annunciation, and response time for an intrusion alarm. If the alarm annunciates at a central security station (outside the company), details should be given as to the sequence of events and the person in the company who is called in response to an alarm.

3.1.7 Medical Alert

There are a number of localized alarm transmitters. These should be supplied to employees who usually work alone in the data center, or anywhere in the organization. They also should be supplied to stationary and roving guards. These systems usually have a unit that can be attached to an employee's belt. In the event the employee becomes ill, or an intruder has entered the premises, the employee only has to hit the surface of the unit. A signal is then transmitted to a central annunciation panel in the building and further transmitted to a central alarm system outside the premises of the building. Guards should be provided with the same system.

Safety and welfare of employees is important to the individual. It also protects the company from any lawsuit that may arise as a result of personal injury or death.

3.1.8 Access to the Data Center (DC)

Absolute access control into the data center, including any operations area, is extremely important. Access must be restricted only to those having a reason to be in the area. Vendor service people (outside service people, field engineers, telephone repair personnel, air-conditioning service people and so on) should be restricted and allowed entry only on a "need" basis—and only with an escort. Organizations with permanently stationed field engineers should prohibit direct access to the machine room from an assigned office. Knowledge of the whereabouts, and the work to be performed should be controlled by data center management or operations personnel, and all access of the machine room should be logged.

In this section, describe the system presently in use to restrict unauthorized personnel access to any data processing environment. Include systems or procedures used to identify data center personnel activities. These will be detailed in this section; lists of authorized employees, data center personnel, executives, vendors, and identification procedures are listed in appropriate Addenda C-5, C-6, and C-7. A sample of the data center sign-in sheet and review procedures is shown in Addendum C-8.

These lists as well as a sample of the sign-in sheet must be prepared during the data collection phase.

3.2 Medical Security

This section, and the items listed in Sections 3.2.1, 3.2.2, 3.2.3, and 3.2.4 briefly describe the emergency medical response procedures and training that are important to the health and safety of company (and data center) employees. Cardiac Pulmonary Resuscitation (CPR) is an important part of employee response training. Data Center employees are potentially subject to high levels of stress as well as electric shocks—both requiring CPR. Stress can also cause minor or severe hyperventilation. First aid and CPR training from hospitals and local Red Cross chapters are important for the protection of lives.

If there are no medical security procedures or trained personnel, planned enhancements should include the requirements for such training.

3.2.1 First Aid Kit

3.2.2 Nurse

3.2.3 General Employee First Aid and CPR Training

3.2.4 EMS (or Fire Rescue Squad)

3.3 Fire Protection (DC)

Fire is probably the most frequently occurring incident of any potential major disaster outside of people problems. Therefore, special attention must be given to the emergency response to this danger. Establishing policies, procedures, and training schedules for personnel is a major responsibility facing corporate and data processing management. The decision whether to fight fires on a manual basis, or to allow employees to vacate the premises and use passive (Halon or sprinklers) methods to be activated in order to extinguish fires is usually based on the ability of on-site personnel to cope with the existing situation. Usually, minor fires are easy to surpress (which mitigates potential extensive damage) by the use of hand-held extinguishing units (Halon, CO_2, and so on). Without previous training, employees should not attempt to fight a fire. Trained employees should be encouraged to put out minor fires, as well as notify appropriate emergency services (fire department, building fire brigade, and so on).

The introductory paragraph in this section should reflect some historical background and a brief description of the fire prevention mechanisms in-place throughout the site, and more specifically, within the data center environs (machine room, print area, check sorter area, paper storage area, electrical cabinets, and other critical areas).

3.3.1 Fire Alarms

Describe the existing methods (equipment) in the building and in the data center that will set off a fire alarm. Water alarms (water resulting from leaks, air-conditioning condensation, overflowing utilities, and burst pipes) should be detected as early as possible. Describe the system procedure that results at the time smoke, heat, or water is detected. This includes the automatic process of power and environmental conditioning shutdown.

3.3.2 Hand-held Fire Extinguishers and Fire Hoses (DC)

Describe in detail and include a diagram (layout) of the location of all types of hand-held fire extinguishers available to personnel. Describe the levels of training given to employees. This should also include a description of the training and use of fire hoses.

Note: All fire extinguishers should be at a height that makes them easy to see and reach. Because of the height of cabinets in a data center, fire extinguishers (as well as floor tile lifters and fire alarm boxes) are usually hidden. Different color arrows indicating the purpose for each fire extinguisher (paper, electrical, chemical, and so on) should be placed at sufficient height for easy identification and location. Suppliers of fire extinguishers will supply color-coded units with appropriate identification/location pressure-sensitive labels describing the use of each extinguisher.

Add any enhancements that have been decided upon and note when these enhancements will be put into operation. These would include but are not limited to:

- Color-coding of extinguishers
- Additional extinguishers
- Training programs

3.3.3 General Fire Safety (DC)

Describe existing procedures and levels of fire protection.

1. Determine whether building and data center wiring conform to national electrical installation and local building codes.
2. List the use of appliances within the building as well as the data center, and note whether or not they are Underwriters Laboratory (UL) approved.
3. Obtain a statement from outside (electrician, fire underwriters, insurance loss control specialists) professionals certifying that the electrical system is properly grounded, and that appropriate surge protectors (or isolation transformers) are installed and are properly rated.
4. Be sure that all exits are properly marked and properly illuminated so that they can be seen even in a smoke environment. Arrows that indicate escape routes should be either put on pillars or hung from the ceiling.
5. Determine whether there is sufficient emergency lighting in work areas as well as escape routes.
6. Indicate whether there is an emergency power-off switch at each exit from the data center.
7. State whether all exit doors that are electrically operated return the latch to a neutral position if the power fails.
8. Establish that existing procedures to evacuate employees have been tested, and that people are assigned accountability responsibilities. (See Addendum C-9, "Evacuation Route." Post this diagram at appropriate locations throughout the floor and data center.)
9. Note whether all emergency exits are checked daily by a supervisor to see that there are no obstructions.

10. Indicate whether periodic fire drills are scheduled and that all employees are required to leave the building, convene at assembly points, and be accounted for.
11. Note whether supervisors of second and third shifts are required to perform the same fire drills and evacuation procedures.

3.3.4 Passive Fire-Fighting Measures (DC)

This section describes the "passive" or automatic fire-extinguishing equipment in the entire building and/or the system(s) in the data center. This would describe and the diagram(s) would indicate the location of sprinkler heads and their method of actuation (wet pipe with fuse links or dry pipe system, either pressurized or flow-controlled). In addition, mention the passive (automatic) fire-supression method used in the data center—Halon 1301 (or another) and whether the Halon system is a single- or a double-shot system—the method to abort the activation, and the time delay between alarm annunciation and actual discharge. Also list the procedures for personnel to follow for evacuation, for power-down (both manual and automatic), and for shutdown of all environmental conditioning equipment prior to release of the Halon (or sprinklers).

3.3.5 Fire Drills (DC)

Each department head or supervisor will be notified by the company safety committee, the building management, or the personnel department of *surprise* fire drills.

Describe in detail the procedures that are anticipated to be implemented within the organization. (See the details in Section 3.3.3.)

3.4 Emergency Response Plan

In the event of an emergency, detailed action plans (emergency response and procedures) have to be established and disseminated among employees. During a crisis, it is necessary that managers and personnel be required to make the minimum of decisions. The action plans/procedures should be detailed for each of the specific emergencies determined to be prevalent (or probable) for the facility.

Each potential emergency should be listed on a separate page; the responsibility for initiating the plan and a list of detailed procedures should also be noted. The entire document should be published in a single document and distributed to all personnel in the organization.

The document should include response/action plans concerning many critical injuries and body-function failures. This would include burns, lacerations, deep cuts, impacts, head injuries, choking, smoke inhalation, heart failure, epileptic attacks, electric shock, hyperventilation, and high allergic reaction (anaphylactic shock). Many local emergency medical services, hospitals, the Red Cross, the American Heart Association, and burn centers have established procedures that can be easily duplicated and incorporated into the Emergency Response Manual.

A sample Emergency Response Manual is available from DIA*log Management, Inc.

If the organization has an Emergency Procedures Manual or an OSHA administrator, review of this material or discussions with the OSHA Administrator or the building maintenance department as to appropriate material for inclusion in the manual is important. Updating material presently in force should be reviewed by a safety committee, the local fire department, and loss control engineers of the company's insurance carrier.

3.4.1 Contents

Probable emergency procedures that should be addressed are:

1. Fire and explosion
2. Tornado
3. Flood (internal or external caused)
4. Earthquake
5. Civil disturbance
6. Bomb threat
7. Power failure (local or areawide blackout)
... plus other threats.

Details on evacuation and procedures for plant, building, and data center shutdowns should also be included.

3.4.2 Emergency Response/Action Coordinator (Building and/or Plantwide)

Depending on the size of the organization (size of the building, number of floors, floor area, number of people, location of people, building separations, and so on), security and emergency response responsibilities would rest with one or more qualified individuals.

List those people and their alternates who would be responsible for initiating an emergency response plan.

3.4.3 Data Center Emergency Response Miniplan

Certain procedures that pertain to the actions of data center personnel are not applicable to those of other general activities. In addition, many data centers are isolated or insulated from other activities or building areas. A separate emergency response plan with additional requirements should be developed for the data center.

3.5 Housekeeping

3.5.1 Data Center Prohibitions

Detail all of the restrictions that are presently enforced within the data center and operations area, and/or any other area where expensive data processing equipment is installed, or where data is processed or stored. For example:

- No smoking
- No eating
- No drinking
- No appliances to prepare food or drinks

3.5.2 General Cleanliness

Rules that must be enforced to keep the data processing area clean and free of waste materials, and the maintenance schedules for the area have to be detailed. Restrictions concerning the types of cleaning apparatus and tools should be designated and distributed to appropriate maintenance personnel (e.g., heavy-duty vacuum cleaners, abrasive cleaners that leave powdered residue, inflammable solvents, spray cans with cleaning agents, or wax that leaves airborne particles and residue).

3.6 Power (DC)

The paragraph should describe the organization's and/or data center's approach to providing clean, uniform, and continuous power. Sections 3.6.1 to 3.6.5 describe the specific levels of power protection (enhancements) that have been designed into the facility to take care of fluctuations/variations (based on historical data) in power. Describe here the overall approach to power taken by the data center.

3.6.1 Isolation Transformer (Required)

The minimum protection required for any computer equipment as well as many other electronic devices is an isolation transformer. Detail here the type and characteristics of the particular isolation transformer. Also detail any other power protection device.

3.6.2 Motor Generator (Optional)

3.6.3 Uninterruptible Power Supply (UPS) (Optional)

3.6.4 Engine Generator (Optional)

3.6.5 Reference Grid (Optional)

3.7 Air-conditioning (DC)

In this section, you will be required to detail the existing air-conditioning (and other environmental-conditioning equipment used in the data center; for example, electrostatic precipitators and humidifiers). A brief description concerning the location of the system, where other components (compressors, air handlers) are located, and the size and type of equipment is also necessary. The descriptions should include whether each is an overhead or underfloor system (or both).

Enhancement may include stand-by backup system; diverting of house air-conditioning at time of failure; location of fresh-air intakes; filtration; and so on. In addition, detection of failures can be accomplished by the installation of a temperature/humidity recorder.

3.8 Maintenance Agreements (DC)

Describe in general the approach to equipment maintenance. See Addendum C-10, where copies of all maintenance agreements should be placed. (*Note:* Periodic verification of old and new maintenance is important to determine if all equipment is covered, and whether the agreement covers equipment that has been abandoned or discarded.

Maintenance agreements are required because they are usually the most definitive list of equipment and peripherals by model, serial number, issue date, issue level, and so on.

3.9 Document Security

3.9.1 Classification

It is important that each organization establish a document classification program. Management should provide the criteria for such classifications and mandate an active, ongoing records management security/retention/classification program. This should include, but should not be limited to storing, shredding, and forwarding of documents.

3.9.2 Storing: Security and Retention

3.9.3 Shredding: Security

3.9.4 Forwarding: Security Methods for Distribution

3.10 Software Security (DC)

Piracy and illegal use, sale, or distribution of proprietary or leased software is one of the most serious breaches in organizations today. Illegal acquisition and sale of in-house programs leads to potential capabilities to invade or abuse existing systems. Civil or criminal suits may result from distribution of multiple (illegal) copies of leased software throughout the company or outside the company, which is a breach of most contractual stipulations.

This section describes the methods in place to protect software. Use Sections 3.10.1 through 3.10.2 to describe the methods presently used in the organization.

3.10.1 User Code/Password Practices

3.10.1.1 System Release Level

3.10.1.2 Application Level

3.10.2 Required Password Change

3.10.3 Other Applications Security Measures

Describe existing security measures presently being enforced. This covers security mechanisms for data libraries, data, terminals, and user restrictions.

Security enhancements that are planned for should be described here. Indicate security procedures, the person(s) responsible for their implementation, and the date(s) of installation.

3.10.4 Pack Lockout

Describe the method for protecting program library pack and the method for access.

3.10.5 Program (Tape) Library Control System (PLCS)

Describe the method used for introducing program changes and procedures for documenting, backing up, and changing (compiling) new object code.

3.10.5.1 Tape Library Management System

Describe the present (commercial or homegrown) tape library management system. (See Section 4.2.6.)

3.10.6 Erasure of On-line Copies of Sensitive Data

Describe the methods and procedures for removing sensitive data from packs, eliminating such data from tapes, and/or preventing access to, or erasure of, sensitive data.

3.11 Datacomm Security

3.11.1 Physical Security of Terminals

Terminals located within the environs of the data center, remote terminals, and/or any terminals that are on-line should be provided with sufficient physical and logical security mechanisms to protect them from abuse, damage, theft, or unauthorized use. The descriptions should cover the physical and logical security measures that are presently in effect, and should include, but are not limited to:

1. Terminal locations and physical security measures taken during and after working hours.
2. Methods used to monitor and record attempts to access the system, from what terminal, and time of access attempts.

3.11.2 Data Encryption Standards (DES) Devices

Organizations who transmit or receive highly classified, trade-secret information or other sensitive data should consider using data encryption. Most terminals and/or modems can be outfitted with standard data encryption "chips" that are commercially available. If such requirements are necessary (or are used) to protect data from illegal datacomm eavesdropping, this should be described in this section. If a data encryption standard has been implemented, the method, less specific details, should be described here. If a limited use of DES is utilized, consideration should be given to standardizing the activity to avoid operations confusion or decisions that have to be made as to whether or not a specific set of data should or should not be protected during transmission.

3.11.3 Leased Lines

Describe the utilization of leased versus dial-up lines as a method for datacomm security.

3.11.4 Dial-back Feature for Dial-up Lines (DC)

All dial-up lines should be provided with security-designed modems that require dial-back verification. These can be installed in place of existing equipment.

Where a less sophisticated method is appropriate, security should require that users desiring access be screened through access-control phones, thus avoiding automatic log-on. Users should leave their phone number, the number will be verified, and a call-back by the data center establishing log-on procedures will be made.

Every effort should be made to avoid *modem tone* output on log-on attempts. Voice response eliminates unauthorized system access.

3.12 Application Development Standards (DC)

The implementation of new applications *must* be controlled from the feasibility study to actual release for production.

In this section you should couple the vulnerabilities with specific controls. Areas covered should include controls as outlined in the matrix for computer controls chart.

1. Transaction origination
2. Data processing transaction entry
3. Data communications
4. Computer processing
5. Data storage and retrieval
6. Output processing

Each of these six should be viewed for areas of:

1. Erroneous falsified data input
2. Misuse by authorized end users
3. Uncontrolled system access
4. Ineffective security practices for the application
5. Procedural errors within the data processing facility
6. Program errors
7. Operating system flaws
8. Communications system failures

There are six basic control categories that should be addressed for each general problem:

1. Data Validation
    a. Consistency and reasonableness checks
    b. Data entry validation
    c. Validation during processing

d. Data element dictionary/directory
   e. Audit and review
2. User identity verification
3. Authorization
4. Journaling
5. Variance detection
6. Encryption

The phases are to be defined as:

1. The initiation phase
2. The development phase
3. The test phase
4. The implementation phase
5. The post-implementation review phase

3.13 Application Support Standards

The same security and implementation procedures that form the base for system development should be implemented for any changes. Audit involvement must be a basic *security* consideration in any change.

3.14 Application Purchase

Purchase, lease, or license of externally developed software packages must be viewed in the same manner as the internal development project. List here the process your organization goes through in evaluating packages and the methodology in establishing criteria for selection.

3.15 Role of the Auditing Staff

The Foreign Corrupt Practices Act of 1977, and other federal, state and local laws and regulations, as well as Generally Accepted Accounting Procedures (GAAP) have established the *requirements* for the control objectives approach to accurate record maintenance and data integrity. These include but are not limited to:

- Appropriate authorizations
- Appropriate accounting classification
- Substantiation and evaluation
- Adequate physical safeguards

The accounting system control objectives benefit not only the security of the software product; they also

- Provide clear management guidance for the detailed design of the application systems.
- Provide auditors with a set of criteria with which to review specific controls in order to ensure that management's policy has been carried out.

The auditor (internal or external) should consider, and it is the responsibility of the plan development group to document here, the basic control objectives in place at the organization.

- Appropriate authorizations
- Appropriate accounting classifications
- Substantiation and evaluation
- Adequate physical safeguards
- Recognition of economic events
- Acceptance of transactions
- Integrity of processing

- Integrity of reports
- Integrity of databases
- Integrity of interface

The auditor's responsibility is to compare the organization's basic control objectives against the components of a fixed procedure that are actually in place. The designer should involve the auditing department in order to ensure that sufficient controls exist, based on the sensitivity of the application.

3.16 Role of the User Department(s)

The user is the motivating base for the development of specific systems and applications. Regardless of the extent of an application and its use within a multiuser environment, it is the responsibility of the user to establish the criteria for the application.

The user should be a participant in a security review meeting where his or her input concerning the controls available and the types of controls that should be installed in the software development process are warranted.

Document here any procedures that are in place relative to the input of the user community during the development process.

4. Disaster Preparedness

This is a general policy statement. Rewrite this paragraph to reflect the policy of the organization.

4.1 File Backup (DC)

Each organization functions in a "unique" environment in performing its operations. Requirements for file backup procedures will vary, depending on time constraints, size and number of transactions, and/or company policy. Of prime consideration is the ability (necessity, obligation, responsibility) to protect the information assets of the company and its interests.

The user (owner) of the software and data has to evolve a file backup strategy in conjunction with the data processing department that will, under an adverse condition, provide the capability to recover within the established time frame to support the organization. This would fall under the structure of a vital records program which has two primary objectives:

1. Protection
2. Reconstruction (immediate or disaster)

Reconstruction addresses the assurance of the continuity of operations in the event of any loss of data, within a reasonable time delay—recognizing that a reasonable time delay is unique to the requirements of each organization.

These vital information management procedures and policies not only include data and software, but include run books (instructions), documentation, equipment configuration and documentation, and so forth.

Document in this section the file backup procedures for each application (production job, database), followed by the organization.

4.1.1 Decision Criteria (DC)

Based on the sensitivity and/or criticality of the software, and more important daily generated data, a backup requirements methodology must be developed. Detail in this section the generation depth and audit trail activities that the company has selected for each application. In most cases the methodology selected should have no relation to cost-effectiveness, owing to the nature of the importance of the data.

4.1.2 Pack Family Usage Rules (DC)

Describe here the pack family (name) and the specific usage.

4.1.3 Backup Strategies (DC)

4.1.3.1 Operating System Code

Describe the generation backup within the data center (number of copies and basic offsite plan).

4.1.3.2 Application Code

Describe the backup (generations) and dump frequency for application code.

4.1.3.3 Databases

Describe the number of generations of the database (on tape).

4.1.4 Dump Schedule (DC)

Chart your present system for the dump schedule.

4.1.5 Source Documents (DC)

The source documents protect the organization by providing the ability to restore the system if the entire database is intentionally or inadvertently corrupted, modified, or improperly used. Off-site and/or other methods (i.e., microfilm, microfiche) should also be noted.

4.1.6 Terminal Floppy Dumps (DC)

It is important that the backup of each transaction entry on on-line terminals should not be overlooked. Off-line activity usually contains one-of-a-kind information, and a backup of these activities should be made at a prescribed schedule and secured off-site in the same manner as other magnetic media and documentation.

Document the established schedule for the backup of diskettes. Users should be responsible for their own backup activities. Although they are the owners of the data, the central information custodial activities should reside with the data center librarian.

4.1.7 Microfiche Reports (DC)

Computer-generated or photo-processed microfilm and microfiche provide an important *security* function for source and specialized documents entered into the computer. If stored off-site on a regularly scheduled basis, the loss of the original document in the event of a disaster is unimportant, as this type of medium provides reconstruction capability.

Document all vital records backup procedures as well as retentions and off-site procedures.

4.1.8 Period-end File (Tape) Retentions

There are two categories of file retention. One deals with the weekly, monthly, quarterly, semiannual, or annual files that are used periodically after the specific period has ended. These files are used for reference, or in the extreme case, for reconstruction of damaged, lost, or corrupted files. The second deals with the archiving of vital corporate data, with retentions established by users, company policy, regulatory requirements, or for other legal conditions.

In this section document all file activity and note whether the data would remain on the system for the specified period in production files or in archival mode. Retentions should be verified and documented by the user, legal department, or financial executives of the organization.

4.2 Off-site Storage (DC)

Document the company's off-site security program, listing:

1. Documentation
2. Disaster/recovery kit
3. Data files
4. Software
5. Hard copy (microfilm, microfiche)

This program should be reviewed in the worst-case condition. All documentation and records (magnetic media and/or other records) should provide the total capability of recovery in the event all on-site records are lost.

4.2.1 Off-site Management (DC)

Document in detail the off-site management of files. Without the proper off-site storage programs, the recovery plan will have a major flaw if proper preparations have not been made.

4.2.2 Off-site Strategy (DC)

See the description for Section 4.2.1.

4.2.3 Off-site Rotation Schedule of Current Tapes

In this section, document the off-site rotation schedule of current tapes. It is important to address the number of generations, the time of day backup is performed, and the time of day when backup is actually removed from the premises.

4.2.4 Off-site Archive of Period-End Tapes

Describe the organization's procedure for the determination of archiving of period records (company policy, industry standards, accounting requirements, government regulatory requirements). Document the present off-site structure used in the organization.

4.2.5 Documents Off-site

List all documents maintained off-site. This list should include *all* vital documents necessary to reconstruct operations in the event of a total (or partial) loss of on-site records, or any incident that will restrict access to the building.

4.2.6 Tape Library Management System (DC)

Describe the present tape library management system (TLMS) utilized by the organization. (*Note:* A sample report should be placed in Addendum D-2.) If the TLMS has been developed in-house, a copy of the latest version of the system (magnetic media/tape) should be stored off-site. If it is a purchased software item, guarantees should be part of the agreement that state, "In the event of a loss of the software, the vendor will supply immediate replacement, and assist in its re-implementation." It is also recommended that the company engage in a third-party escrow agreement concerning the source code availability.

4.2.7 Scratch Tape Management

Magnetic media is a finite resource. The continued reuse of tapes and other magnetic media (disk packs, diskettes) require control in order to ensure the viability of the medium to provide continued performance. A company policy *must* be established as to use, reuse, rotation, and destruction procedures.

Document the present scratch tape management procedures in use. If the present system does not provide adequate protection and control, the new system should be described and listed in the planned enhancements.

4.2.8 Scratch Floppy Disk Management

The ease by which floppy disks can be removed from company premises and the nature and sensitivity of material usually found on this type of magnetic media requires the implementation of high-level security controls.

Methods to erase and/or destroy diskettes should be part of the security procedure.

Document the present and future programs concerning the security and scratch management of diskettes.

4.3 Backup Site Strategy (DC)

4.3.1 General Alternatives

In order to achieve the timely recovery of an affected site, it is necessary to provide for a backup capability. These alternatives are covered in detail in Part 1, Section 3. The plan is based on the actual backup facility being in place, or the assumption that a corporate decision has been made relating to the backup strategy or strategies. The plan should reflect strategies that will be in place perhaps 30, 60, or 90 days hence, if a backup alternative has not been selected.

Document the backup strategy subscribed to or assumed by the organization.

4.3.2 Specific Alternatives

Document the plan methodologies once the selection of the strategies are finalized.

4.3.3 Chosen Alternatives (Period of Time)

After careful consideration and the identification of critical applications and resources required to support these critical applications a decision has to be made concerning how long a period of time can go by without the support of date processing. The alternatives documented under this section cover weeks 1 to 6. The next section (4.3.4) would cover chosen alternatives from week 7 on. Each organization has to develop its own specific *time-based* backup alternative methodology to be documented in the sections indicated.

4.4 Mainframe Configurations

4.4.1 Primary Site

Describe the existing computer configuration at the primary site, and include supporting diagrams and documentation in Addendum D-6.

4.4.2 Backup Site

The backup site configuration is based on the critical application requirements as established by the user community and management policy. Regardless of the strategy or strategies to be used, specific resources are required at the time computer operations are interrupted. Identify the resources at (or to be built into) the backup site that are needed to provide compatibility and capability at the time of a disaster. Detail this configuration in Addendum D-7.

4.5 Datacomm Configuration

Communications comprises one of the most important and vital components in the successful backup recovery activities. Owing to the variables found in most data processing operations relating to the ability to create an on-line, interactive computer utilization, planning for any type of communications failure is, in many cases, more important than the actual loss of the data processing center, or part thereof.

The ability to restore, and the advance planning necessary to provide alternate communications to support the critical user community at the time of a disaster, can be as complex as the entire backup procedure.

4.5.1 Primary Site Communications

Describe the existing communications configuration at the primary data site. Communications network diagrams and descriptions are to be placed in Addendum D-8.

4.5.2 Backup Site Configurations

Detail the communications configuration at (or planned for) the backup site. This description should include information concerning the methodology to be implemented to go around the existing central station, or special internal activities that would complicate reconfiguring the communications lines to the alternative processing site.

Details and diagrams of this communications network and methodology should be placed in Addendum D-9.

4.5.3 Additional Equipment Requirement

Any devices unique to the communications plan should be documented here and in Addendum D-10. This would cover concentrators, multiplexers, special modems, communication conventions unique to the facility or with the telephone central office, microwave or satellite technology, optical fibre, local area networks, direct connect, and so on.

Any unrecoverable communications configurations have to be identified, and alternate topologies have to be created to accommodate the critical on-line applications.

4.6 Emergency Phone Numbers of Employees

4.6.1 Alphabetical List of Phone Numbers

All key personnel (team leaders, alternates, and team members) should be listed in an easily accessible phone directory. This list should contain the office phone number (if the person is at another location), extension (or direct dial number), home phone number, and/or any other method by which the individual can be contacted at the time of an emergency. In addition to key personnel, the phone numbers of all individuals, executives, employees, supervisors, department heads, and so on, should be included on the alphabetical list. Many organizations update and publish an in-house (or organizational) phone directory. This directory should be placed in Addendum D-11.

4.6.2 Calling Tree for Disaster/Recovery Teams

The calling tree for the disaster/recovery action teams requires a different configuration than the alphabetical listing. At the time of a disaster, the action plan PERT diagram identifies those individuals and team members who have to be mobilized to put the plan in place and get backup operations into high gear. The sequence and number of individuals and teams mobilized depend on the initial appraisal of the conditions at the time the disaster has struck. See Addendum D-12.

4.6.3 Calling Tree for All Employees

The calling tree for all employees would include those associated with the data center, as well as those within the user departments. Maintaining accurate employee lists containing employees' responsibilities, addresses (home), and telephone numbers is vital to maintaining the level of continuity of operations to support the activities of the organization.

Important: Home addresses and telephone numbers are confidential, as is the disaster/recovery plan itself. The indiscriminate distribution of this information (or the plan) can have dire consequences. Preplanning as to how confidentiality will be maintained is vital to ensure that distribution is justified. The problem of confidentiality can be resolved by limiting the lists distributed. Key personnel retain segments of the lists (listing those for whom they are responsible), and the calling tree only contains key contacts and their alternates. *Note:* See Insurance Section.

Owing to the requirement of maintaining the privacy and confidentiality of individuals, many personnel departments will not release this information. Therefore, there are four basic questions that have to be resolved concerning the availability of employees at the time of an emergency.

1. Is the employee vital to the backup and/or recovery procedures?
2. Has the employee been informed of his or her part in the disaster/recovery activity?
3. Has the employee agreed to participate in the backup and/or recovery specified by the disaster/recovery action plans?
4. Has the employee (not the personnel department) provided private information such as address and phone number?

In addition, the employee should be required to respond to a simple questionnaire concerning availability and length of time that he or she can be away from home (if necessary). It is also important for the employee to be aware that his or her inability to move to a distant location or to assist at the time of a crisis will not jeopardize his or her position with the organization.

Another important factor is the morale of employees at the time of a disaster. If the working facility has been affected, and company personnel realize they no longer have a place to work, it is the responsibility of management to sustain the confidence level and morale of employees (even though those employees are not involved in the actual backup and recovery processes). Maintaining communication with employees and assuring them that their positions with the organization will go a long way in saving tens, perhaps hundreds of thousands of dollars in retraining a new staff, is essential.

The primary calling tree (employees) is located in Addendum D-13.

Important: Maintaining the calling tree to ensure that all key contacts have up-to-date listings is vital to the entire plan structure. Although this list is separate from the Disaster/Recovery Management Team and the disaster/recovery functional teams, it provides the base for additional support capabilities and specialized skills in the event one or more key personnel become unavailable.

Each person who has been identified to be called in the event of an emergency (primary contacts), as well as those with special skills, must be trained in emergency response procedures. All shift personnel also need to have some emergency training with respect to fire, medical emergencies, and so on. Alternates to the primary contacts should be available. *Note*: Vacation and holiday schedules should be carefully prepared to ensure that critical personnel are always available.

Advance training must be supplied for primary contacts as well as other support personnel and/or clerical, administrative, or general employees so that the extent of plan implementation, training of employees, and periodic maintenance is the only thing that will contribute to a successful outcome of an emergency condition.

The following lists should be maintained:

1. Key contacts (disaster/recovery plan initiation)

2. Primary contacts outside the data processing environment
3. Those who are not necessarily involved with the disaster/recovery procedures

4.7 Vendor Information

Implementing the disaster/recovery plan requires support from areas outside the organization. During the assessment phase, it is necessary to obtain the cooperation of various hardware equipment and supplies vendors who can determine whether or not damaged hardware, equipment, and supplies are recoverable. This determination should be made by professionals relating to their hardware, equipment and supplies, as their recommendations and determinations are usually accepted by insurance adjustors.

Every item of hardware, software, equipment, peripherals, supplies, forms, services, and so on must be listed along with the vendors' (suppliers') telephone numbers, emergency phone numbers, and people that can be contacted at the time of a disaster.

4.7.1 Vendor Addresses, Phone Numbers, and Contacts

Most vendors, especially hardware and equipment suppliers, will exert every conceivable effort to provide an organization with replacements at the time of disaster. They often supply the first-off-the-line equipment waiting to be shipped or divert equipment being transported to another company. In fact, it has been the equipment actually delivered, but not yet installed, that has been transshipped from one company to a damaged site. Additional support can be expected from the vendors, including teams of system engineers and installers and system and application specialists.

Whether you have a single or multiple vendor environment, cooperation on the part of all is one thing you can count on. Nevertheless, discussions with local representatives concerning a vendor's specific policy (written or unwritten) about disaster support is important.

The vendor list should contain complete information on communications equipment.

4.7.2 Vendor Product List

A separate list is to be compiled by vendor, listing each of the products supplied by the vendor. This list should include both permanent and consumable items; it is difficult during the time of an emergency to remember all the items, some of which may be unique to a specific installation. Reference Addendum D-15.

4.7.3 Vendor Support Letters

Although most vendors have a disaster/recovery policy, the policy exists as a verbal commitment, rather than one that is published and distributed to customers. Requests for written commitment by a vendor will, in many cases, illicit only a verbal statement. Smaller vendors, in some instances, have been known to give written commitments. The larger firms have general company policy, but are unable, for legal reasons, to provide a written statement. Of primary concern to vendors is a common local disaster that may affect the computer equipment of numerous organizations.

The inability to obtain definitive commitments from vendors points dramatically toward the need to develop backup strategies that will effectively support the organization.

4.8 Disaster Kit

In addition to providing for backup and recovery strategies and action plans and deciding and planning for the various supplies necessary to continue operations, the data processing activity requires a minimum of supplies in order to:

1. Establish the operation at the backup site.
2. Sustain those activities for a period of three to ten days without concern for obtaining supplies.

The *disaster kit* contains the necessary supplies to support the backup (or even on-site) activities in the event there is a loss on-site of these items. The Supplies Team must consider the highest peak use of supplies during a period and provide, off-site, a disaster kit. The kit should certainly contain items that take a long time to procure or are not normally stocked. It is important to note that if you were to make immediate inquiry of the availability of certain supplies, you will find that quantities required for the kit might not be readily available.

The contents of the disaster kit are listed in Addendum D-17.

4.9 One-Month Supply List

If the disaster has rendered the computer operating environment incapable of performing, then it can be assumed that all the existing inventory of supplies are either inaccessible or they have been destroyed. Preparation of a one-month supplies requirement by the Supplies Team ensures that vendors will be capable of delivering and restoring inventory levels (either at the backup site or at the home site) before the one-month's disaster kit is exhausted.

The thirty-day supplies list should also be drawn up for return to the home site after the disaster conditions and activity have been normalized.

4.10 Forms Books

Where standard (catalog) items are reasonably available and the operation at the backup or alternate site is sustained by the use of these items, extensive cataloging of forms and paper is not necessary. Such supplies will be contained in the disaster kit and on the thirty-day operations requirement list.

This is not usually the case in most organizations and computer operations. The Supplies Team must carefully analyze and catalog (in the format of a forms book) those items that are unique to the organization. This is not restricted to computer operations. Some organizations, such as banks, have numerous forms required in daily operation that are not necessarily used on computers. Companies that have their own in-house forms printing operations must consider stockpiling, off-site, sufficient inventory, since these departments may be located in the same place where the disaster has occurred.

Many organizations keep forms inventory at the supplier and rotate forms inventory periodically, with backfilling going on at all times. These procedures must be documented, and the location of all of these forms and special documents must be listed in a single, readily available reference.

General provisions for office supplies must be carefully reviewed. If an extensive number of operations are affected, regular supply operations may not carry sufficient inventory to satisfy disaster condition needs.

Only a well controlled forms and supplies acquisition program will work at the time of a disaster situation. Samples of all forms, quantity usage, ordering cycles, and lists of primary, secondary, and tertiary suppliers must be listed. The forms book is stored off-site in more than one location.

Important: The Supplies Team should have prepared a complete set of artwork or positives that can be distributed to *any* forms manufacturer.

4.10.1 Facility and Office Space Requirements

In addition to backup capabilities (the place where computer operations can be performed), space is required to house user departments and other personnel involved in computer operations. Executive management will also require facilities from which to operate during the critical period.

A control center has to be established for the Disaster/Recovery Management Team to centralize all backup and recovery activities.

The New Facilities Team will be responsible for maintaining an inventory, or specifically provide for:

1. Space for executive management to use temporarily until the original site is restored.
2. Office space that can be temporarily or permanently occupied in the event the entire facility is destroyed.
3. A control center for the Disaster/Recovery Management Team.
4. Shell space or potential space into which the data processing facility can relocate after operations have been established at the backup site.

4.11 Vital Records Retention

Records retention schedules established on an empirical basis, usually established by the retention schedule of the taxation departments, do not normally meet the overall organizational, industry, regulatory, and legal requirements.

Regularly overlooked in organizations without a prescribed records management policy and program is the establishment and extension of vital records retention into all data processing records and data. It is important that in developing a records management and records retention program, that the daily transaction activity is not ignored. The loss of these documents in the event of a disaster makes it virtually impossible to reconstruct the period of time from the last backup cycle to the interim transactions. This is especially critical in financial institutions, where tens of thousands of transactions occur on a daily basis.

Specific records retention policies are established based on:

- *Organizational policy.* Records are retained based on the nature of the business and the volatility of activity. Those records that establish and prove the legal status of the organization and provide for the economic viability and survival of the business are essential.
- *Legal requirements.* Laws, rules, and regulations established by federal, state, county, and local government agencies, departments, and commissions require the specific retention of all or part of any organization's records. Tax (income, corporate, sales, profit, and so on) records, encompassing laws such as the Foreign Corrupt Practices Act of 1977, Privacy Laws, Right to Financial Privacy, Bank Records Act (OCC, FHLSB, FDIC, CUA, and so on); organizations such as the Equal Employment Opportunity Commission (EEOC), Environmental Protection Agency (EPA), U.S. Department of Commerce (various statistical data); and in foreign countries Company's Acts, inland revenue requirements, customs laws, and so on provide for and establish legal obligations on the part of the company, financial institutions, and companies in interstate and international trade to maintain accurate records of all activities and transactions. Records retention policies and vital records must be reviewed by the organization's legal department.
- *Processing retention.* The user is responsible for establishing a records retention schedule in order to maintain the integrity and auditability of its operations. Consideration must be given to sufficient records retention that will provide for the restart of the processing activity in the event of a disaster.
- *Disaster/off-site records retention.* Considerable preplanning is required to define, establish, and implement a comprehensive off-site storage program that will provide for a sufficient cyclical basis ensuring that resumption of normal activities is possible within a reasonable period of time. The preplanning is important, since the plans must include a cyclical schedule that allows for resumption of the data processing support function as quickly as possible while it maintains the continuity of operations and the integrity of data from the most recent period prior to the adverse incident.

4.12 Software Escrow

Organizations that operate solely with purchased, leased, or licensed software or have incorporated various external software packages must exert extreme care when choosing the vendors with whom they will do business. Although the software industry over the past few years has substantially stabilized, and less companies are going out of business, there are still vendors who occasionally reach a position where they can no longer support their customers.

Recently, contracts with software development firms have included contractual stipulations that provide for a third-party retention of source code. These contracts ensure customer access to the software source code in the event the supplier is no longer in business or can no longer provide update services. This should be a condition for the purchase of all software.

The source code is often retained by legal representatives of the vendor or customer. In some cases, off-site storage facilities provide this type of service in safe-deposit boxes with keys held by a legal representative of the customer. The safe-deposit box can only be opened by a court order stating that the financial or business condition of the vendor is such that it can no longer provide support service.

As part of contractual stipulation, the vendor is obligated to deposit with the legal representative new generation versions of the source code as they are being updated.

4.13 Employee Commitment Letters

Employees involved and responsible for emergency procedures as well as backup operations must be solicited to determine their availability, travel restrictions, limitations regarding time away from home, and so on. The commitments cover, among others, employees involved with computer and noncomputer operations, critical user department personnel, and administrative personnel.

Commitments must be in writing, and sample questionnaires and letters of commitment will follow this section. Each organization may have additional requirements.

See Section 4.6.3, "Calling Tree for All Employees." Responses to the commitment letters are in Addendum D-22.

4.14 Priorities for a Limited Processing Environment

In order to develop effective backup and recovery strategies, it is necessary to establish the requirements of the critical applications and those critical resources necessary to support those applications. Although this plan addresses the worst-case environment, the Disaster/Recovery Planning Team will have identified those threats, hazards, vulnerabilities, risks, and exposures relevant to the data processing operation. Definitions will have been established relating to maximum downtime, and scenarios will have been written concerning other impacts that are less than worst-case conditions.

The user community is required to submit a complete list of critical applications that will:

1. Maintain the integrity and audit trail of all financial and economic records.
2. Maintain the continuity of customer support as well as those functions that would question the credentials or credibility of the organization.
3. Maintain the confidence level of employees.
4. Maintain the cash flow.
5. Maintain those activities that are subject to legal constraints (federal, state, or local laws and/or regulations).
6. Maintain the normal (or close to normal) operating procedures of the organization.

In many instances, it is virtually impossible to consider full computer support services. The backup alternatives are designed and developed, based on critical (high-priority) applications. Management's responsibility is to designate priorities on a mandatory, necessary, and/or desirable basis. Selection of backup strategies is based on the classification of support functions, the period that the adverse incident (interruption) occurs, and the predecision, preplanning activities that have been put in place. These considerations include, but are not limited to, those decisions that have been outlined in Section 4.3 as well as:

- Degradation of services
- Recovery strategies based on internal capabilities
- Ability to revert to manual operations
- Commercial recovery backup site
- Totally redundant in-house facility
- Distributed processing
- Alternative data entry
- Fortress concept with total redundancy
- Stand-by micro- or minicomputer centers (user departments)
- User stand-by provisions

In certain organizations, a single strategy may suffice to accomplish the performance of critical applications and special computer functions, whereas in other situations, multiple strategies may be more appropriate.

Important: Communication plays a vital role in the capability of an organization to be able to backup its user community. The communication environment (local, national, and international) networking requires extensive evaluation during the preplanning activities. The utilization of an off-site backup facility must be totally transparent to the user, and an alternate communications environment, properly planned, will provide stability and continuity to the users. Included in, but not limited to, communications backup for support of critical applications are:

- Alternate networking topology
- Alternate leased lines
- Dial-for-lease and voice-for-data

Selection of critical applications, and ultimately the backup strategy or strategies, is a complex procedure that requires infinite detail planning on the part of management, the user, and data processing communications and backup facilities personnel. The importance of integrating numerous stand-alone, independent PC, and mini- and microcomputer functions should not be overlooked. These separate components of information services provide many of the basic data utilized in the mainframe operation. These activities must be approached in the same priority/critical selection procedures as mainframe activities.

4.14.1 Current Site Operations Schedule

Although the backup strategies are developed based on critical application selection, it is important that a current record of the data processing monthly schedule be available as part of the record stored off-site. This provides operations personnel with a road map of what will be the required performance (critical job schedule) at the backup site, and it will avoid maintaining support services based on an individual's memory or scrambling. If any additions or deletions occur, this schedule should be replaced with the most current schedule as part of the plan maintenance program. It is important to understand that key personnel may not be available at the time of the interruption or disaster, and those performing backup operations require every available resource to maintain processing support.

If possible, this run schedule should be annotated to contain the following information:

- *Applications system (criticality level).* The system's name (user) and programs contained within the application (or specific programs), should be given with as much detail to ensure proper performance during backup operations.
- *Frequency of report or application during the period (one-month).* Indication should be made concerning batch and/or on-line activity and response time.
- *Risk or exposure.* What activity of the company will be affected: legal, customer service, financial, engineering, research and development, vendor relations, public relations, process (manufacturing) failure?
- *Period of acceptable interruption.* How long can the user be without specific support?

The current monthly schedule should be placed on Addendum D-23 or in a separate volume (depending on the extent of the document) and referenced as to its location in Addendum D-23. The schedule should be stored off-site as part of the plan and other documentation.

4.14.2 General Priority Scheme

The approach to establishing a priority scheme includes:

- A schedule of priorities based on critical applications, as discussed in Section 4.14, and on user requirements.
- A schedule of priorities further defined (daily, weekly, monthly, quarterly, semi-annually, or annually), based on the time the adverse incident strikes.
- A schedule of priorities divided up by one or more backup strategies (alternatives).

The chart provides a method of ranking by period and by backup alternative. Further enhancements can be broken down to provide more detail, and a chart can be developed to show priority level (time phased), application system, key system contact, user liaison, processing requirements (resource, CPU time used, performance time), frequency of activity, terminal and/or on-line requirements, new development (existence of new software off-site), any other factors that would contribute to the assistance of backup personnel in performing emergency tasks.

Special conditions should be noted relating to audit department input, audit trail methodology, and unique characteristics of software, hardware, communications, or personnel.

4.14.3 Probable Backup Site Operations Schedule

Based on the previous data compiled in this section, the planning team will establish a backup site operations schedule to submit to the user and management for approval. This schedule should cover *mandatory, necessary*, and *desirable* processing requirements and indicate the backup strategy alternative selected, as well as the time of recovery expected, and so on.

A copy of the probable backup site operations schedule is to be placed in Addendum D-24. If placed in a separate volume, note its location in Addendum D-24.

4.14.4 Application Recovery Plans

The development of individual application recovery plans in sequential recovery order provides the applications software and operations teams with a defined methodology for bringing up each system. A sample form for application recovery is to be put in Addendum D-25.

4.15 Logistical Considerations

Under Section 4.17, the Disaster/Recovery Planning Team will develop the various team assignments; identify the skills and skill levels of personnel involved in the initiation, activation, and performance of the entire plan to recover at the backup site; maintain continuity of operations; and support the organization until the primary (or home site) has been restored and is operational.

Those people involved in each of the teams will require special logistical support. These support considerations include, among others:

- Funds
- Transportation
- Housing
- Meals
- Clothing (and laundry services)
- Entertainment

4.15.1 Credit Cards, Cash, Etc.

Those who will have to travel to the backup site may have to fly or find other modes of transportation. Instantaneous travel reservations, notification of vendors, and local activities of others will require a number of costly procedures. Preplanning the funding of all these procedures is vital to the morale of those participating.

The organization cannot expect individuals to bank the company at the time of a disaster. Policy has to be established and procedures have to be outlined to ensure that individuals are sufficiently prepared to undertake the project at the time of a disaster.

1. *Credit cards.* The most efficient method is to provide personnel who expect to be away from home with company credit cards. If the organization has a firm policy against this procedure, the cards should be held by an executive of the company (preferably off premises, perhaps in a lockbox within the disaster kit, or at the control center in a company safe), and distributed only at the time of the disaster. An agreement should be drawn up by the legal department that makes use of the card outside the scope of approved expenses the responsibility of the individual.

Important: If the organization has subscribed to a commercial backup facility, provision can be made to have funds available at the backup site for any emergency that could arise.

2. *Salary checks.* Overlooked is the problem of distributing and cashing or depositing employee salary checks. The payroll department (or those responsible for preparing and distributing payroll checks) should be consulted as to alternate means of providing employees with payroll if they are away from home for extended periods of time. Direct deposit into employees' bank accounts, permission for a spouse to sign (endorse) checks, cash payments, and so on should be considered and included in the employees' solicitation (availability) letters.

4.15.2 Housing

Housing near (or convenient to) the backup site has to be prearranged. A number of alternates have to be selected, because accommodations vary based on the time of year, location, and so on. Personnel at the backup site are usually aware of the best accommodations in the area and have made such arrangements for other firms. Employees should not be responsible for making their own reservations, nor should they be concerned about payment for these facilities. Prearranged accounts should be established by the company, and all billing for services should be sent directly to the company. If this cannot be arranged, payment should be made by check or credit card.

4.15.3 Local Transportation

The same procedures for providing housing for personnel should be in effect for car rentals or provision of some form of local transportation between hotels and the backup site. Individuals away from home for extensive periods of time need some form of transportation to free them from hotel confinement. Personnel at the backup site can help arrange (prearrange) for transportation.

4.15.4 Office Space

The disaster/recovery plan primarily addresses the backup and recovery of computer operations. Although facilities are provided for the restoration (backup) of computer functions, and operating personnel will have sufficient space at the backup site, the provisions for other personnel involved in computer support activities, user departments (other functional units of the organization), noncomputer operations, and executive management have to be provided for in the event there is a total loss of the organization's facility. Providing a suitable working environment for 25, 50, 100, 300, or perhaps 500 or more people is no simple task. Because of the volatility of real estate in major urban and suburban areas, it is difficult to depend on one or a few sources. In certain areas, organizations are in business to provide completely furnished office space (office centers). These are usually finite resources, and although they may satisfy a limited number of people, continuous activity must be generated to provide major square footage, office equipment, supplies, and communications for the larger number of people that could be affected.

It is also important to consider the location and the type of recoverable office space. Existing personnel may not be able to conveniently travel to newly acquired office space in the suburbs if they have been used to mass transportation systems within an urban area, and vice versa.

At the time of a disaster, it is important to plan for the least disruption and effects on reliable, long-term employees. Recovery time to relocate, if beyond a couple of weeks, may result in a falloff of personnel. Policy has to be established in advance concerning how long the company will continue paying nonperforming personnel. Insurance (see Section 4.16) has to be verified if coverage exists for relocation, employee payroll, continuation of benefits, extra expenses, and so on.

The New Facilities Team, supported by the nondata processing Supplies Team, Equipment Team, Communications Team, and so on, have to constantly research available facilities.

The New Facilities Team has to provide facilities (in the event the affected site is unrecoverable) that will bring the data processing operations back to a permanent site within a reasonable period of time. This is covered in more detail under the charter and assignment of this team (see Section 4.17.13). In general, the physical plant requirements to recover the data processing mission permanently demand readily available:

1. Large, undivided space
2. Raised floor
3. Conditioned power
4. Environmental conditioning (isolated)
5. Security
6. Convenience to existing personnel
7. Facility within budget requirements
8. Readily accessible and installable communications

4.15.5 Control Center

The control center is needed to provide a central location for the Disaster/Recovery Management Team to perform the control aspects of the plan. This can be permanently established in another building of the company within reasonable distance from the primary site.

Other provisions can be made at hotels (using conference rooms, suites, or ballrooms). Care in planning this approach is important, since many conference facilities are reserved in advance.

A primary concern of establishing a control center is the availability of numerous communications facilities (telephone, telefax, teletype, terminal/computer connections, messenger services, couriers, and so on).

4.15.6 Employee Notification by Broadcast (and Other) Media

If standard methods of communication have been destroyed, other alternative methods have to be found to communicate with employees, vendors, suppliers, and customers.

Messages to be used on general public media (TV, radio, newspapers) have to be carefully prepared to ensure that negative implications of announcements are not assumed by the public.

Courier or messengers can be used to send information to personnel, vendors, off-site storage facilities, and so on. Key personnel have to be assigned calling (direct-visit calls) on support personnel.

4.16 Insurance

The present status of insurance within an organization which reflects the capability of the company to economically survive a disastrous event is an area of activity that has to be delegated to insurance experts.

Data processing management, on the other hand, must be cognizant of the potential risks and exposures that exist if the proper coverage is not available. The specific forms and samples of amendments to existing policies represent the new approach of the insurance departments of the organization relative to computers and computer operations. It is important to note that insurance companies have become more aware of the implications and impact of computer failures and disruptions on the overall performance of a business. High deductibles and high premiums have been imposed on those organizations who have taken little or no action in the development of plans to recover from an adverse incident. Those that have taken action are the beneficiaries of lower deductibles and premiums.

Underwriters now provide field personnel with detailed questionnaires concerning the "security" of a facility before they will issue policies.

Insurance professionals, either inside the organization or retained by the company (brokers), have to direct their investigation into the following areas in a manner not normally addressed:

- Property and casualty insurance
- All risk policies
- Business interruption
- Detailed computer and media coverage
- Extra expense

Outside the computer environment, a separate overview has to be performed within the user community. Contacts with vendors have to be reviewed concerning the obligation of the organization to provide insurance on leased equipment.

Insurance companies will review all emergency procedures to see if they comply with the organization's legal obligations and the safety of its personnel. Many insurance companies will invalidate insurance coverage or refuse responsibility for compensating firms delinquent in their responsibilities in this area. Care must be exercised in reviewing insurance agreements to *identify exclusions.*

Insurance companies provide loss control inspections at no charge to the insured.

Correction of deficiencies of identifiable risks can substantially lower premiums.

In today's high-risk environment, insurance companies will not underwrite policies (or will apply huge premiums) for companies who do not

- Have a *tested* disaster/recovery plan
- Store their data, software, and documentation off-site
- Have a qualified backup site

Providing an effective insurance program requires detailed studies and careful execution of insurance worksheets.

Although reconstruction (or rehabilitation) of an affected site can be readily put into dollar terms, the activity at the backup site and other activities that occur during a disaster are far more elusive. Essentially, it is the excess of total operating costs that would incur during the restoration period. Normal operating costs that would have existed at the original site, had there not been a disaster, are deducted from the extra expense. Care must be taken to derive an extra expense cost that would cover all disaster activities.

Self-insurers and those with high deductibles should review their financial position to determine if they can sustain a disaster and absorb the total cost of a disaster.

Open document Sections 4.16.1, 4.16.2, and 4.16.3 for detailing existing insurance coverages. A copy of the organization's insurance policy, primarily relating to data processing (business interruption, casualty and liability, and extra expense) should be placed in this document for reference.

Important: Employee fidelity insurance should be considered for those employees involved in highly sensitive activities, especially those using systems that are involved in money transfers.

4.17 Disaster/Recovery Action Teams

The most indispensable and critical resource at the time of a disaster are *people*, especially data processing personnel who, by virtue of their training, technology, knowledge of the organization, and experience in working with the user community, provide the ability to recover from any adverse incident within the time frame established by the plan.

There are two assumptions that planners have to make concerning personnel:

1. The strategy for recovery assumes that in the event of a disaster, sufficient personnel will (survive) be available to perform the tasks to back up and recover operations.
2. A risk does exist, a disaster may result in the loss of personnel, and recovery may be impossible if such a condition exists.

In addition to internal personnel, contractors, consultants, and vendors (equipment manufacturers, data services, facilities managers, systems designers, field engineers, and personnel obtained from temporary agencies) can provide assistance in the recovery functions.

Teams and Charters

See the Representative Disaster/Recovery Management Team Charter

Although we recognize that few organizations have the personnel to establish the number of teams noted, it is important to understand that all the functions represented by the team assignments must be performed regardless of the number of teams finally established. Therefore, the planning team must set the parameters for the minimum staffing for each activity at the time of a disaster. The number of functions, as well as the number of teams, will be determined by the complexity of the organization and the availability of personnel. (See the list of possible disaster/recovery teams on page 33.)

The team charters and responsibilities are covered in the plan, starting with

Section 4.17.1, "Disaster/Recovery Management Team." This team plus the following teams have been identified as a minimum. Their charter, responsibilities, leadership, team members, management liaison, and disaster/recovery actions are outlined in the master plan.

4.17.2	Transportation Team	
4.17.3	Off-site Storage Team	
4.17.4	Supplies Team	
4.17.5	Administration Team	
4.17.6	Software Team	
4.17.7	Applications Team	
4.17.8	Communications Team	
4.17.9	Operations Team	
4.17.10	Data Preparation Team	
4.17.11	Salvage Team	
4.17.12	New Hardware Team	
4.17.13	New Facilities Team	
4.17.14	Audit Team*	
4.17.15	Public Relations Team*	
4.17.16	User Liaison Team*	

*No team charters appear for these teams.

Note: The numerical system used above, and the numerical identification of the team charters that follow parallel the numerical system of the plan. See the model plan's table of contents.

REPRESENTATIVE DISASTER/RECOVERY MANAGEMENT TEAM CHARTER

4.17.1 Disaster/Recovery Management Team (DRMT)

4.17.1.1 Team Charter
The DRMT is responsible for the overall coordination of the disaster/recovery process. All other teams report to this team at the time of a disaster. In effect, this team succeeds the disaster/recovery planning team, which heads efforts in the predisaster mode. There will be a large carryover in membership.

4.17.1.2 Typical duties of the DRMT
- Coordinate teams
- Secure financial backing from the president and board
- Approve all actions not preplanned
- Give strategic direction

4.17.1.3 Role of the Coordinator
- To be the working head of the team
- To execute all disaster/recovery team decisions via this and other teams

4.17.1.4 Role of the Administrator
- To approve or disapprove all actions with respect to the strategic directions of the firm
- To be the liaison to upper management
- To finance all operations
- To expedite matters through all bureaucracy

4.17.1.5 Membership
Coordinator: [Name of MIS Manager]
[Title of MIS Manager]
Alternate Coordinator: [Name of Data Center Manager]
[Title of Data Center Manager]
Administrator: [Name of V.P. Operations]
[Title of V.P. Operations]
Alternate Administrator: [Name of V.P. Finance]
[Title of V.P. Finance]
Member 1: [Name of Data Center Manager]
Member 2: [Name of V.P. Finance]
Member 3: [Name of Internal Auditor]
Member 4: [Name of Director Security]
Member 5: [Name of User Coordinator]
Member 6: [Choice A]
Member 7: [Choice B]

Expanded Membership: Heads of all other teams

Figure 6-1

REPRESENTATIVE DISASTER/RECOVERY NEW HARDWARE TEAM CHARTER

4.17.12 **New Hardware Team**

4.17.12.1 Team Charter
It is the responsibility of the New Hardware Team to order replacement hardware for that damaged in the disaster. Hardware ordered may not be a one-for-one replacement, since this may be the best time for an upgrade, consolidation, and so on. All areas of hardware are to be dealt with, including:
- Mainframe
- Peripherals
- Data communications
- Terminals
- Micros
- Environmental control equipment

4.17.12.2 Typical Duties by Phase
Phase 2:
- Obtain a list of damaged and destroyed equipment

Phase 3:
- Decide on new hardware
- Order new hardware
- Install and test new hardware

Phase 4:
- Evaluate performance

4.17.12.3 Membership
Coordinator: [Name of Data Center Manager]
Alternate: [Name and Title of Alternate]
Members: [Name of Vendor, Field Engineer]

Figure 6-2

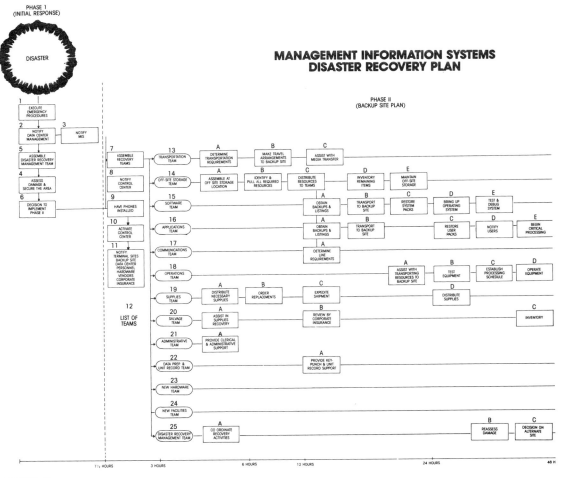

Figure 6-3 Sample PERT diagram.

Effective Disaster/Recovery Plan Model

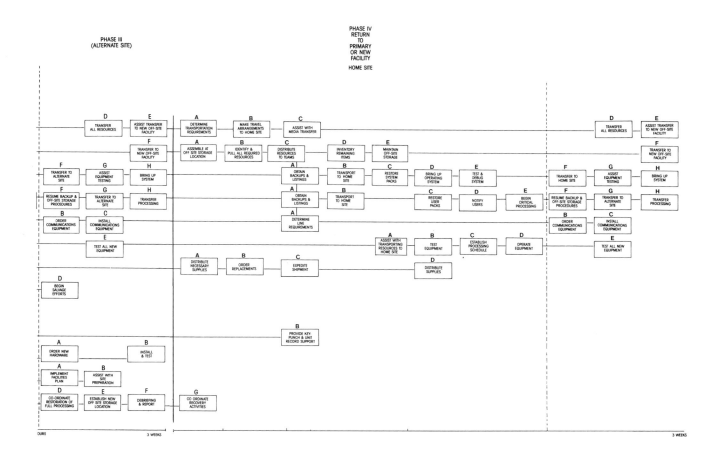

EXAMPLE OF ACTIONS TAKEN BY THE TRANSPORTATION TEAM BY PHASE UPON THE EXECUTION OF THE DISASTER/RECOVERY PLAN

NOTE: The paragraph identification numbers are from the actual disaster/recovery plan.

5.6 **Elaboration of PERT Chart (see the PERT chart on pages 154–155)—Phases 2, 3, and 4 (Team Actions)**

5.6.1 Transportation Team

Phase 2 Actions:

13A Determine Transportation Requirements
Transportation requirements are preplanned, but it is still necessary to reevaluate needs in light of the actual disaster and its schedule. Additionally, some people may be absent and require alternates.

13B Make Travel Arrangements to the Backup Site
The Transportation Team will coordinate the movement of people (based on the backup site plan of operations) to the backup location, control center, or other facilities designated as the operations centers. This requires close coordination with support personnel at the backup site, airline companies, hotels, and so on. *Note*: Special conditions exist during high travel times (summer vacation periods, Christmas holiday, and so on). The Transportation Team will provide alternate methods of moving people as part of their team plan.

13C Assist With Media Transfer
After the initial transfer of media from facility data vaults and/or off-site storage, to operations at the backup site (if available), off-site storage of media must continue in the same manner as operations at the original computer site. Media must be backed up and stored off-site, maintaining every aspect of security and protection.

13D Move Supplies to the Backup Site
The Transportation Team will maintain adequate supplies at the backup site as long as the contingency exists. This is to be coordinated with the Supplies Team.

Phase 3 Actions:
(ongoing)

Phase 4 Actions:

13E Determine Transportation Requirements to Return to the Home Site
The requirements were preplanned, but they require verification.

13F Make Travel Arrangement to the Home Site
As we have indicated, it is necessary for the Transportation Team to coordinate the movement of people (based on their predetermined plan of operations) to the home site, control center, or other location designated as the permanent or temporary computer operations center. This requires close coordination with support personnel at the home site, airlines, and so on. Reservations with airlines should be preplanned based on the schedule provided by the DRMT to cover the personnel being moved back to the home or alternate operating site.

13G Assist With Media Transfer From the Backup Site to Media Storage
The Transportation Team, in cooperation with the Software Team, the Applications Team, and the Off-site Storage Team (14), will provide the necessary secure transportation of all media from the temporary off-site storage to the primary facility. This will also include all operating supplies, documents, equipment, and so on that was utilized at the backup facility. Upon return to the "new" home site, the Transportation Team will continue to provide services to move media (and other required materials) to the off-site facility until the contingency is declared normal. At that time the responsibility will rest with the data processing operations personnel.

Figure 6-4

EXAMPLE OF ACTIONS TAKEN BY THE OFF-SITE STORAGE TEAM BY PHASE UPON THE EXECUTION OF THE DISASTER/RECOVERY PLAN

5.6.2 Off-site Storage Team

Phase 2 Actions:

14A Assemble at the Off-site Storage Location
The Off-site Storage Team should have a complete inventory available of all mandatory (critical, high-priority) applications and the listings of software, database files, and documentation necessary to start recovery procedures at the backup site.

14B Identify and Assemble All Required Resources
If the data center library and media are intact, all accessible media should be withdrawn from the working library to ensure that all software and database media are the most current versions. If this is not possible, then the off-site storage media must be assembled. Only those applications being run at the backup site are required at this time.

14C Distribute Resources to Software Team, Applications, and Team Backup Operations Team
Coordinating its activity with the Transportation Team (through the DRMT), all appropriate media should be delivered to the Software and Applications Teams. Determination as to who should transport the materials and how will be determined by the DRMT.

14D Inventory Remaining Items
On-site: A complete inventory will be taken of media within the data center library (internal vaults) to determine the effectiveness of recovering of necessary and desirable applications, if the media can be salvaged, and what backup requirements are necessary to recovery at the primary or new data center after reconstruction.
Off-site: Continuous operation at the backup site could make media remaining in storage obsolete, depending on the length of time at the backup site. All users should be involved in procedures for keeping data current to the best of their ability under the circumstances.

14E Maintain Off-site Storage
All recovery at the backup site must be continuously backed up and stored off-site (see 13C, Transportation Team). A commercial backup center will only provide media vaults for active daily storage, but these centers are subject to the same "incidents" as the original data center, and off-site storage and frequent cycling is mandatory. Arrangements should be made for this prior to a disaster. If an off-site media storage vault is not available within the vicinity of the recovery site, a determination must be made as to the most appropriate location (or get the personnel at the recovery site to assist) for an off-site storage facility. This is the responsibility of the Off-site Storage Team and the Disaster/Recovery Coordinator.

Phase 3 Actions:

14F Transfer to Original Off-site or (New) Storage Facility
In cooperation with the Transportation Team, user groups, the DRMT, and the Disaster/Recovery Coordinator, the Off-site Storage Team will inventory all software, data, and user activity (nonprocessed) during the emergency. They will verify all inventories with appropriate operations personnel on the Software Team, Applications Team, and Operations Team and provide for the materials, data, and software at the backup center and alternate off-site media vault storage, to be returned to the original off-site security center. [*Note*: Depending on the distance from the backup recovery center (and media storage vault) to the original media vault, it may be advisable for the Off-site Storage Team to provide for periodic updated tapes to be returned to the primary off-site storage facility. This will minimize the recovery activities (return to home site) and provide further protection and security by having at least one additional up-to-date version of all software and data at the original off-site storage facility.]

Phase 4 Actions:
14G Assemble at the Backup Off-site Storage Location
The Off-site Storage Team should be provided with a complete inventory of available applications performed during the contingency and what data is being stored at the temporary off-site facility. The listings of software, database files, and documentation necessary to restart recovery procedures at the home site should be confirmed by the Software, Applications, and Operations Teams.

14H Identify and Assemble All Required Resources
All media within the library of the backup site will be withdrawn from the production library to ensure that all software and database media are the most current versions performed at the backup site. When the contingency operation is to be abandoned, the Off-site Storage Team will be responsible for the safety and security of all media. The Off-site Storage Team will have to coordinate their activities closely with the Software, Applications, and Operations Teams.

14I Distribute Resources to Various Teams at the Backup Site
Coordinating its activity with the Transportation Team (through the control center), all appropriate media should be delivered to the Software and Applications Teams upon their return to the home site.

14J Inventory Remaining Items
On-site: A complete inventory will be taken and supplied to the appropriate teams of all media within the backup facility production library and the temporary off-site storage facility to determine the effectiveness of recovery of necessary and desirable applications at the home site. If the media can be used, what additional backup requirements are necessary to recovery at the home site or new data center after reconstruction?

The Off-site Storage Team will also be responsible for all data residing in the backup facility, and supervise the removal of such data and software once they are assured that proper backups are provided.

The Off-site Storage Team, in conjunction with the Operations Team, the internal auditor, and the Disaster/Recovery Coordinator, will ensure that no information proprietary to the company will remain at the backup site once it has been abandoned. Provision will also be made for the shredding and/or destruction of any hard copy (printouts). All documentation, disaster kit, and so on will be removed and located in their appropriate facilities prior to the normalization procedure of this plan.

Off-site: Continuous operation at the backup site could make media remaining in the home site off-site storage obsolete depending upon the length of time at the backup site. All users should be involved in the procedures for keeping data current to the best of their ability under the circumstances.

14K Renew Home Site Off-site Storage Operations
Although operations at the backup site will be continuously backed up and stored off-site (see 13C), all media will have to be identified and removed to the home site off-site storage facility.

Upon notification by the disaster coordinator that the backup facility is to be abandoned, the Off-site Storage Team will immediately reestablish all off-site storage activities at the home site. Extreme care will be taken to ensure the continuity and integrity of backups during transport and placement into the off-site facility at the home site. The Off-site Storage Team will coordinate its activity with Operations and Audit to ensure that all backup media is identified and properly stored. This team's activity will not be normalized until the home site is functioning properly and all procedures are implemented as they were prior to the disaster.

Figure 6-5

4.18 Other Teams
4.18.1 User Team (brief plan for each user)
4.18.2 Audit Team
4.18.3 Hardware Team
4.18.4 User Liaison Team
4.18.5 Security Team

If the functions listed below are out of the realm of each of the teams activity, consideration should be given to develop a:

4.19 Additional Teams to Be Considered
4.19.1 Third-Party Liaison Team (coordinate activities with other organizations interacting with data center operations)
4.19.2 Public Relations Team
4.19.3 Alternate Strategies Coordinating Team
4.19.4 Legal Team
4.19.5 Crisis Management Team (Management Level)

5. DISASTER/RECOVERY ACTION PLANS

The disaster/recovery action plans and PERT chart developed in this section of the plan deals with a worst-case disaster condition. Lesser incidents can be acted upon by utilizing subsets of the overall disaster/recovery action plan. This section provides the road map of what actions are to be taken and by whom (team), the time in which actions are to be taken, and when they (or specific functions) are to be completed—based on the selected time requirement to establish operations at the backup site.

The PERT (disaster/recovery team action plan) flow diagram is divided into four phases:

Phase 1 Emergency response plan
Phase 2 Implementation of the disaster/recovery plan backup site initiation
Phase 3 Backup site operation and recovery process
Phase 4 Normalizing disaster mode . . . returning to the home site

Each of the PERT (disaster/recovery action plan) flow diagrams will be unique to each organization and will be developed based on each company's specific strategies and requirements. Many of the activities represented on the chart will take place in parallel by the individual teams, coordinated by the Disaster/Recovery Management Team.

Phase 1: Emergency Response Plan

The emergency response plan is initiated on the first indication of an adverse incident: the fire alarm goes off, a bomb threat is reported, a fire is seen prior to an alarm being annunciated, an intruder is observed or the alarm goes off, the system crashes, and so on. The procedures developed for *action* and incorporated in the emergency response plan are initiated. Depending on the time of day and the number of people (and who) on the premises, the actions taken are based on the judgment of the most senior official available.

All emergency procedures should be designed to:

1. First insure the safety and welfare of personnel and property
2. Protect property and/or mitigate the effects of any adverse incident
3. Inform the appropriate emergency agency
4. Inform senior management
5. Cover every anticipated or probable disaster

Phase 2: Implementation of the Disaster/Recovery Plan Backup Site Initiation

The DRMT is responsible for assessing the damage resulting from the adverse incident, determining whether the recovery of the affected site will exceed the required recovery (restoration) time, and declaring the disaster. At this time the plan is initiated, and all the notifications, control center actuation, and backup sites are notified of mobilization of disaster/recovery teams, vendors, suppliers, etc.

At the first stages, and depending on the number of people available, as many activities as possible are initiated in parallel. Within the time frames established by the PERT chart, the Disaster/Recovery Management Team will direct personnel in the most effective manner until all the teams have been mobilized and put into action.

Phase 2 action plans are designed to initiate and direct all the activities necessary to establish the backup site and other operating facilities that support the organization's activities, based on the "predecision" strategies provided for in the disaster/recovery plan.

The instructions, action plans in Section 5 of the plan, should describe in detail the duties of each of the teams and they should correspond with the PERT flow diagram. Phase 2 can be considered complete for those involved in this activity when the backup site is operable and on-line, providing the users with critical application support service information processing.

Parallel with initiating operations at the backup site, other teams, such as the New Facilities Team, the New Hardware Team, and the Salvage Team, are starting up recovery operations of the primary/home site. Other departments, if affected, will be required to initiate their plans to resume operations.

Phase 3: Backup Site Operation and Recovery Process

Phase 3 action plans for the recovery teams mentioned in phase 2 is an extension of their restoration activities to reconstruct, repair, or relocate the data processing operation. Subplans developed by on-site or off-site users who will be affected by the damage that has occurred to the primary site (on-site users) and who do not have access to their area, will have to initiate their miniplans in coordination with the DRMT. Off-site (on-line) users, whose data processing support has been affected, will have to initiate their action plans to perform tasks, capture data, and be available to go on-line when the backup site is operational. The User Liaison Team is extremely valuable at this point, because it will ensure that all the functions are performed during the time of interruption and maintain contact with all users, without interfering with the activity of bringing up the backup site.

During phase 3, the major aspect of operations of most of the teams who were involved in phase 2 and have been relieved once operations are resumed, is to cooperate with the DRMT to recover, restore, or construct a permanent site for the data processing operation and/or other organizational units that have been affected by the disaster.

Phase 4: Normalizing the Disaster Mode . . . Returning to the Home Site

The PERT diagram team operations extend beyond the restoration of services at the backup site. Insurance settlements, salvage operations, and acquisition of new hardware and facilities should have started, and should be fairly well completed by the middle or end of phase 3. If full recovery requires the complete reconstruction of the home site, it is impossible to estimate the time of reoccupancy of a new facility. If a commercial site is used, there is a contractual requirement that the organization release the site after six weeks. Therefore, at the initiation of the disaster phase 2, determination must be made by the DRMT and corporate management as to what strategy will be implemented to ensure a site (semipermanent or permanent) for data processing operations.

Phase 4 action plans and procedures are far more complex than those initiated at the time of a disaster. Normalizing a disaster mode encompasses the restoration of all activities at the new site, with new hardware, and bringing systems and data up to date with what has been run at the backup site, deferred applications, degraded activities, and establishing on-line activities to the satisfaction of the auditing department.

The DRMT will be far more active, over a longer period of time, as work proceeds in restoration of the home site facility.

During phases 2, 3, and 4, one of the most important aspects of backup, recovery, and restoration lies with the Communications Team and the provisions (preplanning activities) established to ensure that communications can be rebuilt and restored during each phase to service all the critical and normal user activities and requirements.

Review

The actual master plan contains sufficient details concerning all of the team responsibilities during phases 1, 2, 3, and 4. The further enhancement of these action plans are the responsibility of the Disaster/Recovery Plan Preparation Team, and the subplans on how each of the teams will perform these tasks are to be written by the specific teams and placed in Addendum E-3. The Disaster/Recovery Management Team has the responsibility of reviewing each of the subplans and ensuring that the activities will properly interface with the concepts and dynamics of the plan.

Sections 5.2 to 5.9 detail the action plans from phase 1 to phase 4. See pages 154 to 155 for details on how to elaborate the team activities when the disaster mode is invoked. *Note:* If additional teams are added, the disaster/recovery plan preparation team will detail the activities of the new teams through all four phases, and the specific teams will have the responsibility of preparing the team subplans.

The chart on pages 154–155 shows the process used in developing the disaster/recovery action plans.

6. Training for Disaster Recovery

The development of the plan using the team approach is understandably an advantage over other methods in preparing any business plan. Those involved have firsthand knowledge of each plan element, and the function and operation of the plan in the event of an adverse incident. These are the individuals that created the plan and its components. Yet, the plan has little or no value to the organization if the knowledge is not transferred to those who are responsible for the activities to be generated at the time of an unexpected or undesirable event. Pulling the plan "out of mothballs" and trying to implement it at the time of a disaster is useless.

The plan has extensive value when used as a "position" reference. The ability to perform this automatically at the time of the disaster requires a training program of everyone involved in the action plans.

Assuming that distribution of the plan is sufficient is courting disaster. Few will read it, less will understand the implications or requirement, and all will probably make it a shelf decoration. For the plan to work—to ensure that the organization in total will perform correctly—there must be a training program. Resources, both people and funds, must be appropriated. A plan that cannot be implemented could cause excessive time in initiating various elements of the plan, and perhaps thousands of dollars in lost revenue.

Part of the training program includes meetings with the teams to discuss their preparation of their individual miniplans. Additional training can be performed with various groups in a classroom environment. Test disaster, controlled recoveries, and drills covering various emergency responses (fire drill, bomb threat evacuation, and so on).

The DRMT and the Disaster/Recovery Administrator/Coordinator should develop a training program and review sessions with sufficient frequency in order to maintain currency.

6.1 Recovery Team Training

Individual teams are responsible for training personnel within the scope of their miniplans. The team assignments, and the ability of individuals to perform their tasks, in some cases without supervision or direction, is vital to the successful restoration of operations at the backup site.

The Disaster/Recovery Administrator/Coordinator is responsible for overseeing all training activities and maintaining the currency of training exercises.

6.2 All Employee Training

To ensure that every employee within the scope of the plan receives sufficient training and understands the principles of emergency response, disaster/recovery initiation, action plans, and recovery procedures, specified training sessions should be a requirement. This section addresses the various training programs to be performed by the Disaster/Recovery Administrator/Coordinator, fire safety instructors, evacuation supervisors, and so on.

Insurance companies will be glad to make presentations on various activities, fire safety, use of fire extinguishers, proper evacuation and accountability procedures, bomb threat procedures (local police community liaison officers are usually available, the FBI will perform instruction in certain companies, especially those engaged in highly classified government contracts); local fire departments will teach people how to use various types of fire extinguishers.

All employees, and more important, those employees involved in the strategic planning and action plans, must be aware of all the concepts and details relating to disaster/recovery planning, especially those actions that relate to their specific team assignment.

The Disaster/Recovery Administrator/Coordinator will be required to design, document, and implement the training program. In disseminating material to various participants, a copy of management's letter endorsing the program is vital to the success of any restoration actions. The objectives of the training plan, the training program schedule, and the training to be performed should be listed and detailed by group.

New personnel policy needs to be written concerning training and assignments. Training should include, but should not be limited to:

- Company policy on prevention and security
- Emergency procedures
- Team assignments
- Disaster/recovery philosophy and strategies
- Plan testing
- Plan monitoring and updating

6.3 CPR and First-Aid Training

In Addendum C-8.1 list those individuals on each shift who have passed a CPR course given by Red Cross or local EMS services. Also list in C-8.1 those individuals who have had Red Cross or emergency first-aid training.

7. Validation and Testing

Critical decisions regarding the effectiveness, adequacy, and timely performance of each aspect of the action plans developed must be made by qualified personnel (an EDP auditor, for example) and must be based on a reliable overview of a complete "reasonableness" review, static testing, and actual "dynamic" testing. This type of certification review provides technical information, whereas the certification provides the structure needed to make such critical decisions.

7.1 Validation

Since the plan and the planning process addressed anticipated problem areas (assumptions) and defined the needs for strategies, action plans, and specialized skills, it is necessary for the planners to perform, in conjunction with the audit department, a high-level "quick" review of the entire plan in order to gain an understanding of the problems, strategies, concepts, and issues involved. This review should provide that the task conform to:

- Placing boundaries on the effort
- Partitioning (dividing) the work to be performed at the time of a disaster
- Identifying the critical areas or areas of emphasis
- Certifying that procedures will validate the plan as a functional document

See certification checklists, Addendum G-1.

7.2 The On-site One-Application Test

This section outlines the anticipated test procedures taken to perform an in-house "static" test.

7.3 The On-site Complete Test

This section establishes the parameters for performing a complete in-house "static" test of the entire system. Selection of the timing for the test is to be made by the Disaster/Recovery Administrator/Coordinator with users' concurrence.

7.4 The Backup Site Test

Although the plan may be considered functional on paper, it is impossible to have a true understanding of how successful every function can be performed until the entire off-premises text is performed. "Dynamic" backup site testing establishes the criteria for molding the plan into a perfect or nearly perfect functional document.

7.5 The Surprise Test

Although many elements of the plan may prove functional on a series of preplanned tests, it is impossible to calculate how people will function at the time of a disaster. Therefore, it is important that at least once-a-year each organization perform a *surprise test* actually simulating a disaster incident. One of the most important aspects of this test is the debriefing operation performed after the test, and modifications made to the plan based on "failure points."

7.6 Test Schedule

Until the backup site is in operation, it is obvious that off-site tests cannot be performed. The test schedule referenced in Addendum G-3 can be established to perform in-house "static" tests until the backup facility is contracted for (that is, a commercial site) or an in-house facility has been developed.

7.7 Problem Log

When performing tests at the home site ("static" tests) or at the backup site, it is important to document every aspect of the test. This should be performed by the auditing department. A review of each activity should be held at the end of each test to address problem areas and modify the plan accordingly.

8. Plan Update/Control

Plan maintenance involves more detail than simple changes that result from personnel changes or responsibility reassignments. The general business environment and user activity is far from a static condition. New systems, applications, and business conditions require continuous monitoring. The changes must be reviewed with regard to their effects on the performance of the plan at the time of any adverse condition.

Reference Addendum H-1 for those items that have to be periodically reviewed to ensure that the plan is current. Although this list was established to cover the similarities found in data centers and organizations, some organizations may find it necessary to add items that are singular to the specific company.

In overview, the following items should be considered as part of the update and control of the plan as conditions change over time:

1. New systems programs and their equipment requirements
2. Identification of critical applications as systems activity changes
3. New users and/or applications
4. New backup procedures
5. Changes in documentation
6. Changes in organizational policy
7. Changes in file or data retention
8. Modification of buildings and areas that would affect emergency response procedures
9. Changes in personnel
10. New equipment acquisitions (computer and noncomputer operations)
11. Changes in communications
12. Reviews with personnel at the backup site

Management has the responsibility to periodically review the updates and verify that the plan has remained current with business activities. The validation activities performed by audits should be included as part of the plan's maintenance, update, and control.

8.1 Update Log

A section should be set aside as part of the plan (Addendum H-1) to log all changes to sections of the plan. Copies of rewritten sections should be sent to all "full" plan holders. Where changes affect individual teams, each should receive revisions. All details of the distribution should be logged.

8.2 Approval

The Disaster/Recovery Administrator/Coordinator is responsible for obtaining prior approval of any changes to be incorporated in the plan. The approval is the responsibility of the Disaster/Recovery Management Team. Distribution of any changes cannot be made until such approvals have been made and documented.

8.3 Formal Review (Once a Year)

Regardless of the number of review meetings held during the year, an annual review of the entire plan is important. It should include the determination of the status of all prevention and security procedures and performance results of tests. This review should also cover anticipated changes and future plans of the company and their effects on the performance of the plan at the time of an emergency or adverse incident.

8.4 Checklists

The disaster/recovery planning team (or the individual assigned the task), should review the plan checklists in Addendum H-2 every month. Those items covered under "Monthly" have to be documented. During the first few months after the completion and implementation of the plan, those items identified as planned enhancements must be verified with regard to status; all progress should be documented.

9. Plan Distribution and Security

In this section the Disaster/Recovery Administrator/Coordinator will document the distribution of the plan. Section 9.1 will contain the actual distribution list (total plans and/or sections) to be maintained. The Disaster/Recovery Administrator has the additional responsibility of physically viewing how the security of the plan is maintained. (See Section 9.5.)

9.1 List of Recipients

Record the distribution of the disaster/recovery plan (and/or individual sections if, in the event an employee is discharged or resigns, evidence is available as to who has possession of the plans and how they can be retrieved).

9.2 Screening Information

Prior to general and/or restricted distribution of the plan, the security department, as well as those involved in the records management program (classification of documents), should review the plan for sensitive information. Verification of the distribution list should be made by the security department as to what and to whom full documentation (or need-to-know) should be made and documented.

9.3 Serialization

All distributed copies of the plan (whole or partial) should be serialized; the document number should be placed on the distribution list (see Section 9.1).

9.4 Termination or Transfer of Copy Holder

The security officer, Disaster/Recovery Administrator/Coordinator, and the personnel department should have available a list of all plan (whole or part) holders. In the event that an individual is discharged or transferred, it is the responsibility of the personnel department to inform the security department to retrieve the plan from the individual prior to his or her leaving the premises.

9.5 Copies

Strict policy has to exist within the organization relating to the copying of all or any section of the plan. Each person receiving a copy (or section) of the plan should be obligated to sign a confidentiality and nonreproduction agreement (see Section 9.7). Company policy concerning reproduction of all (or portions of) the document has to be outlined on the first page of the plan.

To prevent unauthorized copies, the distributed copies should be printed on dark red paper in black ink. Most copiers see red as black, so copy sheets will come out black, with little or no legibility of typed matter.

9.6 Physical Security of the Plan

The security officer and/or the Disaster/Recovery Administrator/Coordinator has to verify that each plan (whole or part) holder has taken appropriate steps to place the document in a secure location.

9.7 Security and Confidentiality Agreement

See Addendum I-1.

10. Planned Enhancements

During the development of the plan, the disaster/recovery plan preparation team will develop a list of projects that require attention to ensure that reasonable steps have been taken to prevent damage or abuse to and the security of the facility, the operating environment, the equipment, the software, and the data. These are to be listed in Addendum J-1, "Planned Enhancements." The DRMT and the Disaster/Recovery Administrator/Coordinator will obtain authorization relative to the items listed and assign individuals to perform these tasks. Each of the items listed should have the name of individual(s) assigned to the project and the anticipated date of completion. The DRMT and the Disaster/Recovery Administrator/Coordinator will monitor the progress of these activities. The first list is by no means ever complete, and as the program progresses through the organization, additional items may be identified and new projects assigned.

POSTSCRIPT TO THE DISASTER/RECOVERY PLAN

Table of Contents Review

The writing of the plan at the workshop by no means completes the project. On the contrary, it is only the beginning. The teams have to complete their specific mini-plans, the user departments as well as all personnel involved in the strategies and action plans will have to be trained, planned enhancements will have to be completed, and the plan will have to be tested and maintained.

The DRMT, with the support of senior management, should encourage participation by every functional unit of the organization, and encourage suggestions and further enhancements to make the project more effective throughout the organization.

11. Addendum Review

Note: Addenda start with C. This allows open Sections A and B in the event special conditions exist at the organization.

Addendum C-1

Data Center Layout (DC)

Obtain a layout of the data center. If a layout does not exist, one should be created, preferably drawn to scale. The layout should show where all exits and entrances are located, how doors open (in or out), and all posts in the room. The legend should indicate the height of the floor and the dimension of the drop ceiling from the room's true ceiling. The floor grid should be detailed as close as possible. The following items should also be indicated:

1. Each piece of equipment (computer hardware). Give each item a number, and list on the side of the drawing the machine model and its serial number.
2. The location of the input power supply box.

3. The location of the input communications box.
4. The location of the emergency power-off switch(es).
5. The location of emergency air-conditioning switch(es).
6. The location of the power transformer(s).
7. The location of the emergency fire alarm box.

This becomes the master drawing. A number of copies of this drawing should be made:

1. One drawing (layout) to show where all sensing (detection) devices (heat, smoke, water) are located (indicate in legend whether in ceiling or under the raised floor).
2. One drawing (layout) to show where all fire extinguishers (and type) are located. It should also show where emergency plastic covers are kept to throw over machines in the event of a water flow. This drawing should also indicate where Halon discharge heads (or sprinkler heads) are located.

These drawings (layouts) serve a number of important purposes:

1. They can be used to mount on the wall or bulletin boards to acquaint and instruct personnel with the various safety features and emergency systems located in and around the computer room.
2. Preparing a complete drawing (layout) of the computer room provides another method of inventorying items, hardware, and peripherals and serves as a double check against vendor maintenance lists and contracts. Some data centers have installed special devices that are not mainframe vendor supplied; these usually get overlooked.
3. They are excellent references for the New Hardware, Salvage, Insurance, and New Facilities Teams and vendor engineers to assist in reconstructing the facility in the event of a disaster.

Note: The original layout for the data processing room can usually be obtained from the building engineers, architect, or facilities management (if space is leased).

Addendum C-2

Building Layout (DC)

Physical security of the building is the first step to providing internal security for the entire organization. The plan should show the layout of the floor (if a single-story building), or a layout of each floor the organization occupies if a multistory building.

The layout should indicate all entrances and exits, the way doors open, emergency exits, and doors that have "crash bars" (one-way exit doors). The types of doors (wood, steel, sheetrock, composition board, fire-rated, and so on) should also be noted. Loading docks and security of loading areas, and all emergency egresses, fire escapes, and so on, should be indicated as well. Locate and determine if existing fire hoses will reach all points on the floor (getting around dividers).

If external video security cameras are used, note where they and the monitoring station(s) are located.

The areas for car parking (in, under, or alongside the building) should be noted as well as the security enforced to prohibit illegal car entry.

Locate and identify all external air-conditioning fresh-air units, and the intake and types of fitters used.

Locate and indicate which side of the building power and communications lines come in.

If the building is a multitenant building, identify who is above, below, and alongside the data center and critical areas of the operation.

Addendum C-3

Area Layout (DC)

If the business (and/or the data center) is located in an urban area (within the business district of the city), a layout of the area should encompass a few blocks surrounding the actual building.

If located in a combined residential/commercial/industrial area, it would be advisable to produce a layout that would cover an area of about one-quarter to one-half mile circumference around the building.

If located in an industrial area, it should cover the same area as previously indicated [or depending upon access(es) to the area, and what could affect the access to the area or directly affect the building].

If located in a suburban or rural area, a greater area should be represented in the layout.

In all cases, the following should be indicated:

1. Parking areas for employees
2. Dangerous or critical areas that nighttime workers would have to enter or pass through to get to and from their cars
3. The location of power and communications centers (central offices)
4. Fire department and police department
5. Hospitals and EMS centers
6. Main highways
7. Railroad lines (especially those that carry freight)
8. Airports and flight patterns (commercial or general aviation)
9. Critical or dangerous manufacturing activities (propane filling plants, paint factories, chemical factories, dangerous substance storage facilities, and so on)
10. High-powered microwave or radar devices
11. Rivers and streams
12. Company buildings and facilities that are likely to have or have had demonstrations or civil disorder
13. Locations where crowds are likely to congregate for meetings, demonstrations, or any other purpose (for example, a racetrack or a baseball stadium)
14. Traffic patterns; traffic flows; rates of people, car, and truck traffic

Addendum C-4

Emergency Call List (DC)

Intrafacility

This list should contain all emergency numbers (and/or extensions) for internal activities (both daytime and night emergency numbers).

1. Security
2. Facility management, building management
3. Personnel department
4. Internal emergency medical services (inplant, hospital, nurse, and so on)

Disaster/Recovery Management Teams

This list should contain the names of individuals assigned to the DRMT and all other disaster/recovery teams; the team leader's phone number (internal and home emergency number); the name of the alternate team leader; and any special emergency assignment [field engineer, insurance department (internal)], and so on.

It is advisable that the internal extension, special phone numbers (internal), home phone numbers, and call annunciators (beepers) be tested here for each of the critical personnel.

Emergency Services

Use national emergency number 911 where applicable.

- Fire department
- Law enforcement agencies: police, sheriff, state police, secret service, FBI, private security companies, central alarm companies, military police, state guard, and the national guard

Cellular Phones

The growth of cellular phones, and their ability to function when other forms of voice, fax, or teletype communications have failed, make them an ideal adjunct for maintaining communications at the control site, etc.

Addendum C-5

Card Key Holders (DC)

In this addendum, prepare a complete list of authorized persons who can access the computer operations area or the actual machine room. This gives data processing management the opportunity to determine who should be allowed access to the area or the machine room. By listing all existing access cards, for those with access codes (key pad) entry, management can determine the advisability of individuals entering the restricted area without having a requirement to do so.

This procedure should be implemented into other highly sensitive user areas of the organization.

The machine room or data processing areas should be completely off-limits to those persons who are not actually employed in the area. If development work is performed in the data processing area, systems analysts and programmers should be prohibited from the machine and/or operations area, and if it is necessary for them to enter the facility, it should be done by escort.

New PC and micro-based access systems restrict entry into the data processing building/area or machine room based on time. For example, an individual's access card will only work during the time he or she is expected to be in that work area.

Personnel policies should include maintenance of lists of those who have access to sensitive areas; upon an individual's discharge or resignation, cards should be retrieved, or if key pads are used, the code should be changed immediately.

Another column may be added to the chart showing restricted access times.

Plan maintenance includes the monitoring of access restriction changes and approval.

Addendum C-6

Authorized Entrants (DC)

This list should contain all persons other than employees of the firm who have authorized access to the facility.

Procedures such as proof of identity, phone verification with vendor or supplier, preregistration procedures, and so on should be in place as part of the access entry authorization policy. It is also recommended that escort requirements exist for one-time vendor access. Field engineers, service personnel, and other maintenance service personnel should be required, at minimum, to produce an ID prior to entry; the facility should also maintain a sign-in log. The sign-in log *must* be reviewed daily by the manager of the data processing center.

Procedures should be established with the vendor or supplier to notify the data center manager of any reassignment, discharge, or resignation of personnel. It is recommended that a sign-in or registration sheet be part of the entry procedures to the building as well as the data center.

Addendum C-7

Authorizing Management (DC)

The responsibility of access authorization should lie with someone with specific authority in the organization. This authority should be controlled by a senior executive of the organization, as well as MIS management and data processing operations (data center management). Authorization for access or denial of access is the total responsibility of this group. Policy is to be established by senior management, and all changes must be in writing. Maintenance responsibility of this list has to be authorized in addition to the actual authorization procedure.

Addendum C-8

Data Center Sign-in Sheet (DC)

A sample of the actual sign-in sheet is to be placed in this section. If the organization does not have a sign-in form, the one presented here might fulfill the organization's requirements.

Addendum C-9

Evacuation Route (DC)

Using one of the drawings (layouts) obtained for placement into this addendum (floor plan—data center or building layout), the evacuation route from the area should be designated. In addition to placing the evacuation diagram into this section, a blowup should be made, the route(s) identified in red arrows and lines, and mounted in prominent areas in the data center as well as in the hallways.

Addendum C-10

Copies of Maintenance Agreements (DC)

Upon completion and identification of all hardware, peripherals, and auxiliary equipment within (or used by) the data center, a complete check should be made against all vendors' and suppliers' maintenance agreements. Suppliers would include such items as tape and disk pack purchases, forms, and so on, as well as items that normally are not covered by contractual procedures (although they should be!).

Software agreements should also be included, and terms of leasing, purchase, or franchising must be verified. Insurance coverages, liability, source codes, and so on have to be clearly defined as to responsibility and liability.

Actual copies of the maintenance agreements are to be placed in this addendum. In addition, letters of agreement between the company and vendor are to be placed here.

These maintenance agreements, as well as letters outlining specific authority, liability, and so on, will be needed by the insurance company, adjusters, New Hardware Team, Communications Team, New Facilities Team, and Applications Recovery Team, among others, at the time of a disaster.

Addendum D-1a

Off-site Agreement (DC)

Commercial Facility

A copy of the off-site storage agreement and procedures outline is to be placed here. The terms, methods, and security procedures for tapes (or disk pack) going off-site have to be detailed. Those who have authority for access or giving instructions should also be listed. Access codes used to place orders at the off-site facility should be placed in a sealed envelope, and access to the instructions should be kept top-secret until those who are authorized to retrieve tapes (or their alternates) are available.

The plan preparation team should have reviewed all the procedures prior to the final plan preparation, and all the procedures and security details should have been documented.

The commercial off-site facility should provide a copy of their specific security procedures, and these details should be put into this section.

If microfilm and/or other vital documents as well as documentation are placed off-site, the procedures for retrieval should also be outlined in this section.

In-house Off-site Program

Providing off-site storage as an in-house activity from one company location to another poses major security problems that must be addressed.

Strict policy concerning schedules, personnel clearances, corporate policy relating to records management, access to tape storage, handling procedures, and so on all require development to ensure the protection of information assets and their availability at the time of a disaster.

The details of these plans and activities require more attention as well as strict adherence to policy and procedure. They must be carefully developed and implemented by data processing center management.

PC and Minicomputer Backup and Storage

The same procedure developed for the mainframe processing activities must be extended to noncentralized computing operations—more specifically to PC, mini-, and microcomputing activities throughout the organization. It is the responsibility of MIS management to take the position that *all* information of the organization has to be protected regardless of where it is performed. Off-site storage is part of the entire information management program; the total maintenance and continuity of information pertains to PCs as well as mainframes.

Each user department should develop a procedure for backup and off-site storage of information.

Addendum D-1b

Vendor Information (DC)

This addendum covers all the details, authorizations, and identification requirements.

Addendum D-2

Tape Library Report (DC)

If the data center uses one of the tape library management systems, an up-to-date schedule should be set up in an addended volume to the plan and referenced here. Any modification to the tape library system or routine should replace previous versions.

Addendum D-3

Backup Site Alternatives (DCI)

A single backup alternative in light of the multiple applications that have been considered critical may not be practical. A number of alternatives have to be developed in order to accommodate as many users as possible during any short- or long-term computer interruption.

Addendum D-4

Backup Site Agreement (Initial) (DC)

Place a copy of the backup agreement in this addendum. This letter represents the primary backup contract, whether an agreement with:

1. A commercial facility
2. A cooperative agreement between two organizations
3. A co-op formed by two or more companies
4. A service bureau
5. An agreement between two data centers of the same organization

Addendum D-5

Backup Site Agreement (Secondary)

Place a copy of any further agreements made with any organization(s) that will support various applications not covered by the primary backup site agreement.

Addendum D-6

Primary Site Mainframe Configuration (DC)

A complete inventory of the data center (machine room) as well as any equipment used by systems analysts, programmers, operations personnel, and so on should be listed here. Verify the configuration against the data center layout. If maintenance contracts and/or sales contracts cover all equipment, peripherals, and so on as part of the entire inventory, these may be put here as well.

Addendum D-7

Backup Site Mainframe Configuration (DC)

This list should contain the entire hardware configuration available at the backup site. This configuration should provide sufficient minimum resource to cover critical applications.

Addendum D-8

Primary Site Datacomm Configuration (DC)

List the necessary communications resources to support critical applications and the resources available at the primary site.

Addendum D-9

Backup Site Datacomm Configuration (DC)

List the necessary communications resources to support critical applications and the resources available at the backup site.

Addendum D-10

Required Ancillary Equipment (DC)

List all data center (and backup site) ancillary equipment necessary to support backup operations.

Addendum D-11

Home Phone Numbers (DC)

List for the DRMT, backup and recovery team the members' names, titles, home addresses, home phone numbers, work phone numbers, and work area numbers. In addition, this list should contain the names of suppliers and individuals who have been put on special teams. For example:

1. *Salvage Team.* Insurance company agent, public adjuster, and field engineer.
2. *New Equipment Team.* Vendor representative and field engineer.
3. *New Facilities Team.* Real estate agent, architect, and engineer.

Addendum D-12

Calling Tree for Disaster/Recovery Teams (DC)

Set up the procedure necessary to reach all the individuals and their alternates assigned to various teams. The calling tree should start with one person calling two people calling four people, in the order of criticality.

Addendum D-13

Calling Tree for Other Employees (DC)

The administration team (or other assigned team) will have to contact all employees of the organization to inform them of the status of the workplace, and where employees should report to work in the event the organization's facility has been extensively damaged or destroyed.

The same type of *calling tree* should be established as performed with the disaster/recovery teams: one calls two, two calls four, four calls eight, and so on.

Addendum D-14

Vendor List (DC)

List all vendors supplying hardware, equipment, peripherals, supplies, and communications. If possible, each vendor should provide a letter representing that its best efforts will be applied in the event of a disaster. In addition, list the company phone number, the name of the sales contact, and shipping source and contract; also obtaining an emergency phone number is extremely advantageous.

Addendum D-15

Vendor Product List (DC)

A separate list by vendor and product should be prepared. This assures the continuity of the quality and compatibility of products to data center use. Some organizations combine this list with the actual vendor list.

Addendum D-16

Vendor Support Letters (DC)

Include here letters obtained from vendors who provide critical resources or supplies indicating their participation and support at the time of a disaster.

Addendum D-17

Disaster Kit

The disaster kit should contain all the supplies necessary to start up operations at the backup site. *Important reminder*: Sufficient tape supply should be in the disaster kit. When the database is loaded, a dump copy should be made and immediately sent off-site; tapes will be needed for regular backup.

Addendum D-18

One-Month Supply List (DC)

The data center operations personnel have to compile a list of items necessary to support off-site backup operations for a period of at least one month. This kit may be placed at the supply company, prepaid. The disaster kit may be located at the off-site storage location and picked up at the time the off-site tapes are retrieved.

Addendum D-19

Vital Records Retention (DC)

If the data center operates retention of files under a library management system (UCC-ONE, name to be added), then a copy should be sent to the legal department to verify the legal retention of vital records. This would include, but is not limited to:

- Original corporate documents
- Source documents
- Other hard copy records
- Magnetic media
- Microfilm or microfiche

Addendum D-20

Escrow Agreement

If the organization purchases leases, has purchased leases, or franchises software and its contracts specifically exclude the vendor's providing the organization with the source code, an escrow agreement should be negotiated and the source code deposited with a third party (off-site storage facility, bank safe-deposit box) with signatory requirement by an officer of the court in the event the software company can no longer support the organization.

Addendum D-21

Purchased Software (DC)

List all purchased software by product, application, vendor, purchase date, release level, type of support, purchase price, and annual fee. Any special conditions concerning purchased, leased, or franchised software should be amended to this list.

Addendum D-22

Disaster/Recovery Plan: Confidential Employee Questionnaire

People become singularly the most important resource at the time of a disaster. When forming the emergency response and all other disaster/recovery teams, it is important to identify those individuals who can or will be available for extended-hour service, as well as those who can leave their normal area and go to a backup site for a lengthy period of time. This questionnaire establishes the availability of personnel, the length of time they can remain away from home, whether they will require sustenance, the need for home care of minors, aged parents, or relatives, and transportation requirements. It also determines whether the employee is willing to use personal credit cards on behalf of the company. Questionnaires must be kept strictly confidential.

Note of concern: During a common area disaster like an earthquake, tornado, or flood, most people will be more concerned with the safety and welfare of their families. Organizations must recognize that at the time of such a catastrophic event, every organization will be in the same position. Police, state, or National Guards will usually be available to secure areas to prevent looting and vandalism and impose martial law. It often takes a few days before the area affected will return to any form of normalcy and businesses can start assessing the status of the damage and recovery. The human resources department should be enlisted to develop such a questionnaire to ensure that it conforms to federal, state, and local equal employment and labor laws.

Addendum D-23

Current Site Operations Schedule

On some occasions when disaster strikes, the people who are familiar with the normal daily performance may not be readily available. A copy of the daily, weekly, monthly (and special activities), operations schedule has to be prepared (and kept current) to ensure that when a disaster strikes *anyone* can maintain the established operations schedule. *Note*: Replace this page with a printout of the data center operations schedule. This must be continually updated as production schedules change.

Addendum D-24

Probable Backup Site Operations Schedule

When selecting the critical applications and the necessary resources to support them, it may be necessary to degrade service support. Although it is difficult to determine what the schedule will be at the time of a disaster, owing to the event occurring at various times, it is important to at least establish a minimum operations schedule for the backup site operations. Recognize that even for these (critical) applications, there might be:

- A reduction of service response
- Strategies developed to return to manual activities
- Complete withdrawal of all services
- Strategies that have been preplanned to accommodate the mandatory/critical applications at the backup site
- Other strategies that deal with *necessary* and *desirable* applications

Addendum D-25

Application Recovery Plan Worksheet

A sheet should be filled out for each critical (mandatory), desirable, or necessary application. Within this form, the status of the application and resources should be defined in order to understand what will be required to recover and maintain specific applications. Those applications that might be interfaced (integrated) should be clearly documented.

Addendum D-26

Copies of Policies and Annual Premiums

The risk management department should provide copies of all insurance policies that cover insurance coverages pertaining to data processing activities. *Important*:

Policies should show coverages for terminals, PCs, micro-, and/or minicomputers. Even though they do not directly relate to the mainframe, these terminals will be required as part of the backup recovery program and should indirectly or directly be under the control of data processing management. If properly prepared, the forms to be found in the disaster/recovery data collection manual may be used in place of copies of the policies. Here it is important that copies of the policies be stored off-site for reference by the Salvage Team in their negotiation with insurance companies.

Addendum E-1

Inventory of Damaged Items

The extent of the damage (or the level of destruction) to the building and/or the data center will dictate the methods of inventorying (or assessing the damage) to the equipment, hardware, facilities, power, communications, and so on. In cases where there has been extensive damage to the facility (building and/or data center), the authorities will usually deny access until the area is considered *safe*.

If ceilings have fallen, walls have collapsed, and so on, it will take considerable time to clean up the area before an inventory can be taken. Here is where floor plans, area layouts, and equipment lists are invaluable in identifying what should be where, and what has actually survived in one fashion or another.

The form here represents just the first stage of evaluation that the Salvage Team will prepare in conjunction with public and/or insurance adjusters. If the damage is of a lesser degree, inventorying of items, equipment, and the like is relatively easy. Vendors and suppliers should be involved so that a true picture of repair or replacement time (and cost) can be made.

It is advisable that all calculations be independent of those from the insurance company adjuster. This is important since insurance personnel will attempt to come up with a lesser or more reasonable cost and time; the effort of the organization should be directed to an accelerated activity regardless of cost.

Addendum E-2

Time to Replace Hardware—Assuming Emergency Expedite of Order

The layout and equipment lists should contain all the items necessary to recover the home site. Organizations, in some instances, have an excellent opportunity at the time of a disaster and as part of the recovery process to upgrade the system. This has to be preplanned. The New Hardware Team has the responsibility of keeping current with new versions and potential software changes that may be necessary.

The chart that appears in this addendum should be prepared prior to the occurrence of any adverse incident.

The preparation of purchase orders in advance for all major operating equipment and components will avoid major time delays. Copies of purchase agreements should be obtained from vendors or suppliers. Each organization should obtain, if possible, a letter of commitment from vendors. Consider the condition that an organization has over a long period of time accumulated an extensive number of terminals. Purchasing these over a long period of time, in small quantities, usually poses no problems. Requiring a vendor to supply a large number in a short period of time may not be feasible. This has to be carefully reviewed with the vendor to determine what strategies and alternatives are available in the event of a disaster. Obviously, the condition becomes more critical with the mainframe, where a release level must be matched. Another difficulty is obtaining replacements for old or obsolete equipment.

If the organization has opted to prepare a shell site, and will depend upon vendors to respond quickly, it is far more critical to have every item researched and documented prior to a disaster.

Addendum E-3

Team Plans

Section 4.17 documents teams by title, their charters, and typical duties by phase. (Also see pages 36–41.) Section 5 documents the *action plans* by specific phase, and by specific team activity during the time of a disaster. Although much of the "activity" is documented here, and the PERT flow chart is prepared in Section 5, each team will be required to detail how it will accomplish the specific tasks assigned to it based on the time constraints established in the action plan (PERT diagram). These are the *miniplans* necessary to expand the action items to actual performance. Further discussion on the actions required by the user will be covered under separate documentation at the end of the addenda. User plans should interface with the data processing center plans, and they could be amended to the master plan.

Addendum G-1

Certification Checklists

The management control of prevention, security, and disaster/recovery planning can be achieved by a *certification/accreditation program*. There are major benefits to this approach to control. The processes suggested here (certification/accreditation) can help protect against fraud, illegal practices, mission failures, and embarrassing "leaks," as well as mitigate the effects of a disastrous event and, more important, legal action. Testing and maintaining the plan only provides one level of confidence concerning the established policies, practices, and plans. What this process provides is a way of keeping managers from being "surprised" by problems within the data processing operation and applications. Certification/accreditation in the area of prevention, security, and disaster/recovery planning is only one aspect of the approach to this assurance methodology to satisfy that the computer operation functions within its defined performance, security, quality, and reliability requirements.

The process does not end with the first certification/accreditation project. It is necessary to continue the assurance/credentials approach by recertification/reaccreditation of the plan.

The checklists provided in this addendum are guidelines directed primarily towards those responsible for performing computer security certification/accreditation and those responsible for establishing certification and accreditation programs:

- Senior executive officers
- Computer security managers
- Users
- Disaster/Recovery Administrator/Coordinator
- EDP auditing personnel

This program benefits the organization by improving management control over computer security and increasing awareness of computer security throughout the organization. The Certification Checklists are available through DIA*log Management, Inc.

Addendum G-2

Threat Evaluation Form

Section 2.3 of the plan discusses the steps to be taken to achieve an understanding of the threats, hazards, and vulnerabilities as well as the "risks" to which the data center and/or the user community are exposed during normal business activity. Although we feel the major aspect of risk assessment is resolved utilizing baseline security, the authors of the plan may want to specifically highlight key "risks" and to discern what is significant from what does not merit contingency provisions. The form in Addendum G-2 provide a uniform method of documenting such risks.

```
┌─────────────────────────────────────────────────────────────────────┐
│                          FORM G-2                                    │
│                       FOR ADDENDUM G-2                               │
│                     Threat Evaluation Form                           │
├─────────────────────────────────────────────────────────────────────┤
│                                                                      │
│  Threat Name _____  Threat frequency _____       │
│  Description  _____ High/Low   │
│               _____         │
│                                                                      │
│  Potential or Risk    ☐ High  ☐ Medium  ☐ Low                        │
│  Measures and Countermeasures—Description                            │
│  (This should cover prevention, security, and disaster/recovery      │
│  strategies and action plans.)                                       │
│  _____    │
│  _____    │
│  _____    │
│  _____    │
│  _____    │
│  _____    │
│  _____    │
│  _____    │
│                                                                      │
│  Rating                                                              │
│  Capability of Controls    ☐ High  ☐ Medium  ☐ Low                   │
│  Rating                                                              │
│  Quality                                                             │
│  Measures    ☐ Above required  ☐ Sufficient  ☐ Insufficient          │
│  Impact                                                              │
│  ☐ Destruction  ☐ Disclosure  ☐ Modification  ☐ Denial of service    │
│  Prepared by: _____  Date: _____      │
│  Reviewed by: _____  Date: _____      │
└─────────────────────────────────────────────────────────────────────┘
```

Addendum G-3

Testing Schedule

Establishing test procedures to determine if the procedures, strategies, and action plans will work in the same manner as they were conceived requires careful planning. Deviating from plan components will not provide the feedback to determine whether the plan needs modifications.

Staging static tests should be done in small segments, as each phase of testing has been debugged and modifications have been made. In-house tests can be performed on equipment by merely shutting down the system and bringing it back into operation by loading software and the database. From this point, tests can be performed by dropping specific applications and using off-site software and backup tapes to bring up the specific program.

Testing calling trees and mobilizing teams at the control center can check time and performance. Testing of emergency procedures is absolutely vital.

Critical dynamic tests are those performed at the backup site utilizing off-site facilities, bringing up the system and one or more on-line users. Before these tests are performed, care should be taken to ensure that the in-house static tests have been carefully documented and modifications have been made, as off-site tests are expensive and time consuming.

At least two announced off-site tests should be made before a surprise test is performed. Results of surprise tests or emergency drills require extensive debriefing. The plan will need to be updated to correct those areas or activities that have not performed satisfactorily. All tests should include use of the on-line network.

Emergency Plan Tests

Building management, security administration, and other functions (areas) of the facility, main office, and so on, should plan various emergency incidents and test procedures. Some organizations have numerous emergency services (police, fire department, emergency medical services; in critical industries, federal agencies will participate in preplanned programs).

Monitoring Tests

The internal auditing department should be involved in designing and monitoring various plans. Once the plans have been tested sufficiently, and have been accepted for full implementation, the internal auditing (or EDP auditing) staff should be responsible for the monitoring of all subsequent tests. In addition to monitoring tests, the internal auditing department can be made responsible for establishing a variety of tests of different magnitude, especially during the first year. Senior management should participate in and be notified of all test procedures.

Addendum G-4

Problem Log

During all test procedures, and in the event of a disaster, it is imperative that every condition that interfaces with the smooth performance of the plan procedures be documented; modifications and resolutions made on the spot to correct the condition are to be documented as well.

Addendum G-5

Characterization of Test

All key personnel involved in any test or actual *under-fire* activities should prepare a personal evaluation of the performance. These reports will be part of the activities evaluation and add to the future modifications necessary to update the plan.

Addendum H-1

Update Log

All changes to the plan are to be documented. A suggested form is presented in this addendum which contains:

- Plan sections affected
- Description of change
- Add/delete/modify suggestions
- Date
- Approved by

Plan Maintenance

The maintenance of the disaster/recovery plan can be segmented into 6 categories:

- Time
- Equipment
- Facilities

- People
- Applications
- Data

Changes that occur over periods of time that will affect the recovery capabilities and performance of any support center or user will be the ability of the plan to keep current with continuous revisions as well as keeping systems programming synchronous with hardware, software, applications, and data. It is just as important to maintain and update the list of people affected (names, responsibilities, addresses, phone numbers, and vacation schedules of key participants).

In addition, it is necessary to conduct periodic reviews of off-site storage procedures, methods, and practices of maintaining documentation, new requirements for the disaster kit, as well as thirty-day supply requirements, changes in forms, changes in key suppliers, new suppliers, and so on.

Emergency procedures have to be periodically reviewed to determine if changes in procedures have to be made.

The basic checklist provided in the master plan establishes a schedule for the disaster/recovery group (administrator/coordinator or the DRMT). In addition to modifying (or making required changes to) the plan, it is necessary to communicate the changes and updates to those groups (teams) that have the responsibility to perform the specific tasks. Conversely, it is the responsibility of the individual teams to review their action plans and require performance and notify the Disaster/Recovery Administrator/Coordinator of any changes.

Senior management, through the auditing department, should be periodically (specify time periods) reviewing the existing plans utilizing a recertification/accreditation process. They should then report to management the status of the plan.

Addendum H-2

Maintenance is an absolute necessity. The following outline provides the mechanism to ensure that all aspects of the plan are properly maintained in an organized manner. Volatile elements such as personnel are reviewed more frequently than the backup site agreement.

DISASTER/RECOVERY PLAN—PERIODIC UPDATE QUESTIONNAIRE	
Date: _____ Quarterly: Jan/Apr/Jul/Oct Semiannually: Jan/Jul Annually: Jan	
Administrative Planning	
Period	**Question**
Monthly	Terminations: a. List all terminations. b. Have all keys and documentation been collected?
Quarterly	Has the disaster/recovery plan test schedule been met?
Quarterly	Have planned enhancements to the physical security system been made as scheduled?
Quarterly	Has the documentation off-site been checked?
Quarterly	Have the off-site requirements and contents been confirmed.
Semiannually	Has the disaster/recovery plan been tested?
Semiannually	Are the telephone lists up to date?
Annually	Has the backup site agreement been reviewed?
Annually	Has the off-site storage agreement been reviewed?
Annually	Are all purchased software products under a software escrow agreement?

Annually	Is the latest version of products in the software escrow?
Annually	Has the retention schedule been reviewed?
Annually	Has the backup site configuration been modified?
Annually	Has the computer insurance been reviewed?
Annually	Has the maintenance agreement been reviewed? (Are all units covered?)
Annually	Are employee commitment letters current?
Annually	Have new applications been added, and have all disaster/recovery implications been handled?
Annually	Is the disaster/recovery plan up to date?

Physical Security and Environmental

Period	Question
Monthly	Has the training schedule been met?
Monthly	Are all alarms in working order?
Monthly	Have physical security breaches occurred? (What preventative actions were taken?)
Monthly	Have new employees been instructed in fire safety?
Monthly	Are all emergency exits operable?
Quarterly	Are all locks in working order?
Quarterly	Has a fire drill been held?
Quarterly	Are the first-aid kits full?
Semiannually	Are all fire extinguishers in working order?
Annually	Have all keys been changed?
Annually	Has the power quality been reevaluated?
Annually	Have air-conditioning requirements been reevaluated?

Software Security

Period	Question
Monthly	Have passwords been changed?
Monthly	Was the system log reviewed for security breaches?
Quarterly	Has old media been discarded?

Addendum I-1

Security and Confidentiality Agreement

Each employee of the organization should be required to execute a security and confidentiality agreement to protect the information resources and trade secrets of the organization. The legal department should be consulted to develop such an agreement to ensure that the rights of individuals and the organization are protected. (See the sample document, "Security and Confidentiality Agreement," on page 182.)

Addendum J-1

Planned Enhancements

During the development phase as well as the testing and maintenance aspects of the plan initiation and execution, the plan's developers will recognize that certain enhancements to the physical and logical security will be required. This addendum lists those "planned enhancements," the data for anticipated installation and/or completion, the authority for such enhancements; and the individuals who will be responsible for their implementation. (See the sample form, "Planned Enhancements," on page 183.)

Addenda K-1 through K-12

Disaster/Recovery Plan Monthly Checklists

In addition to the periodic checklist that appears in Addendum H-2, the disaster/recovery planners may want to develop a monthly checklist that will cover the volatile aspects of the plan. These monthly checklists, fashioned after the example in Addendum H-2, plus unique conditions to the organization, make it easy to send out and receive changed information in a uniform manner.

**FORM FOR ADDENDUM I-1
SECURITY AND CONFIDENTIALITY AGREEMENT**
(Sample Document)

I _____

employed by _____

located at _____

have reviewed the confidentiality provisions of the organization's prevention, security, and disaster/recovery plan, and herewith agree to the following:

1. At no time will I divulge to any other individual the contents of the document(s) provided to me, unless such release of information is directly related to the performance of my responsibilities, and shall limit the dissemination of information contained in the referenced document to the specific authorization list.

2. At no time will I reproduce, transmit, or communicate any information covered by this confidentiality notice, or any other information that is considered confidential, secret, or proprietary to the organization and/or the plan herein referenced, to any other employee, contractor, or consultant (or any agency of the government) without the express written permission of an authorized officer of the organization. All authorizations *must* be in writing.

3. I understand that I am subject to disciplinary actions, or discharge in the event of any breach of security or confidentiality herein described, or established in other provisions, policy, regulations, or communications of the company. The decision of the company to invoke disciplinary measures as herein described is final, and cannot be waived by any member of the company.

Accepted by: _____

Date: _____

_____
Authorized Signature

Date: _____

FORM FOR ADDENDUM J-1
PLANNED ENHANCEMENTS
(Sample Form)

Description	Responsible Party	Proposal Date	Approval Date	Start Date	Completed Date
Add Smoke Alarms			9/1/89 (est)	2/11/89 (est)	3/30/90 (est)

Addendum L-1

Backup Facility Requirements and General Information

Each commercial (or in-house) backup facility should prepare a requirements and general information document: in effect, a plan within the plan. This should contain details on how to interface and interact with the backup site at the time of a disaster. It should also list what facilities such as hotels, restaurants, auto rental, and so forth are available. Details should also include use of ready rooms, cold sites, and control center operation, just to name a few.

Addendums L-2 through T

The items discussed in Addendum L-1 are detailed in these addendums and should be filled in to detail items listed. (See the Plan Table of Contents in front of Part 2.)

Addendum U

Procedures for Reporting and Obtaining Personal Property and Clothing Damaged or Lost

Employees usually have a number of personal items in their place of employment. Each employee should be required to list these items, in advance, and the list should be kept in personnel files, off-site. Employees whose items of clothing may be damaged during the time of the disaster, where the employee may be required to perform services in a fire, water, or environmentally affected site, should be entitled to receive replacement of, or compensation for, such clothing items as long as they were damaged in the performance of assigned duties.

Other problems such as use of credit cards, distribution of payroll, and family care are documented in Addenda U-2 through U-5.

Addendum V

Substance Abuse

This addendum should contain a Confidential Employee Agreement stating the employee (data center) has not or will not use or bring onto the premises any illegal substance or alcohol.

PART 3

GETTING THE JOB DONE

1.1 HOW TO GET THE JOB DONE

No one individual in any organization wants to, or can, undertake any more projects than is needed. This especially pertains to the development of a disaster/recovery plan that covers the broad range of activity and is as complex as an effective disaster/recovery plan for data processing and/or any other functional business unit(s) of the organization.

As professional consultants in this field, the greatest obstacle we have found in getting organizations involved in disaster/recovery planning is obtaining management support and finding the group of people who can be assigned to the project. Historically, the project is usually laid at the doorstep of data processing. And although the data processing plan requires the technical input of experts in this area, the plan should be motivated, administered, and controlled at the corporate management level, since it is a *business survival plan, not a technical plan.*

Over the years, in performing seminars and workshops on disaster recovery and business interruption planning, we have found it necessary to continually modify, retrofit, and develop new procedures and strategies on the methodology of "getting the job done." Developing a plan that deals with assumptions and probabilities is, to say the least, difficult to convey even to the most sophisticated audience. Getting these groups to sit down and do the job becomes almost impossible.

Some participants of our seminars and workshops have attempted to develop effective disaster/recovery plans for business units or data processing on their own. Others have investigated using or have used outside consultants. Yet other groups have invested considerable amounts of money in the purchase of mechanized, "cookbook, fill in the blank spaces" plans and have tried to make them work. Some have been successful; others have relegated these documents to bookshelves.

Approaching the project in the methods just described lacks a number of success keys, but each possesses its own rationale for failure. Attempting a plan on your own usually fails because plan developers usually work in a vacuum with organizational tunnelvision. They are not aware of, or do not have available to them, the state-of-the-art technology that provides all the components that create an implementable, testable plan, and they seldom have the cooperation from any other areas of the organization. Time also becomes a substantial barrier. Regardless of how motivated the planning group is with respect to the plan, communicating motivation throughout the company, and establishing the effective disaster/recovery planning program as a high-level priority is a major task. What starts out to be an energetic, dynamic, well intentioned group, is soon reduced to a bunch of frustrated, disillusioned individuals. Without participation and contribution from the entire organization, primarily senior management support, the continuity of the project is substantially diluted, and the project becomes paralyzed.

Consultants usually have a better shot at getting the job completed. Less internal resources are utilized, but cost involved is high. An unknown quantity are the extensive hours and substantial participation of organizational groups necessary to provide organizational background. Even with maximum contribution, these plans lack the organizational culture, "intelligence," "motivation," and "substance" of plans developed by in-house personnel. Further, the final product does not reflect the organization's personality and culture, and the final document has a little than better chance of staying alive. Without the further use of consultants for training, testing, and maintenance, the document loses its "staying" power quite quickly.

The plan development methodology presented in this handbook, provides the best of two worlds. *One, the material contained in this handbook effectively becomes the organization's consultant* providing the planning concepts and technology; and two, with the Disaster/Recovery Administrator/Coordinator acting as the seminar leaders, and with total participation of the organization, the plan has a chance of getting completed. With the participation of company departments, including data processing, the plan will usually remain effective and alive. Training and maintenance is easier, since the individuals who are involved in writing the plan will be responsible for testing and maintaining it and have a vested interest in its success.

1.1.1 The First Step

The Disaster/Recovery Administrator and Coordinator *must* become totally familiar with the contents of this handbook: its methodology, planning concepts, processes, and procedures. It is important that all the information contained in Part 1, "Effective Disaster/Recovery Concepts and Requirements," be read and understood in as much detail as possible. Learn as much of the internal workings, the interfacing, and interactions of every department and function of your company as possible. The Disaster/Recovery Administrator and Coordinator will become the resident experts—the "instructor/consultant" at the forthcoming company seminars and workshops on disaster/recovery planning. The material presented in this handbook and the plan itself have been written for the worst-case occurrence of a catastrophic event. Events of lesser severity can be handled as a subset of the master plan, and at the appropriate level of the occurrence. Participants must be conditioned to the worst-case situation scenario.

1.1.2 The Second Step

Review in detail every section of the plan as outlined in the Table of Contents Review in Part 2.

The Disaster/Recovery Administrator and Coordinator should be knowledgeable about the concepts of each of these sections and the intent of each paragraph of the plan.

1.1.3 The Third Step

Review the concepts of the interactive sessions and prepare the lecture (presentation) material necessary to provide instruction and obtain responses from the participants.

1.1.4 The Fourth Step

Test the session out with other individuals assigned to the project before you present a full participation seminar or workshop. This should be a limited meeting, rather than an actual seminar or workshop. At this time, it might be appropriate to involve a limited number of other user departments that will have a direct interest or can provide suggestions and recommendations to enhance the seminar or workshop as well as the actual plan.

1.1.5 The Fifth Step

Schedule the program as follows:

1. Preplanning session (two to three days)
2. The data collection/decision-making period (six to eight weeks)
3. The plan preparation workshop (four to five days)

Once the commitment is made to proceed with the project, the dates are established, and the departments and participants are notified, no changes should be allowed to be made to the schedule. The project must be treated with the highest priority; the people involved should be notified that scheduled dates must be adhered to, and that the program cannot be deferred or delayed. It is the responsibility of senior management and the Disaster/Recovery Administrator and Coordinator to create the urgency that the plan development deserves.

UNIT 1: PREPLANNING SEMINAR
Program Procedure

Phase 1: Preplanning

1.1 In order to begin writing the disaster/recovery plan, it is important that all the items listed below be in a complete form.

1.1.1 All profile documentation completed by the company and submitted to the Disaster/Recovery Administrator/Coordinator. This document should list all participants, the scope of the project for each concerned, and background data on each group preparing a plan.

1.1.2 Selection of the organizational disaster/recovery steering committee.

1.1.3 Selection of the Disaster/Recovery Management Team.

1.1.4 Selection of the Disaster/Recovery Administrator.

1.1.5 Selection of the Disaster/Recovery Coordinator.

1.1.6 Determination of what department will administer the plan.

1.1.7 The scope of the plan. Those departments other than data processing that will be involved in the development of the plan.

1.1.8 The depth to which they wish to involve other organizational units. Will each of these critical units be developing its own plans at the workshop, or is this a secondary activity after data processing develops its plan?

1.1.9 Other organizational units, if any, that will be developing plans simultaneously (at the workshop). List these departments, and note who will be on the planning teams. See the individual and company profiles.

1.1.10 A list of groups that are not at the same site but which will participate. Where are they located?

1.2 Other information and decisions can be derived from answering the following questions.

1.2.1 Has the company determined what its backup strategy will be? Where will it be located?

1.2.2 What are the critical services that have to be supported (data processing), and what are the critical resources necessary to support these critical applications?

1.2.3 What is the company's present procedure for off-site storage? Where is its off-site storage facility? Give a general background of its present activity.

1.2.4 Where is the company's command or control center to be located?

1.2.5 What has the company done about alternate communications, alternative networking?

Figure 7-1

UNIT 2:
THE PREPLANNING SEMINAR/WORKSHOP

The Preplanning program (seminar/workshop) will have to be developed by the Disaster/Recovery Administrator and Disaster/Recovery Coordinator to suit specific organizational requirements and culture. It is important that all participants understand the concepts of the plan, the terminology used, the common interests, management's commitment, and how the plan will be put together for each participating group.

The Preplanning session, which will take two, perhaps three days, provides participants with the guidelines in the plan development methodology as well as the structure of the plan. The Preplanning session should include, but is not limited to:

2.1 Defining the Project

Defining the project calls for an overview of the concepts of effective disaster/recovery planning; the general (or detailed) scope of the project; the participation of organizational units or individuals; management's commitment; resources required, including personnel and funding; the objectives of the plan; and the assumptions under which the plan is developed and the action plans are created.

2.2 Interactive Seminar/Workshop Sessions

During the Preplanning seminar, each of the departments, components, users, and computer and noncomputer operations should be required to participate in a number of interactive presentations.

For the Disaster/Recovery Administrator and Coordinator, these interactive sessions may be the first time they are exposed to the overall workings, functions, and performance of the organization. It is without a doubt, their first opportunity to completely understand the interrelationships and interactions between departments, divisions, and operating units of the company.

The interactive sessions should also provide the first overview as to the scope of the plan, by identifying what management and the users understand are their critical applications and the critical success factors that support the attainment of organizational goals. At the public seminar/workshops performed by DIA*log Management and the authors, these interactive sessions provide a wealth of information to assist the instructor/consultants in learning about the organization; the same should occur at your sessions. This is vital to the actual plan preparation portion of the program.

The authors believe that few organizations exist where the disaster/recovery plan itself can be limited solely to the data processing operation. Unquestionably, the data processing department has become one of the most critical success factors in the business performance. Yet, without the ability of the user department to perform, and in many cases the derivation of information that flows from noncomputer operations and that ultimately resides in the computer, the overall business function can substantially come to a grinding halt. Therefore, the development of disaster/recovery plans that include corporate headquarters, all of the administrative and executive operations, noncomputer operations, divisions, users and so forth, has become as important as the plan developed for data processing.

The interactive sessions provide the "hook," the understanding of organizational relationships ensuring that the critical success factors are properly addressed and incorporated into the overall planning scheme.

Each participating group has a specific responsibility in addressing the subject matter covered by these interactive sessions.

2.3 List of Interactive Sessions

- Organization and disaster/recovery planning
- Scope of plan
- Cost of downtime
- Maximum allowable downtime
- Affects and impacts on critical success factors
- Selection of critical applications
- Risk assessment—measuring dependency
- Baseline security
- Personnel security
- Facilities evaluation
- Internal controls
- Insurance
- File backup strategies—audit trail—off-site storage
- Communications
- Organization for disaster/recovery planning and implementation
- Organization for disaster/recovery planning and execution
- Selection of disaster/recovery teams and team assignments
- Backup alternatives
- User backup and recovery strategies
- Maintenance—testing—training
- Recovery procedures and post-recovery considerations
- Data collection
- Establishing organizational policy relating to information management and information processing
- User and business unit disaster/recovery planning considerations
- Establishing, manning, occupying, and operating a disaster/recovery control center
- Continuity of corporate management in the event of a major catastrophe

Figure 7-2

UNIT 3: DESCRIPTION AND CONCEPTS OF INTERACTIVE SESSION FOR THE PREPLANNING SEMINAR/WORKSHOP*

3.1 Organization Structure (Corporate and Data Processing)

This opening session assists the Disaster/Recovery Administrator and Disaster/Recovery Coordinator in gaining useful knowledge about the internal workings and functional operations of the corporation and the specific department. Each participating group will first present an organizational chart showing the relationship of its specific function, its relevance to the organization and data processing. Each group then presents an organizational chart of the data processing operation as it relates to its utilization. (The participant team members should be prepared to illustrate this chart on an overhead transparency film. This should be prepared prior to the session. If there is an existing organization chart, it can be copied onto transparency material to save time during the session.

3.2 Facilities Evaluation

Understanding the potential threats, hazards, and vulnerabilities (risks) to which the organization, the user, and the data processing department are exposed is extremely important in developing a *prevention* and *security* program. Most organizations view their operations strictly on an internal productivity basis, and overlook potentiality lethal conditions that could seriously interfere with the normal operation of the organization or the data processing environment.

Participants will prepare transparencies showing an area plan of the general external building activity, traffic flow patterns, surrounding buildings, potentially hazardous industries, and critical or dangerous activities that may affect access to the area in the event of a disaster. The second diagram should show the location of the user in relation to the internal building activity. The third should show the data center, the accessibility of the machine room by outsiders, security provisions, fire precautions and prevention, and so on. After each facility description by the participant, the Disaster/Recovery Administrator and/or Coordinator should critique the existing procedures and security systems as well as make recommendations as to the improvements necessary to protect the environment. The purpose of this exercise is to acquaint all participants with the *facility evaluation methodology* (see the company and data processing facility evaluation checklists) that must be undertaken during the data collection phase of the plan development program.

3.3 Cost of Downtime/Maximum Allowable Downtime

Part of the justification for the development of a plan is the potential losses to an organization that result from its inability to perform normal business functions. Each participating group will be asked to evaluate its business functions and determine (in a broad view) what the organization will lose, not only financially, in the event there is an extended interruption of data processing activities or any critical function of the company.

The objective of this session is to specifically determine how the organization will deteriorate in its business performance as time passes without any one department's support or the lack of information processing. More definitively, what is the impact on the organization and business continuity in the event the data processing operation is lost? For example, how will the loss of information affect the decision-making capability of management? How will strategic, tactical, and operational decision making be affected? Cost of downtime should be viewed on a multidimensional basis.

- Financial loss
- Loss of cash flow
- Operational stoppage
- Customer relations and satisfaction
- Legal/regulatory compliance
- Data integrity
- Loss of business continuity
- Loss of business opportunities
- Breach of contracts
- Loss of competitive position

Time plays an important factor in the cost of the downtime scenario. In viewing the critical applications and the critical success factors of the organization, a decision has to be made as to *when* these functions must be recovered in order to maintain the continuity of business performance. Each participating group has to determine the maximum allowable downtime for each of the critical applications.

3.4 Internal Controls

This review/interactive section discusses the factors shaping the control environment and the criteria for building a strong external control environment. Here we will discuss the problems of computer systems operations, system development, on-line processing, access control, and so on.

Internal controls are based on management's policy and standards for the protection of the business. These *controls* or control objectives provide the organization with the security that ensures profit, effective organization performance, and efficient operations. The lack of controls leads to adverse incidents and creates an environment that will not meet management's business objectives. Although controls are vital to the information processing activity, organizations should not overlook the controls necessary to perform normal business (noncomputer) functions.

The subject of internal controls cannot be taken lightly. With better than 90 percent of computer disasters generated by people, controls and more specifically internal controls within the computer applications are vital not only to the success of the business performance on a day-to-day basis, but to the reduction of possible and probable business discontinuity resulting from a computer failure.

Again we have to look at the critical business functions and critical success factors that can affect the ability of the organization to stay in business. These include, but are not limited to:
- Accuracy of information
- Timeliness of information
- Continuity of business performance
- Prevention of fraud and embezzlement
- Security and welfare of personnel
- Compliance to legal requirements
- Fulfilling of contractual obligations (written or unwritten)
- Efficient, cost-effective business operations
- Positive public image
- Continuity of cash flow
- Profitability
- Continuity of general administrative functions and services
- Continuity of financial activities (collections, payments, credit rating)
- Maintenance of accurate financial statements
- Maintenance of audit trails and auditability of financial and business functions
- Maintenance of competitive position
- Sustained ability to take advantage of all business opportunities
- Maintenance of satisfactory labor relations

Each organization can probably add a significant number of other control objectives appropriate to specific business activity.

What must be carefully appraised is how each level of a disaster, from a minor inconvenience to total destruction of a facility, will affect business discontinuity, and how the implementation of controls can substantially reduce the level of disaster as well as the impact on the organization. A facilities evaluation interactive session should be held; its emphasis should be on the physical protection (security) of the facility and the data center. In this session, the organization should also direct its efforts to the logical or internal system controls. This starts with the writing of specifications for a new application up to and including its implementation as well as the security provisions of existing applications. In order to assist the Disaster/Recovery Administrator and/or Coordinator in giving the participating group a better view of how the interactive session is performed at a public seminar/workshop, the authors have included a copy of the actual material covered during the interactive session, and how these sessions are performed.

3.5 File Backup Strategies

If the production system fails, or it no longer exists owing to a major catastrophe, is your organization capable of recovering the database from the instant of failure? Most organizations do not consider the implications of poor file backup procedures.

This interactive session reviews each organization's procedures and helps you determine the effectiveness of present backup procedures as well as provide you with the methodology for establishing such a process within the organization.

File backup procedures vary greatly from organization to organization. In order to emphasize the importance of proper file backup procedures, it is important to understand what you are doing today, and how the program can be improved in the future. This is accomplished by using the "scenario" approach during the interactive session. It becomes quite dramatic if the users are participants at the same session.

3.5.1 Proposed Scenario for Session Review

A major data processing disaster occurs at 3 P.M. on Thursday afternoon. The plan is executed, and the data processing operation moves to the backup site. The User Liaison Team notifies the users that the best recovery that can be achieved is the backup taken on Wednesday, at midnight, and that all data entered from 8 A.M. Thursday morning until 3 P.M. will have to be reentered. After the howls and screams, a decision has to be made as to what entries have to be made in order to bring the system current after 3 P.M. The basic problem is that too few users are disciplined to the extent that data input is logged by time.

A coordinated effort between the data center and the user community should establish a file backup procedure, as well as the methodology to ensure the recapture of data that can be lost at the time of a disaster.

File backup and off-site storage goes beyond just data and should include:
- Object code
- All current versions of system software
- Application software
- Files and vital records
- Know-how manuals and standards
- Documentation

In fact, anything that is vital to the successful, on-time, current recovery of the system should be included.

3.6 Backup Alternatives

When the system crashes, a disaster occurs. Where can your organization go in order to establish computer support services necessary to maintain the continuity and support services for the organization? The participants will discuss their specific requirements for backup. The organization will have to formulate its backup strategies based on its critical application requirements.

*A complete list and review of all interactive sessions listed in Figure 7-2 is available from DIA*log Management, Inc.

Figure 7-3

SUGGESTED AGENDA FOR THE THREE-DAY DISASTER/RECOVERY PLANNING SEMINAR

Agenda
Program starting time (unless otherwise informed): 9:00 A.M.
Coffee breaks: 8:30 A.M.   10:30 A.M.   3:15 P.M.
Lunch break: 12:30 P.M.

Day 1's Program

- Introduction to program
- Introduction of participants
- General overview of program and disaster recovery
- Parts of the disaster/recovery plan
  - Prevention planning
  - Security planning
  - Emergency response planning
  - Disaster/recovery planning
- The seminar/workshop manuals and additional handouts
- Introduction to disaster/recovery planning
- Slide presentation reviewing the program
- Your organization
- Data processing operation
- Interactive session: Establishing organization policy
- Scope of disaster/recovery plan
- Interactive session: User and business unit disaster/recovery planning
- Identification of critical systems/applications
- Cost of downtime
- Maximum allowable downtime

Day 2's Program

- Continuation of cost of downtime (interactive session): "How Long Can You Stay in Business?"
- Risk assessment methodology/baseline security
- Physical Security (presentation)
- Company and data center facilities evaluation
- Facility evaluation
- Insurance
- Personnel considerations
- Continuity of corporate management
- Emergency response plan

Day 3's Program

- Classifying the crime: "Dictionary of Computer Crime"
- Interactive session: Internal controls
- Interactive session: File backup strategies
- Vital records management
- Records management program: Computer and non-computer
- Off-site storage
- Care and handling of magnetic media
- Backup strategies (handout)
- Teams and action plans
- Operating a control center
- Testing
- Data collection: "Information and Decisions"
- Introduction to the four-day plan preparation workshop
- Corporate plan: Concepts and overview
- Evaluation of the program
- Management reports (handout)

Figure 7-4

FOUR-DAY DISASTER/RECOVERY PLAN PREPARATION WORKSHOP

Four-Day Disaster/Recovery Plan Preparation Session Procedures

The Disaster/Recovery Administrator/Consultant will introduce the methodology for the plan development as an opening introduction. A review will be made as to the mechanics of manicuring the existing master plan to meet the individual organization's specific requirements. The Disaster/Recovery Administrator/Coordinator will briefly review the table of contents and highlight various paragraphs throughout the plan. The work assignments will be distributed to indicate where participants should be at the various periods of the four days.

1.1 Program Procedure: Plan Preparation Workshop

1.1.1 Day 1: Plan Preparation Review

A number of the organizations and/or individuals who are participating in this program may not have attended the Preplanning seminar/workshop. In order to ensure that each group is aware of the plan preparation methodology and the plan components, it is advisable to hold a review session. Effectively, this will be a complete review of the table of contents of the master plan(s) with the entire group.

1.1.2 Days 2 to 5: Plan Preparation Workshop

Each of the participating groups will write its own business interruption preparedness plan (disaster/recovery plan) using the table of contents review section of the book as a guideline. *Note:* Each day, and during the day, the disaster/recovery team leader will review the progress and critique the material that has been completed.

1.1.3 Use of PCs at the Workshop

It has been our experience that the plan preparation teams who use PCs at the workshop to enter data as sections of the plan are completed, become more enamored with the entry and fall short of their mark on a daily requirements basis.

Although there is an advantage to having the daily work entered (so that the plan preparation team can review its work), no team will be allowed to use any computer equipment at the workshop unless it is used in the following manner:
1. Computer equipment can only be installed in a room other than that where the teams are working.
2. A separate room (or a number of units in the same room) will be provided for data entry personnel and equipment for each group.
3. The group must provide a data entry clerk (word processing keyboard operator) to operate the equipment. No team member should be permitted to do the data entry unless it is done after hours.
4. Equipment is to be supplied at participants' cost.
5. Page assignments requirements are as follows:
   Day 1: From 1. to 2.6 of the Table of Contents
   Day 2: From 3. to 4.16.3 of the Table of Contents
   Day 3: From 4.17 to 5.13 of the Table of Contents
   Day 4: From 6. to completion of Validation and Certification (G-1) and all Addenda

1.1.4 Workshop Manuals
The workshop coordinator should provide each of the participants with a copy of this book prior to the session.

1.1.5 Data Collection
The participants have received the disaster/recovery data collection list applicable to their responsibilities during the disaster/recovery preplanning workshop. Each data collection team will be given a general overview of the materials and decisions necessary to write the plan. Part 6 should be referred to as a guide to the data collection team(s) to assist them in obtaining the correct materials in order to properly document the plan.

Figure 7-5

2.1 THE PLAN PREPARATION WORKSHOP

2.1.1 Creating a Workshop Environment

If you want to get the project done, there a few simple rules to follow.

2.1.1.1 The Planning Group

The planning group (team) should be a small group of people representing a cross section of the organization, especially those involved in and assigned to disaster/recovery duties, auditing, critical users, and at least one representative from senior management (Disaster/Recovery Administrator).

2.1.1.2 Schedules

Meetings, especially the first introduction to the project, must be scheduled well in advance to ensure that all participants will be available. Subsequent meeting schedules should be set during the first session and not varied.

2.1.1.3 Location of Meetings

If possible, every effort should be made to hold meetings (seminars and workshops) outside the normal business environment. This is not to say that they cannot be held in large conference rooms, where a classroom type of environment can be created.

A room at a conference center or hotel would be preferable. It is important that the people involved be away from their areas and not disturbed while the session is going on.

What you want to create is a comfortable working atmosphere. The facility, if off premises, should be close enough to the place of work so that people can easily travel back and forth as they are needed or assigned. Or, if materials are needed, they can be easily obtained in a short period of time, without disrupting the sessions.

2.1.1.4 Room Environment

The room or conference area should be conducive to work. Tables should be set out in a classroom fashion. Those running the sessions should have their own table and a podium.

An overhead projector should be set up for use by the leader and participants. Window blinds should be closed to avoid any distractions. Smokers should be separated from nonsmokers.

2.1.1.5 Workshop Room Layout for the Plan Preparation Program

Depending on the number of groups that will be developing plans, it is necessary to have a workshop environment that is conducive to work, with sufficient room around work areas and tables to hold materials. The following sketch shows the type of layout we have found most useful. We suggest that the workshop be held in a hotel conference room, since this provides the type of facilities we recommend. The hotel should be a short distance from the company's location, if possible.

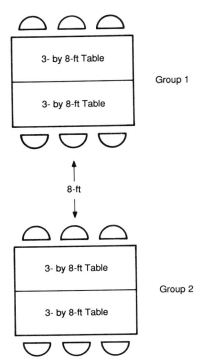

Figure 7-6 Typical 4-day disaster/recovery plan preparation room layout.

2.1.1.6 Materials Required for the Workshop

The following materials are required to assist workshop participants in the rapid completion of their project.

- At least one black or white board with liquid crayons
- Slide projector
- Overhead projector with three to four packs of various colored markers for overhead cells
- Two (although one may be sufficient) packages of 100 transparent overhead cells
- A three-hole punch
- Six to twelve empty binders (three-ring)
- Five hundred sheets of prepunched, ruled paper
- Scissors (one pair for each group)
- Rulers (one for each group)
- Scotch tape (one for each group)
- Fine-line pens (not ballpoints; two to three for each group)
- One dozen yellow highlighters

It would be advisable to provide a location where a copy machine is available, since the groups will have to reproduce a number of documents to be placed in their plan.

We suggest that the instructor/consultant(s) be housed in the facility where the workshop is being held. Many of the groups may require additional after-hours work to be guided by or overseen by the instructor/consultant(s).

2.1.1.7 Breaks

When setting up the agenda, two coffee or soft-drink breaks and lunch should be scheduled. The company should supply the food and beverages for both. If the participants have to fare for themselves, we have found them drifting away and taking excessive time from the session. It is therefore difficult for the leader to reconvene sessions.

Coffee and soft-drinks should be available throughout the sessions, especially the four-day program, where no breaks are planned.

2.1.1.8 Agenda

Prior to each meeting of the planning group, the Disaster/Recovery Administrator and Coordinator should prepare an agenda for the meeting as well as individual work assignments. Each meeting should review the previous assignments and how they are progressing. This gives everyone the opportunity to understand how the program affects his or her specific activity, even though the actions of one group are not specifically applicable to others.

By giving out prepared work assignments, you can maximize the meeting time by not having to dictate assignments and have people take notes. The agenda should include only those items to be covered for the session. The Disaster/Recovery Administrator/Coordinator should allow about two hours for each session; the first meeting, however, is usually a full-day session.

A typical start-up agenda might run as follows:

1. Call the meeting to order
2. Review the activity and assignments of the last meeting (ask for comments)
3. Review uncompleted items
4. Review new assignments
5. Hold discussions
6. Adjourn the meeting

2.1.1.9 Review of Completed Assignments

During the initial planning stages, there is a great deal of material the Disaster/Recovery Administrator and Coordinator will convey to the participating group. Assignments will be made, and participants are expected to return to the next session with the information and materials requested. Each participant must submit his or her information in writing, preferably with a summary sheet covering the details. Prior to subsequent meetings, the Disaster/Recovery Administrator/Coordinator will review the completeness of the material submitted and make information requests if it is not complete. Do not waste time at meetings to review material unless it is pertinent to the meeting itself, and cover only a small group of subject at the shorter meetings. If the Disaster/Recovery Administrator and/or Coordinator finds it necessary to obtain more detailed information from a specific group, a separate meeting with that group should be held before attendance at the larger meeting.

2.1.1.10 A Step-by-Step Approach

In today's work environment, it is impossible to get busy executives to commit to lengthy meetings, or take on work assignments that would substantially interfere with their daily responsibilities. Phasing in various components of the plan, within a reasonable period of time, will assist participants in doing a better job as well as help the Disaster/Recovery Administrator/Coordinator and the plan preparation team put together a more effective disaster/recovery plan.

Each of the subjects, assignments, and so on should be planned and documented in advance. If there is material, memos, files, or submissions that have to be reviewed, do your homework in advance. Do not waste time in reviewing material during the sessions.

Work out contributions and solutions to problems on a one-on-one basis. Do not take the entire group's time to solve a problem in a specific area, unless the problem concerns the entire group.

PART 4

INSTRUCTOR'S GUIDE

TYPICAL SEMINAR AGENDA

Time
1. Opening of session
2. Introduction to the program
3. Background of disaster planning in the corporation
4. Purpose of the seminar/workshop

Break (15 min.)

5. Review of the workshop flow diagram and PERT diagram
6. Introduction to the book on contingency disaster/recovery planning
7. First review table of contents of the disaster/recovery plan

Lunch Break

8. Second pass-through of the table of contents
9. Review of the organizational chart (interactive)
10. Start details of the plan (third pass of master plan)

OVERVIEW

The instructor's guide represents the general outline recommended for presenting the material on disaster/recovery planning to the various groups within the organization.

The Disaster/Recovery Administrator/Coordinator and/or the session leader should become familiar with the details in the information section of this book, as well as the table of contents review. The material presented in this book is intended to provide participants with a structured process in developing the plan.

DAY 1: MORNING SESSION

I. *Opening of session*
   A. *Introduction of participants* (see the profile at end of this manual). Prior to the workshop, each participant group received and should have completed the organization and individual profile (see Appendix 1) and returned it to the Disaster/Recovery Administrator/Coordinator (or session leader) before the company seminar/workshop. The Disaster/Recovery Administrator/Coordinator should review the information on each group's profile before the workshop commences. If the profile is brought to the workshop, the Disaster/Recovery Administrator/Coordinator should review the document prior to opening the session. (It is important for the Disaster/Recovery Administrator/Coordinator to familiarize himself or herself with each of the individuals, the activity of their departments, and the procedures and operations of their company.)

   The Disaster/Recovery Administrator/Coordinator will confirm that each of the participants described in the profile is in attendance, and if there has been any change, the new individual(s) should make out a profile sheet.

   The Disaster/Recovery Administrator/Coordinator, or the head of each group (team), should be identified and all questions pertaining to that specific component group should be directed to that individual.

   Each group should then be asked to identify themselves by name, title, company, responsibility, and reason for attending the program.

   B. *Review of daily schedule* (create overhead cell). The program starts at 9 A.M. (unless otherwise announced).

   Breaks will be at approximately 10:30 A.M. and 3:15 P.M.

   Lunch will be from 12 P.M. to 1 P.M. (this may vary from day to day). On day 3, check with each participant concerning airline schedules. Normal closing for day 3 will be about 4:30 P.M.

   C. *Homework (reading) assignments*. There will be a number of reading assignments of specific sections of the book. The Disaster/Recovery Administrator/Coordinator should request that the overhead cells for each of the interactive sessions be completed during regular seminar hours. In the event that time is running short, these should be completed as a homework assignment.

   D. *Interactive sessions and questions*. It is important

that the Disaster/Recovery Administrator/Coordinator discuss the various interactive sessions in as much detail as possible. Each group, as part of the organization and individual profile, has had the opportunity to review the concept of the interactive sessions.

Each of the groups should be encouraged to participate and interact between each other. The Disaster/Recovery Administrator/Coordinator will act as a moderator.

II. *Introduction of the program.* After completing the administrative details, review the work and activities of the next few days. Discuss the necessity for the preplanning program and how it prepares the group for the plan preparation workshop. Discuss also the plan preparation methodology. *Important*: The Disaster/Recovery Administrator/Coordinator must emphasize the fact that the more the participants contribute—the greater the information about the company, components, methods, and procedures on how they perform their activities—the more definitive the disaster/recovery plan will be.

III. *Introduction to the workshop book and materials*
   A. *First pass through the table of contents.* This pass is designed to acquaint the participants with the format and the major groupings of the master plan components. It is important to review the grouping methodology of the plan. Indicate that most of the static information is contained in the first section of the plan, but that the volatile information is maintained in the action plan section and the addenda.
   B. *Instructions for the Disaster/Recovery Administrator/Coordinator.* Each component of the plan, like all other business plans, has a purpose. The Disaster/Recovery Administrator's first review of this material will establish the tone of session. As leader of the project, it is necessary that you review the concept of each of the paragraphs in the table of contents review. This outline follows the table of contents review in Section 2. The first pass should be a generalized review.

Table of Contents Units

1. Introduction
   1.1 Purpose of the Plan
   1.2 Plan Postscript
   1.3 Management's Letter of Commitment
   1.4 Information Security Policy
   1.5 Personal Liability of Officers
2. Preplanning and Assumptions
   2.1 The Disaster/Recovery Planning and Data Collection Team
      2.1.1 Charter
      2.1.2 Members
   2.2 Scope
   2.3 Risk Assessment
      2.3.1 Probabilities of a Disaster
      2.3.2 Disaster Scenarios
         2.3.2.1 Sample Scenarios
      2.3.3 Costs of Downtime
      • Maximum Allowable Downtime
      2.3.4 Assumed Maximum Downtime (business units)
      2.3.5 Worst-Case Assumption
   2.4 Other Manuals
   2.5 Role of the Disaster/Recovery Management Team
   2.6 Plan Initialization

DAY 1: AFTERNOON SESSION

IV. *Second pass of the table of contents.*
   A. *Scope of the plan* (Section 2.2). This is one of the first *decision data collection items*. Each participating group will have to determine the scope of its own plan. Under the scope of activity, specific responsibilities (beyond or internal to data processing) and the number of data processing operations involved will need to be determined.
   B. *Organizational structure* (interactive session). Discuss the flow of information and user recovery. Review team membership, assignments, and responsibilities as they relate to the data center's disaster/recovery plan, and also review how they work with other company plans to be developed. An important part of teams is the participation of other departments. (See Interactive Session 4; refer to Table of Contents, Section 2.1.)
   C. *Review downtime/cost of downtime.* This is a general review of how to look at the maximum downtime the organization can tolerate based on critical applications. The Disaster/Recovery Administrator/Coordinator will introduce the interactive session that will be conducted and a homework assignment on how the individual departments and functional units would view and conduct business in the event of extended outage, as well as how long they can perform their services without computer support.
   D. *Other manuals.* Discuss other manuals that are necessary. At this segment, a discussion should be held relating to the emergency response manual, personnel safety, and so on. (See page 20 of the table of contents review manual.)
   E. *Risk assessment* (Section 2.3). The Disaster/Recovery Administrator/Coordinator will review the risk assessment section identifying the threats, hazards, vulnerabilities, risks, and exposures. This starts with the simplest form (list of histori-

cal and/or probable disasters). The Disaster/Recovery Administrator/Coordinator should emphasize the problems relating to a risk analysis program: time and cost. The Disaster/Recovery Administrator/Coordinator should evaluate the need for such a program and what are management's objectives in requesting a formal risk analysis. Discussions should be held relating to the comparison of maximum downtime and the effects of computer interruption as well as identifying the threats, hazards, vulnerabilities, risks, and exposures in determining some basic numbers relating to annual loss expectancy (ALE).
F. *File backup methodology.* This is the first overview on file backup, use of library, production tapes, off-site storage, a tape library management system, forms books, vital records, and the data processing department.
G. *Role of the Disaster/Recovery Management Team.* Discuss who is part of the team and their responsibilities. See table of contents review (Part 2).
H. *Plan initialization.* A brief discussion should be held about the initiation of the plan when the company goes into a disaster mode.

DAY 2: MORNING SESSION

V. *Continue the review of baseline security.* The Disaster/Recovery Administrator/Coordinator may opt to select a number of units to meet the needs of the participating companies, including the following. (See Part 10, "Baseline Security.")
  A. The laws and legal implications (a representative of the corporate legal department should make this presentation)
  B. Individual privacy
  C. Low building privacy
  D. Physical security
  E. Emergency plans/discipline and control
  F. Control areas (data center, user areas, floppys)
  G. Universal use of badges
  H. Isolation of sensitive jobs
  I. Equipment replacement
  J. File backup
  K. Planning

VI. *Maximum downtime* (interactive session). Discuss with each company (division or group) the approach it should be taking in determining the maximum allowable downtime and the cost of downtime. Each group will prepare an overhead cell detailing the degradation of activities based on time, cost to company, and performance of support activity. The group should then discuss its findings with the rest. One of the important aspects of this exercise is to allow all the groups to view each other's functional losses to determine if there are other considerations that have to be made in assessing maximum allowable downtime and how quickly support services will be required to restore critical activities. This unit also assists disaster/recovery management to present a realistic picture relating to the effects of a major computer outage. (See Figure 7-2 on page 187.)

VII. *Prevention/security.* Prevention and security are an important part of the disaster/recovery planning process. The primary objective of the entire program is to *prevent* a disaster. Each of the participating groups should view its facility and operations objectively. The result of this exercise and that of introducing baseline security concepts is that each group will discover the potential vulnerabilities of its operation and take appropriate and positive action in incorporating prevention and security measures. (See the facilities evaluation interactive session following this list.)

Table of Contents Units
3. Prevention/Security
3.1 Physical Security
3.1.1 Layouts
3.1.2 Locks
3.1.3 Guards
3.1.4 Cameras
3.1.5 Emergency Telephone Numbers
3.1.6 Intrusion Alarms
3.1.7 Medical Alert
3.1.8 Access to the Data Center
3.2 Medical Security
3.2.1 First Aid Kit
3.2.2 Nurse (Internal Medical Staff)
3.2.3 General Employee First Aid and CPR Training
3.2.4 EMS (or Fire Rescue Squad)
3.3 Fire Protection
3.3.1 Fire Alarms
3.3.2 Hand-held Fire Extinguishers and Fire Hoses
3.3.3 General Fire Safety
3.3.4 Passive Fire-Fighting Measures
3.3.5 Fire Drills
3.4 Emergency Response Plan
3.4.1 Contents
3.4.2 Emergency Response/Action Coordinator (Building and/or Plantwide)
3.4.3 Data Center Emergency Response—Miniplan

Note: The second day incorporates the greatest number of interactive sessions, with each (especially the facilities evaluation session) taking the longest period of time. Consideration should be given to starting the session at 8:30 A.M.)

  3.a Prevention/Security
    3.a.1 Prevention Planning
      An in-depth review is to be made on how

> prevention planning can reduce the possibilities of a disaster. Further expansion on the development of a functional emergency response plan. Each organization should be polled as to its level of emergency preparedness, the number of tests of emergency procedures (announced or unannounced) and the problems of low-level versus high-rise buildings.
> 3.a.2 Review of Section
> 3.a.3 Use of Facility Evaluation Checklist

The more interactive the session becomes, the more the Disaster/Recovery Administrator/Coordinator can contribute to this area of activity. There is a normal reluctance on the part of workshop participants to expose the weaknesses of their operation. Each group should be encouraged to talk freely about what it can do to protect people, the facility, and the general computer environment—both physically and logically. See the facilities evaluation interactive session that follows.

DAY 2: AFTERNOON SESSION

VIII. *Facilities evaluation* (interactive session). Depending on the length of time to complete the baseline security and the downtime interactive sessions, this unit may come at the lunch break. If there is at least one-half hour before lunch, each of the groups should begin preparing its overhead cells showing the area where its building is located, streets, traffic patterns, fire department, hospital, telephone central station, police department, any other emergency services as well as any other facility that represents a potential threat to the building. A second cell should show the data center, floor traffic, elevators, water, smoke, heat detectors, evacuation routes, fire extinguishers, and so on. The program can then resume promptly after the lunch break. At many sessions, the authors have requested a photo essay showing the same details as requested in the floor plan cells. This provides an outstanding visual opportunity to critique security provisions.
  A. *Application security and standards*. The Disaster/Recovery Administrator/Coordinator will discuss the relationship of application development and on-line production to prevention and security measures required to protect the system.
IX. *Internal controls* (interactive session). The Disaster/Recovery Administrator/Coordinator, using the sample overhead cell on a typical payroll system, will request that each group identify specific security provisions they would incorporate in each phase of the payroll system. The session leaders will then list under each of the phases their specific recommendations. This exercise emphasizes the need for stricter control from the initiation of an application design to its implementation.

Upon completion of the Facilities Evaluation Interactive Session, continue presentations and discussions on the following subjects as they relate to the plan.

Table of Contents Units
3.5 Housekeeping
3.5.1 Data Center Prohibitions
3.5.2 General Cleanliness
3.6 Power
3.6.1 Isolation Transformer
3.6.2 Motor Generator
3.6.3 Uninterruptible Power Supply (UPS)
3.6.4 Engine Generator
3.6.5 Reference Grid
3.7 Air-conditioning
3.8 Maintenance Agreement
3.9 Document Security
3.9.1 Classification
3.9.2 Storing: Security and Retention
3.9.3 Shredding: Security
3.9.4 Forwarding: Security Methods for Distribution
3.10 Software Security
3.10.1 User Code/Password Practices
3.10.1.1 System Release Level
3.10.1.2 Application Level
3.10.2 Required Password Change
3.10.3 Other Applications Security Measures
3.10.4 Pack Lockout
3.10.5 Program (Tape) Library Control System (PLCS)
3.10.5.1 Tape Library Management System
3.10.6 Erasure of On-line Copies of Sensitive Data
3.11 Datacomm Security
3.11.1 Physical Security of Terminals
3.11.2 Data Encryption Standard (DES) Devices
3.11.3 Leased Lines
3.11.4 Dial-back Feature for Dial-up Lines
3.12 Application Development Standards
3.13 Application Support Standards
3.14 Application Purchase
3.15 Role of the Auditing Staff
3.16 Role of the User Department(s)
4. Disaster Preparedness
4.1 File Backup
4.1.1 Decision Criteria
4.1.2 Pack Family Usage Rules
4.1.3 Backup Strategies
4.1.3.1 Operating System Code
4.1.3.2 Application Code

                4.1.3.3 Databases
        4.1.4 Dump Schedule
        4.1.5 Source Documents
        4.1.6 Terming Floppy Dumps
        4.1.7 Microfiche Reports
        4.1.8 Period-End File (Tape) Retentions
    4.2 Off-site Storage
        4.2.1 Off-site Management
        4.2.2 Off-site Strategy
        4.2.3 Off-site Rotation Schedule of Current Tapes
        4.2.4 Off-site Archive of Period-End Tapes
        4.2.5 Documents Off-site
        4.2.6 Tape Library Management System
        4.2.7 Scratch Tape Management
        4.2.8 Scratch Floppy Disk Management
    4.3 Backup Site Strategy
        4.3.1 General Alternatives
        4.3.2 Specific Alternatives
        4.3.3 Chosen Alternatives

DAY 3: MORNING SESSION

X. *Disaster preparedness* (interactive session). This section of the program deals with the basic requirement of the "three-legged stool."

   A. *Off-site storage/retentions*. Discuss the present methodology for backup and off-site storage; determine the correct procedure for backing up software, data, vital records, source documents, documentation, and so on. Methodology for determining whether the correct versions and the proper number of generations are stored off-site should be discussed.

   It is also important to point out that file retentions must be in compliance with the law (federal, state, local) and conform to industry standards. Also discuss the classification of data and documents.

   1. Review methods of off-site storage.
      (a) In-house—another location in building
      (b) Off-site—another location of the company
      (c) Off-site—at another company
      (d) Off-site—at a bank vault
      (e) Off-site—at a commercial site
      (f) Methods of shipment
      (g) Time of day and backup procedures
      (h) Risk time
   2. Cover topics dealing with file backup.
      (a) Three types of backup: reload, disaster recovery, archive
      (b) Strategies
      (c) Alternatives
      (d) Scratch tape management (count passes and clean)
      (e) Impact if bad tape is in off-site storage
      (f) Review of TLMS
      (g) Documents that should be off-site
      (h) Forms book
      (i) Scratch management
      (j) Scratch floppy disk management

XI. *File backup strategies* (interactive session).

XII. *Backup site strategy* (interactive session). The Disaster/Recovery Administrator/Coordinator will review all the backup alternatives. This is a three stage presentation.

   A. *Determination of critical applications.* (Use overhead of method of determining critical applications.)
   B. *Selection of backup strategies.* The backup alternatives/strategies available, and the use of multiple strategies categorized as mandatory, necessary, and desirable. TIME OF DISASTER (See "Backup Alternatives.")
   C. *Interactive portion.* Based on critical applications, maximum allowable downtime, and recognition of the various aspects covered in the cost of downtime and customer relations, each company will discuss the potential (probable) backup strategy or strategies they would recommend to management. Based on the selection of backup alternatives/strategies, how would each organization's machine configuration change from the primary site to the backup site? Continue discussion from Table of Contents Review.

Table of Contents Units

4.4 Mainframe Configurations
    4.4.1 Primary Site
    4.4.2 Backup Site
4.5 Datacomm Configuration
    4.5.1 Primary Site Communications
    4.5.2 Backup Site Configurations
    4.5.3 Additional Equipment Required
4.6 Emergency Phone Numbers of Employees
    4.6.1 Alphabetical List of Phone Numbers
    4.6.2 Calling Tree for Disaster/Recovery Teams
    4.6.3 Calling Tree for All Employees
4.7 Vendor Information
    4.7.1 Vendor Addresses, Phone Numbers, and Contacts
    4.7.2 Vendor Product List
    4.7.3 Vendor Support Letters
4.8 Disaster Kit
4.9 One-Month Supply List
4.10 Forms Book
4.11 Vital Records Retentions
4.12 Software Escrow
4.13 Employee Commitment Letters

XIII. *Datacomm backup strategies (presentation)*. This presentation will review the various methods (alternatives/strategies) to recover critical on-line services in the event of a disaster. Each organization will be asked to discuss its present datacomm on-line activity (in-house/external) and how each would or could affect recovery on critical applications.

Table of Contents Units

4.14 Priorities for a Limited Processing Environment
    4.14.1 Current Site Operations Schedule
    4.14.2 General Priority Scheme
    4.14.3 Probable Backup Site Operations Schedule
    4.14.4 Application Recovery Plans
4.15 Logistical Considerations
    4.15.1 Credit Cards, Cash, Etc.
    4.15.2 Housing
    4.15.3 Local Transportation
    4.15.4 Office Space
    4.15.5 Control Center
    4.15.6 Employee Notification by Broadcast (and Other) Media
4.16 Insurance
    4.16.1 Business Property Insurance
    4.16.2 Extra Expense Insurance
    4.16.3 Business Interruption Insurance
4.17 Disaster/Recovery Action Teams
    4.17.1 Disaster/Recovery Management Team (DRMT)
        4.17.1.1 Team Charter
        4.17.1.2 Typical Duties of the Disaster/Recovery Management Team
        4.17.1.3 Role of the Coordinator
        4.17.1.4 Role of the Administrator
        4.17.1.5 Membership

XIV. *Insurance* (presentation). Review present status of organization insurance directly related to computer operations.

XV. *Disaster/recovery action teams* (presentation).
    A. Discuss the elements of team assignments.
        1. Team charter
        2. Typical duties of the Disaster/Recovery Management Team
        3. Role of the coordinator
        4. Role of the administrator
        5. Membership
    B. Review activities of the various teams. This unit is extremely important because it helps each group establish the parameters for team assignments. The instructor/coordinator should refer back to the organizational charts and scope of the plan to direct the Disaster/Recovery Administrator in the formation of teams, and from where the team assignments are to derive.

Table of Contents Units

    4.17.2 Transportation Team
        4.17.2.1 Team Charter
        4.17.2.2 Typical Duties by Phase
        4.17.2.3 Membership
    4.17.3 Off-site Storage Team
        4.17.3.1 Team Charter
        4.17.3.2 Typical Duties by Phase
        4.17.3.3 Membership
    4.17.4 Supplies Team
        4.17.4.1 Team Charter
        4.17.4.2 Typical Duties by Phase
        4.17.4.3 Membership
    4.17.5 Administrative Team
        4.17.5.1 Team Charter
        4.17.5.2 Typical Duties by Phase
        4.17.5.3 Membership
    4.17.6 Software Team
        4.17.6.1 Team Charter
        4.17.6.2 Typical Duties by Phase
        4.17.6.3 Membership
    4.17.7 Applications Team
        4.17.7.1 Team Charter
        4.17.7.2 Typical Duties by Phase
        4.17.7.3 Membership
    4.17.8 Communications Team
        4.17.8.1 Team Charter
        4.17.8.2 Typical Duties by Phase
        4.17.8.3 Membership
    4.17.9 Operations Team
        4.17.9.1 Team Charter
        4.17.9.2 Typical Duties by Phase
        4.17.9.3 Membership
    4.17.10 Data Preparation Team
        4.17.10.1 Team Charter
        4.17.10.2 Typical Duties by Phase
        4.17.10.3 Membership
    4.17.11 Salvage Team
        4.17.11.1 Team Charter
        4.17.11.2 Typical Duties by Phase
        4.17.11.3 Membership
    4.17.12 New Hardware Team
        4.17.12.1 Team Charter
        4.17.12.2 Typical Duties by Phase
        4.17.12.3 Membership
    4.17.13 New Facilities Team
        4.17.13.1 Team Charter
        4.17.13.2 Typical Duties by Phase
        4.17.13.3 Membership
4.18 Other Teams
    4.18.1 User Team
    4.18.2 Audit Team
    4.18.3 Hardware Team
    4.18.4 User Liaison Team
    4.18.5 Security Team
        4.18.5.1 Team Charter
        4.18.5.2 Typical Duties by Phase
        4.18.5.3 Membership

    4.19 Additional Teams to be Considered
        4.19.1 Third Party Liaison Team
            4.19.1.1 Team Charter
            4.19.1.2 Typical Duties by Phase
            4.19.1.3 Membership
        4.19.2 Public Relations Team
            4.19.2.1 Team Charter
            4.19.2.2 Typical Duties by Phase
            4.19.2.3 Membership
        4.19.3 Alternate Strategies Coordinating Team
            4.19.3.1 Team Charter
            4.19.3.2 Typical Duties by Phase
            4.19.3.3 Membership

5. Disaster/Recovery Action Plans
    5.1 Action Plan Overview
    5.2 Definition of the Disaster/Recovery Phases
        5.2.1 Phase 1: Emergency Response
        5.2.2 Phase 2: Backup Site Activation
        5.2.3 Phase 3: Recovery of Home Site
        5.2.4 Phase 4: Return
    5.3 The Master PERT Chart for Disaster/Recovery Operations
    5.4 Elaboration of the PERT Chart (Phase 1) (See sample overhead cell on action plans when disaster strikes)
        5.4.1 Disaster Strikes
        5.4.2 Emergency Response Procedures Overview
            5.4.2.1 Incident Reporting
            5.4.2.2 Fire
            5.4.2.3 Water
            5.4.2.4 Power Failure
            5.4.2.5 Environmental System Failure
            5.4.2.6 Weather and Natural Phenomena
            5.4.2.7 Sabotage (Causing Denial of Service)
            5.4.2.8 Other Special Conditions
        5.4.3 Notification of the Disaster/Recovery Management Team (Box 3)
        5.4.4 Disaster/Recovery Management Team Meeting
        5.4.5 Securing the Area
        5.4.6 Assessing Damage
        5.4.7 Determining Downtime Estimate
        5.4.8 Decision to Declare the Disaster
    5.5 Elaboration of PERT Chart (Phase 2)
        5.5.1 Assemble Teams
        5.5.2 Activate Control Center
        5.5.3 Notify Other Parties
    5.6 Elaboration of PERT Chart (Phases 2, 3, and 4)
        5.6.1 Transportation Team
        5.6.2 Off-site Storage Team
        5.6.3 Software Team
        5.6.4 Applications Team
        5.6.5 Communications Team
        5.6.6 Operations Team
        5.6.7 Supplies Team
        5.6.8 Salvage Team
        5.6.9 Administrative Team
        5.6.10 Data Preparation and Unit Record Team
        5.6.11 New Hardware Team
        5.6.12 New Facilities Team
        5.6.13 Disaster/Recovery Management Team
    5.7 Cancellation of Disaster/Recovery Mode (Phase 4)
    5.8 Disaster/Recovery Postmortem (Phase 4)
    5.9 Detailed Team Plans

6. Training for Disaster/Recovery
    6.1 Recovery Team Training
    6.2 All Employee Training
    6.3 CPR and First-Aid Training

7. Validation and Testing
    7.1 Validation
    7.2 The On-site One-Application Test
    7.3 The On-site Complete Test
    7.4 The Backup Site Test
    7.5 The Surprise Test
    7.6 Test Schedule
    7.7 Problem Log

8. Plan Update/Control
    8.1 Update Log
    8.2 Approval
    8.3 Formal Review (Once a Year)
    8.4 Checklists

DAY 3: AFTERNOON SESSION

XVI. *Discussion of four phases of the disaster/recovery plan* (presentation). The Disaster/Recovery Administrator should review each of the four phases and describe the activity the plan preparation team will be doing during the four-day session. A sample of the blank flow chart should be exhibited (overhead cell). Review team assignments (Section 19, "Data Collection").

XVII. *Presentation on training.* Discuss the plan training methodology to be implemented.

XVIII. *Presentation on validation and testing.*
   A. The session leader will review the various methods of testing the plan.
      1. Application by application (in-house).
      2. Full test of reload of system using on-site production and software tapes (in-house).
      3. Full test of reload of system using off-site tapes.
      4. One or two critical application tests (back-up site).
      5. Full critical application test (backup site).
      6. Total disaster/recovery procedures.

Table of Contents Units

9. Plan Distribution and Security
  9.1 List of Recipients
  9.2 Screening of Information
  9.3 Serialization
  9.4 Termination or Transfer of Copy Holder
  9.5 Copies
  9.6 Physical Security of the Plan
  9.7 Security and Confidentiality Agreement
10. Planned Enhancements
11. Addendum Review
  C-1 Data Center Layout
  C-2 Building Layout
  C-3 Area Layout
  C-4 Emergency Call List
  C-5 Card Key Holders
  C-6 Authorized Entrants
  C-7 Authorizing Management
  C-8 Data Center Sign-in Sheet
  C-9 Evacuation Route
  C-10 Copies of Maintenance Agreements

XIX. *Plan update methodology* (presentation).
XX. *Plan distribution and security* (presentation).
XXI. *Planned enhancements* (presentation). Refer back to risk assessment and the implementation of a baseline security program. During the data collection phase, the disaster/recovery team will have to determine the various prevention and security enhancements that are to be implemented and incorporate them into the plan as planned enhancements, if they have not already been completed prior to the four-day workshop.

Table of Contents Units: Addenda
D-1a Off-site Agreement D-1b Vendor Information D-2 Tape Library Report D-3 Backup Site Alternatives D-4 Backup Site Agreement (Initial) D-5 Backup Site Agreement (Secondary) D-6 Primary Site Mainframe Configuration D-7 Backup Site Mainframe Configuration D-8 Primary Site Datacomm Configuration D-9 Backup Site Datacomm Configuration D-10 Required Ancillary Equipment D-11 Home Phone Numbers D-12 Calling Tree for Disaster/Recovery Teams D-13 Calling Tree for Other Employees D-14 Vendor List D-15 Vendor Product List D-16 Vendor Support Letters D-17 Disaster Kit D-18 One-Month Supply List D-19 Vital Records Retention D-20 Escrow Agreement D-21 Purchased Software D-22 Disaster/Recovery Plan: Confidential Employee Questionnaire D-23 Current Site Operations Schedule D-24 Probable Backup Site Operations Schedule D-25 Application Recovery Plan Worksheet D-26 Copies of Policies and Annual Premiums E-1 Inventory of Damaged Items E-2 Time to Replace Hardware—Assuming Emergency Expedite of Order E-3 Team Plans G-1 Certification Checklists G-2 Threat Evaluation Form G-3 Testing Schedule G-4 Problem Log G-5 Characterization of Test H-1 Update Log H-2 Contingency Backup Plan—Periodic Update Questionnaire I-1 Security and Confidentiality Agreement J-1 Planned Enhancements

PART 5

ORGANIZATION AND PARTICIPANT PROFILE

Important: In order to make your organization's participation in your forthcoming effective disaster/recovery planning seminar/workshop truly productive, the Disaster/Recovery Administrator requests that you prepare the following materials for the seminar.

1. Please review the descriptions of the interactive sessions in Section 9 of this profile.
2. If there is a request for material to be prepared in advance of the three-day preplanning seminar, please bring it to the seminar.
3. Included with this profile are the descriptions and outlines of the communications and insurance interactive sessions. Because of the limited time, the number of participants at the workshops, and the extent of material required in these two subsections, *please* complete these prior to the seminar. For the insurance session, a review by your insurance carrier or broker would be sufficient.
4. The organization structure interactive session requires each participating organization to prepare an organizational chart showing the interrelationships of all activities in the company, from senior management to the operating departments.
5. In addition to the organizational chart of the company, please prepare an organizational chart of the data processing operation. Under each category, put in the number of people involved in each function.

The following are preseminar assignments:

1. Organization and Participant Profile/User Questionnaire
2. Insurance
3. Personnel
4. Organization and Disaster/Recovery Planning

STATEMENT OF CONFIDENTIALITY

The information provided by each participating organization group, and/or individual in this document is proprietary to the organization. The Disaster/Recovery Administrator will not disclose the contents of this document to any person not involved in, or acting as an instructor/consultant in the preparation of, an effective disaster/recovery plan seminar/workshop either in public or with other divisions without the express written permission of an authorized representative of the participating group. Participation in the workshop's interactive sessions does not waive the participant's rights to total confidentiality. Once the session is completed, if requested, the Disaster/Recovery Administrator will return this document.

Thank you,

PREPLANNING WORKSHOP: ORGANIZATION AND PARTICIPANT PROFILE
The Contingency Planning Process

During the forthcoming effective disaster/recovery preplanning seminar/workshop, material will be presented by the Disaster/Recovery Administrator and Coordinator and many sessions will be held that will require interaction by the participants. Author's Note: In over 60 workshops conducted (with over 160 organizations participating), only a few organizations have selected not to participate in one or more interactive sessions. This was due to the extreme confidentiality and sensitivity of their organization's operating procedures. The organizations were primarily government agencies and/or defense contractors.

During the effective disaster/recovery plan preparation workshop, the Disaster/Recovery Administrator and Coordinator will rely on the information obtained during the preplanning workshop to assist each company in conceiving and constructing, piece-by-piece, unit-by-unit, paragraph-by-paragraph the strategies and action plans for an effective disaster/recovery plan. During the period prior to your attending the program, we would request that you review the material contained in this profile and determine those areas of interaction in which your organization will participate. Please check off boxes indicated.

These interactive sessions provide a unique forum where one group learns how another group, division, or corporate component functions, and others have the opportunity to view your operation in a constructive manner. You will find that in preparing the material requested, those areas sensitive or confidential to your group can be readily protected, and that the details can be generalized to a sufficient degree so that there can be no security breach. The completed copy of this document should be mailed to the Disaster/Recovery Administrator prior to the scheduled session. It is preferable that he or she receive the material well in advance of the session in order to help plan a program that is beneficial to your group and others participating.

About Your Organization

As participants in the contingency disaster/recovery planning seminar/workshop, your division or group has recognized that disaster/recovery planning is an important part of your overall company's normal, formal business planning activity and has achieved a high management priority. Representation from various functional areas of the company further exhibit an understanding that this *plan* requires the participation and cooperation of virtually every functional operating unit of data processing and the user community, and it is not limited to, or is, a data processing plan created solely for data processing operations.

Users represent the *base* of the entire planning process. They have the prime responsibility for maintaining the continuity of business functions in the event of loss of data processing support capability or vice versa and, therefore, must be participants in this planning process.

Data Collection/Decision Phase

Numerous decisions have to be made during the preplanning session, the data collection phase and the plan preparation, implementation, testing, and maintenance phases of the plan. Each decision must be based on agreements between management, the user, and the data processing department. It is the user who is ultimately responsible for performing the *business functions* for which he or she will be held accountable, whether there is a partial or total failure of data processing facilities, or backup computer services are not available.

Therefore, in responding to the following questions and preparing materials for the forthcoming three-day disaster/recovery preplanning seminar, information should be obtained (if appropriate) from all departments, functions, components, and levels of the organization (computer and noncomputer activities) that will be ultimately affected by the loss of computer support. Most important, when the actual plan is written, it must include the subplans of all others that fall within the *scope* and *assumptions* and *action plans* of the contingency disaster/recovery plan. Failure to include or involve all functional, operational, and administrative plans will result in incompatible strategies and action plans. Similarly, at the time of an adverse incident, multiple claims may be made on the same limited resources without definitive strategies and determinations as to critical corporate requirements.

IMPORTANT—IMPORTANT—IMPORTANT—IMPORTANT—IMPORTANT

Complete this document as soon as possible and return to:

Name _____

Title _____

Dept. _____

Address _____

City _____ State _____ ZIP _____

If there are any questions concerning the information requests, please feel free to contact the Disaster/Recovery Administrator: Phone (000) 000-0000.

If you cannot complete this document in sufficient time to submit it prior to the session, bring this document to the workshop. It is *your* program—it is *your* plan. You can and have the opportunity to fulfill the goals and objectives of your organization's contingency disaster/recovery plan. A little effort on your group's part now will make the seminar/workshop participation and plan preparation extremely effective and rewarding.

Fill out one for each participant.

1.0 Workshop Participants

1.1 Please list the name of each participant below; give title and specific assignment with respect to the disaster/recovery planning process.

    1.1.1 Company name _____

    1.1.2 Name _____ Title _____

        Organizational responsibility _____

        _____

        _____

        _____

        _____

    1.1.3 Disaster/recovery assignment or responsibility _____

        _____

        _____

        _____

        _____

    1.1.4 Will the disaster/recovery assignment be ☐ permanent   ☐ temporary?

    1.1.5 Please provide a brief profile about each individual:

        _____

        _____

        _____

        _____

        _____

        _____

Note: One copy has been provided in this package. Make additional copies of this form if necessary. Do not use one form for more than one person.

2.0 Your Company/Division/Group/Department

2.1 Give a brief description of all business activities (type of business: financial institution, insurance company, manufacturer, distributor, service organization, consumer or industrial products or services, etc.).

_____
_____
_____
_____
_____
_____
_____

2.2 Describe (or list) the various divisions, subsidiaries, components, and so on.

_____
_____
_____
_____
_____
_____
_____

2.3 How many employees are in the company? _____

2.4 How many people are dedicated data processing personnel? _____

2.5 Do you have a full-time disaster/recovery or security administrator? ☐ Yes   ☐ No

2.6 If yes, is he or she on the team that will attend the workshop? ☐ Yes   ☐ No

2.7 Has management fully endorsed this program? (If available, include a copy of management's letter, company policy statement, and so on.) ☐ Yes   ☐ No

2.8 Do you anticipate any problems in obtaining the resources to maintain and test the plan once it is completed?
☐ Yes   ☐ No
If yes, please describe. _____
_____
_____

2.9 Do you presently retain consultants in your organization? ☐ Yes   ☐ No
If yes, what are their areas of expertise or discipline? _____
_____
_____
_____

2.10 Have you appointed a Disaster/Recovery Administrator? ☐ Yes   ☐ No
Who? _____
Will that person be represented at the seminar/workshop? ☐ Yes   ☐ No

2.11 Have you appointed a Disaster/Recovery Coordinator? ☐ Yes   ☐ No
Who? _____

2.12 Has the Disaster/Recovery Management Team been selected? ☐ Yes   ☐ No

Who? _____ Title _____

Who? _____ Title _____

Who? _____ Title _____

Who? _____ Title _____

Who? _____ Title _____

3.0 Computer Operations

3.1 What computer vendor are you presently using? (If you have a multivendor mainframe operation, please list.)

VENDOR	MODEL	CPU/M	DISK
1.			
2.			
3.			
4.			
5.			

3.2 How many on-line terminals are using the mainframe resource? _____

3.3 Please describe, in general terms, the data processing communications network. (If possible, attach a simple schematic.) The material will be reviewed at the interactive session on communications. Complete this document and bring it to the workshop. Owing to the extent of information required, and the time to complete the material, this should be done in advance of the seminar.

_____

_____

_____

_____

_____

3.4 Are you presently storing your backup tapes off-site? ☐ Yes   ☐ No

Where? ☐ In-house   ☐ Commercial facility   ☐ Other division of the company   ☐ Other building of the company

Other (describe) _____

_____

3.5 What other items (such as documentation, run books, etc.) are stored off-site? _____

_____

3.6 Do you have a tape library management system? ☐ Yes   ☐ No

If yes, describe. _____

_____

4.0 Seminar/Workshop Expectations

As this is the first time you will be participating in a disaster/recovery preplanning and plan preparation workshop, we would appreciate your expressing your views and expectations prior to, at the opening sessions, and at the end of the preplanning session. Although the preplanning program has a fairly defined format and presentation, you can recognize that some of the material presented at any seminar/workshop will appeal to some individuals of your group, and other subjects to others. The interactive sessions provide each organization with a unique opportunity for all company participants to be involved. In a multi-divisional environment (many organizations in different or even similar business activities at the same session) it is even more difficult to satisfy everyone at all times. Although the generic information presented is vital to the preparation of the plan, the instructor/consultant will do everything possible to tailor the preplanning program to each company's specific objectives and expectations. *Your cooperation is necessary to accomplish this task.*

Your response to the following questions will help the Disaster/Recovery Administrator develop a seminar course action that will transmit basic information and provide your organization with individual response and attention to your specific organization's requirements.

The information given on this form is for the use of instructor/administrator instructors/consultants and will be held in the strictest confidence. It will not be discussed, distributed, copied, or disseminated in any manner without the express written permission of the organization. Copies may be made in blank form by you to distribute to other members of your group.

4.1 What is the intended scope of your plan? (See Section 6.0.)

Check one or more.

- ☐ Data processing only
- ☐ Involve users
- ☐ Subsidiaries
- ☐ Corporate headquarters planning
- ☐ Divisions, branches, and so on
- ☐ Other _____

4.2 Who in your organization initiated the disaster/recovery plan project?

Check one or more.

- ☐ Management/president vice president
- ☐ Controller/financial
- ☐ Internal auditor
- ☐ External auditor
- ☐ DP/MIS management
- ☐ Other _____

4.3 Have you attempted to develop a disaster/recovery plan without assistance? ☐ Yes ☐ No

4.4 Do you presently have one or more of the following? (See Section 8.9.)

1. Corporate policy manual for data processing ☐ Yes ☐ No
2. Company or data processing emergency response plan ☐ Yes ☐ No
3. Standards and procedures manual for data processing ☐ Yes ☐ No
4. Data center personnel policy manual ☐ Yes ☐ No
5. Data center operations manual ☐ Yes ☐ No
6. Data center users guide ☐ Yes ☐ No

4.5 Have you or any of your associates attended any seminar, workshop, or presentation on disaster/recovery planning?
☐ Yes   ☐ No

If "yes," please elaborate. _____

_____

_____

4.6 Do you consider the following to be an important part of your responsibility in developing this plan? [Check the appropriate box(es).]

☐ Prevention   ☐ Security   ☐ Emergency response

4.7 Which would you consider important subjects to be covered at the preplanning workshop? (Please rate in order of importance; do not number those you feel are not important.)

_____ Detailed review of the table of contents of the disaster/recovery plan

_____ Prevention

_____ Security (physical)

_____ Security (logical)

_____ Legal implications

_____ Insurance

_____ Backup alternatives

_____ Off-site storage

_____ File backup procedures

_____ Maximum downtime for recovery/cost of downtime

_____ Communications

_____ Internal controls

_____ Facilities evaluation

_____ User involvement

_____ Critical application review

_____ Application recovery plans

_____ Disaster/recovery team development

_____ Postdisaster concerns

_____ Disaster/recovery action plan review

_____ Personnel policies and standards

_____ Corporate policies and standards

For those you have *not* selected, please take two or three items and indicate why they are not important to your organization.

1. _____

_____

_____

2. _____

_____

_____

_____

3. _____

_____

_____

4.8 Briefly describe what you and your team expect to achieve at the disaster/recovery preplanning seminar workshop.

_____

_____

_____
_____
_____
_____
_____

4.9 Can the planning team participating in the program discuss all aspects of the business activity of the company?

☐ Yes   ☐ No

4.10 If we require more participants from your company to help write the plan during the disaster/recovery plan preparation workshop, are they available? Who might they be?
(*Note:* This may be more appropriate after completion of the preplanning program. This may also be discussed with you during the program. In any event, consider those individuals who might be advantageous to have. List by title.)

Name _____   Title _____

Name _____   Title _____

Name _____   Title _____

Name _____   Title _____

4.11 Are there any subjects that you consider important to your organization that you would suggest we might add (regardless of the fact that you might think we will cover it anyway)?

1. _____
2. _____
3. _____
4. _____
5. _____

(Continue list below if required.)

4.12 Commentary

Please provide us with any commentary concerning your perceptions, expectations, objectives, and so on that you or your team feel appropriate for your organization.
Do not limit the scope to data processing disaster/recovery planning.

_____
_____
_____
_____
_____
_____
_____
_____
_____
_____
_____
_____
_____
_____
_____
_____
_____
_____

5.0 **Administrative Information (Optional)**
Please provide us with the following information.

5.1 Who informed you that you would be participating in this program?
_____
_____

5.2 How long from the time you were aware of the disaster/recovery planning seminar/workshop did your business unit make the decision to participate?

_____ Week(s)   _____ Month(s)   _____ Year(s)

5.3 Who was responsible for making the final decision to attend this program?

Name _____ Title _____

Name _____ Title _____

Name _____ Title _____

5.4 Did you investigate any other method of preparing a plan? ☐ Yes   ☐ No

If yes, who or what method(s) was investigated?
_____
_____

Why were they abandoned? _____
_____

5.5 Why did your group select this program?
- ☐ Cost
- ☐ Methodology
- ☐ Uniqueness
- ☐ Good Salesmanship
- ☐ Other Organization's Experience
- ☐ Time Restrictions
- ☐ Speed of Plan Preparation
- ☐ References
- ☐ Attended Presentation
- ☐ Other _____

5.6 What are your feelings about homework reading or other assignments that might be required at the seminar?
_____
_____

5.7 If the sessions had to run later than 5 P.M. or start earlier than 9 A.M., is there any problem in your attending?

Before 9 A.M. ☐ Yes ☐ No

After 5 P.M. ☐ Yes ☐ No

5.8 Do you think advance distribution of some seminar/workshop material would assist your group in being better prepared for the preplanning sessions? ☐ Yes ☐ No
If yes, what would you like to know about the program? (Please go through the rest of this profile before you answer this question.)
_____
_____
_____
_____

6.0 Scope of the Plan

Please list on separate sheets, those organizational divisions (functions, branches, components, and so on) that will participate or be involved in the development of business interruption preparedness plans. A separate form should be completed for data processing.

Use one sheet for each operational unit.

6.1 Name of Unit _____

Location _____
_____

6.1.1 Type of operation, activity, or service _____
_____

6.1.2 Name of participants

Name _____ Title _____

Name _____ Title _____

Name _____ Title _____

Name _____ Title _____

Name _____ Title _____

Name _____ Title _____

6.1.3 Is this operation on-line function? ☐ Yes ☐ No

6.1.4 Do they operate any computer equipment other than the company's central computer? ☐ Yes ☐ No

Explain _____
_____
_____
_____
_____

6.1.5 Is this a critical organization function? ☐ Yes ☐ No

Make additional copies of this form if required.

7.0 Critical Services (Operation Described in Sections 6.1–6.6)

List in order of priority.

(Unit name _____)

7.1 Describe the critical operations, services, and so on that this operational unit performs: _____
_____
_____
_____
_____
_____
_____
_____
_____
_____

7.2 Describe briefly the critical resources (computer and noncomputer operations) needed to support this activity. Include space, power, equipment, and supplies. Do not detail; use general descriptions that relate to their overall operation.
_____
_____
_____
_____
_____
_____
_____
_____
_____

8.0 Preplan Preparation Decisions

8.1 Do you have any idea of the type of disaster/recovery team activities that will be involved in disaster/recovery activities? (The team title should describe the team responsibility. Review with your group the type of actions you consider important in backup and recovery. Responses to this will help the Administrator and Coordinator determine the group's understanding of disaster recovery.)

(This form does not have to be filled out if you so desire.)

Team Name

1. _____
2. _____
3. _____
4. _____
5. _____
6. _____
7. _____
8. _____
9. _____
10. _____
11. _____
12. _____
13. _____
14. _____
15. _____
16. _____
17. _____
18. _____
19. _____
20. _____

8.2 List the top 10 risks you feel are the greatest exposures (these can be physical, logical, manmade, or natural).

1. _____
2. _____
3. _____
4. _____
5. _____
6. _____
7. _____
8. _____
9. _____
10. _____

8.3 Maximum allowable downtime. Have you determined how quickly you have to recover from any interruption due to a disaster? Select the most critical operations (applications) and list them by recovery time (hours, days, weeks, and so on).

Operation/Application	Maximum Recovery Time
1.	
2.	
3.	
4.	
5.	

(Does this correlate with the information in Section 7.0, "Critical Services"?)

8.4 Cost of downtime. Please consider how your organization's services would deteriorate. What would be the costs relating to computer downtime and their impact on company finances, customer service, business opportunities, and so on?

8.5 File backup strategy. Briefly describe you present file backup strategy (source documents, object, job, database, transactions, and so on).

8.6 Off-site storage strategy. Briefly describe your off-site storage strategy. How many generations are off-site?

_____
_____
_____
_____
_____
_____
_____
_____

8.7 Vital records strategy. Please answer the following questions.

    8.7.1 Does your organization have a vital records management program? ☐ Yes   ☐ No

    8.7.2 Are all source documents kept until all accounting controls are completed and verified? ☐ Yes   ☐ No

    Where? _____
_____

    8.7.3 Is there a records management manual? ☐ Yes   ☐ No

    8.7.4 Are all documents classified as to sensitivity? ☐ Yes   ☐ No

    8.7.5 Is there a legal file retention schedule? ☐ Yes   ☐ No

8.8 Employee participation in disaster/recovery. Have you polled employees to determine if they will be available to go to a backup site in the event of a disaster? ☐ Yes   ☐ No

8.9 Company manuals. Are the following manuals available?

1. Emergency procedures manual	☐ Yes	☐ No
2. Personnel policy manual	☐ Yes	☐ No
3. Vital records policy	☐ Yes	☐ No
4. Records management program manual	☐ Yes	☐ No
5. Data center employee job descriptions	☐ Yes	☐ No
6. Application documentation	☐ Yes	☐ No
7. System development standards	☐ Yes	☐ No
8. Data center users guide (manual)	☐ Yes	☐ No
9. Data center operations manual	☐ Yes	☐ No
10. System software documentation	☐ Yes	☐ No
11. Computer hardware and peripheral documentation	☐ Yes	☐ No
12. Internal audit procedures manual	☐ Yes	☐ No
13. Forms book	☐ Yes	☐ No
14. Security policy and standards manual	☐ Yes	☐ No

Note: If there are any other documents that are important to the operations, procedures, and policy of the company, please list them:

1. _____
2. _____
3. _____
4. _____
5. _____

9.0 3-Day Disaster/Recovery Preplanning Seminar—Interactive Sessions
Commentary on Interactive Sessions

Many of those who attend the preplanning seminar will be totally unfamiliar with the concepts and the complexity of developing a disaster/recovery plan. The objective of the interactive session is to direct thought processes into areas and activities that parallel the development of the plan, but deal in each case with specific business issues. A review of the table of contents of the plan, correlating the appropriate interactive session with the specific plan requirement, will tend to direct participants to the type of contribution necessary to achieve the workshop objective, the development of a viable, effective disaster/recovery plan. All of the material needed for the development of the interactive sessions can be found in the previous sections of this handbook.

9.1 Organization and disaster/recovery planning. This interactive session teaches the participants how to look at their organization from the perspective of its structure, interactions, and intra- and interrelationships concerning the use of and dependency on information processing activities. The instructor/consultant presenting the material should emphasize individuals' accountability and responsibility and the areas of critical business functions that could be seriously affected by an unexpected crisis or disaster. The discussion should be divided into two sections:
1. The primary organizational structure
2. The makeup and reporting responsibilities of data processing

The next aspect of organizational structure deals with the organizations that have to be developed in order to fulfill the responsibility of developing an effective disaster/recovery plan. This requires the establishment of a series of "groups" that will be involved in the following:

9.1.1 Organization for plan development. The organization for plan development involves numerous individuals and functions. The preliminary steps in developing the plan require numerous administrative activities as well as a great deal of information, and therefore it is important that senior management (or a key senior executive) be directly involved in the establishment of this organization for plan development.

The following departments or individuals should be part of this organization: Disaster/Recovery Coordinator, Internal Auditing, user departments, technical support from data processing, security, an external consultant (if required), and vendor representatives.

9.1.2 Organization for plan execution. This organization is the group "designed-in" to the plan, and is slightly different than the organization required for the creative and decision aspects of plan development.

This organization comprises, in addition to the Disaster/Recovery Management Team, line operations and specialist groups. This team will be in charge and perform all of the "action plans," coordination, and administration of the backup and recovery programs.

9.1.3 Organization for plan testing and maintenance. While it is expected that many aspects of the plan will be tested, most organizations cannot undertake the testing of the entire plan, requiring the simulation of a total disaster. Some portions of the plan will be tested statically, others dynamically to ensure that the concepts and procedures will fulfill the disaster mode requirements. The organization for testing the plan will involve the entire organization established for plan execution, as well as other participants within the company. The important aspect here is the understanding of the roles each of the individuals and groups will play at the time of a disaster, and the verification of the state of readiness of the organizations and plan components.

Plan maintenance should be assigned to the Disaster/Recovery Coordinator, with reporting back to the Disaster/Recovery Administrator.

9.2 Scope of plan. The scope of plan interactive session should be used to analyze the critical interactive users and applications that interface and are totally dependent on data processing. In the true sense of the scope of plan we deal with those functions and departments that would be involved if the plan is anticipated to go beyond the data center. The participants are taught how to look at their organization charts, and those that will be required to develop plans, in order to maintain the critical functions and business performance continuity. The primary thrust for plan development should be the data processing operation, but critical users who support vital aspects of business performance must be considered part of the planning process as well.

The scope of plan interactive session is one of the critical components of the planning process, as it begins to identify the critical activity areas, critical business functions, and ultimately the critical applications.

9.3 Cost of downtime/maximum allowable downtime/affects on critical success factors. This interactive session covers the technology of analysis of the organization's critical information processing systems and applications, the degradation of business performance, and depletion of support services. Each of the groups participating in the seminar should document how their operation(s) would degrade, how the services they perform would be affected by the loss of computer support, and determine how long they could sustain business activities without computer resources. As part of this interactive session participants should list the critical success factors of their operation and define how a disaster will impact on the ability of their group or the organization as a whole to fulfill its mission.

Once it is determined how the organization's activities will be degraded, a cost or "loss" factor can be applied in terms of business, customer satisfaction, profits, research and development, etc.

9.4 Selection of critical applications. Once the maximum allowable downtime has been determined, and the critical success factors documented, the group(s) can then move to the selection of those critical applications that support the base requirements of the information processing system. It is important to note that the time of day, the day (date), the day of the week, the week of the month, the month of the year, and the period of the year (quarter, end-of-year), substantially change the criticality of applications. There is no way to foresee when a disaster will strike. Selection of critical applications is dependent on normal business activity, and adjusted at the time of the adverse incident. Final decisions concerning the selectivity of critical applications, and what aspects of the business will be supported at the time of diminished processing capability, is that of management. Each of the groups participating will have to rate or prioritize their applications in order to determine the backup resources necessary to maintain continuity of performance. The selection of critical applications leads to the determination of data processing backup alternatives and user "stand-by" provisions.

9.5 Backup alternatives (backup strategies). The objective of this interactive session is to assist the planners in selecting the most appropriate backup strategies (alternatives), and user stand-by provisions. The selection of a backup (recovery) site is the key, as well as the basic success factor in building a valid disaster/recovery plan. Throughout the handbook, various methodologies have been presented to guide the instructor/consultant (Disaster/Recovery Administrator and Coordinator) in devising such strategies with the data processing and user communities. Contingency-disaster/recovery planning is about protecting information assets through the ability of the company to maintain the continuity and integrity of the flow of information.

The strategy (or a series of strategies) built-in to the disaster/recovery plan should be devised to satisfy the business and corporate critical success factors and organizational mission in utilizing information assets and information processing resources in protecting the viability of the business. The seminar interactive session provides a unique forum for developing a corporate consensus on the strategies to be employed at the time of a disaster. Once this consensus has been reached and the selection of the backup strategies achieved, the organization can proceed to develop a workable, implementable, testable, and maintainable plan.

9.6 Baseline security techniques. The baseline security technique was conceived to provide planners and risk management with a set of standards in order to eliminate repetitive, arduous, time-consuming risk analysis (risk assessment). The baseline security technique, documented in this handbook, provides the guidance for the verification or requirement for specific prevention, security, and controls within the data processing environment. The importance of these baseline security elements evolved from a study of the risks, threats, and vulnerability that could affect a data center. From this a set of prerequisites have been established that provide the base for an organization's security program.

Using the "standards" established, or in reverse, performing a risk assessment based on specific requirements, the planners can implement a viable security program. The checklist found at the end of the sample interactive session on baseline security techniques indicates how your organization measures up to these requirements.

9.7 Personnel considerations. The potential threat of intentional/unintentional disasters caused by people is real. Disclosure, modification, destruction, and abuse of information or information resources is present within any organization. It is necessary for managers to be constantly on the alert regarding employee behavior, attitude, and job performance. This interactive session highlights the various requirements of managers and supervisors to detect, mitigate, or eliminate any aspect of employee dissatisfaction. This interactive session should be presented by personnel professionals and/or psychologists who have been trained in identifying problems before they materialize.

9.8 Facilities evaluation. Identifying physical threats to a particular facility or environment is part of the prevention program to be initiated within the organization. Risk assessment or use of the baseline security technique approach is immaterial. The important part of the project is for personnel attending the session to be able to view their particular environments and identify the potential and/or real risks and exposures. The presenter should guide the group in examining their facilities and operations from without and from within in terms of the ability of anyone or anything to breach the security and/or safeguards installed to protect the environment.

9.9 Internal controls. The protection of information assets stored and/or accessed in information processing systems is the primary objective of any security program. The entire subject of internal controls requires far more effort and time than can be expended during the preplanning session. The important objective of this interactive session is to alert data processing professionals and users as to the necessity and requirements for internal controls. The failure of implementation of internal controls and security software packages is that they are usually user-driven. Excessive internal controls and security measures create a user-unfriendly computer utilization environment. The presenter has to dispel these concepts and drive home the importance of such security measures.

9.10 Insurance. This interactive session should be designed to review existing insurance, as well as evaluate the insurance requirements for disaster/recovery planning. There are many considerations for funding activities during an emergency condition that normal insurance coverage does not address or insure. The participants should review the criteria for increased insurance that will be applicable at the time of a disaster.

9.11 Emergency response planning. The ability to respond quickly to the many adverse incidents that could impact on the data center or user community will in the long run minimize the extent of damage and reduce downtime. Emergency response planning based on the identifiable threats to the facility, data center, or user community is vital to maintaining the continuity of activity. The interactive session should review existing emergency response plans (if any), and assist in developing proper procedures for emergency conditions.

9.12 File backup strategies. A computer, regardless of its availability before or after a disaster without source code, application software (latest versions), or data is of no value. Many organizations still view off-site storage programs as superfluous, and have yet to institute acceptable plans to ensure their ability to recover systems after a catastrophic event. It is the object of this interactive session to review existing procedures, define the losses of data that will occur, identify the ability to recover lost data from source documents, and provide an organizational or user methodology for file backup procedures and off-site storage program.

9.13 Teams and action plans. This interactive session gets to heart of the planning process: the development of the various teams that will be required to function during a disaster condition, their responsibilities, and action plans. This training session should provide an overview as to how all of the teams will function: how they will direct backup activity, user stand-by provisions, recovery strategies, and post-disaster activities. The participants will be required to identify the teams necessary to perform the action plans, the structure of their teams, and how they will initiate their actions and conclude their mission, all within the time constraints established by the maximum allowable downtime.

9.14 Communications. Because of the complex and varied communications (voice, data, fax, teletype, internal and external specialized comms), no interactive session can provide the details or guidelines for recovery at the time of a disaster. Each of the participating groups will have to view their communications requirements, their present activity versus disaster alternatives, and communications topology. Participants should be able to review and document existing communications activity, and specialists will be required to design alternative strategies.

9.15 Organization for disaster/recovery planning and implementation. This interactive session is presented and used to develop the organization(s) that will develop and implement the disaster/recovery plan. As part of the process, every aspect of authority, group responsibility, and plan development process should be reviewed and documented. This is part of the report that will go to management for approval.

9.16 Organization for disaster/recovery planning and execution. Every aspect, concept, and implementation of the strategies, procedures, and stand-by provisions require a highly responsive and efficient organization in order to ensure that the plan is executed properly at the time of a disaster. Each of the action plan teams are appointed, and given the responsibility of coordinating their individual activity with other teams and within the overall scope of the plan.

9.17 Selection of disaster/recovery teams and team assignments. This interactive session is used to prepare the phase assignments for each of the action teams. These assignments include what has to be performed, how each team will accomplish their task, how they will coordinate and interface with other team activities, what their assignments are after the backup site is on-line, and what responsibilities and performance will be required during the recovery (return to home-site, disaster normalization, and post-disaster periods).

9.18 Maintenance/testing/training. What are the requirements for plan update and maintenance? Who is responsible? How will the information flow to those responsible? These are the questions this interactive session addresses. In addition to plan maintenance, which is also dependent on plan testing, the presenter should review the various testing procedures and determine the most effective methods of testing for each of the groups participating. Each of the teams, users, and groups will also develop their training programs.

9.19 Recovery procedures and post-disaster/recovery considerations. During the disaster mode the performance of business activities has been abnormal in relation to the standards and procedures of the organization. Before normal operating procedures are implemented (disaster normalization), all of the operations and activities that were performed prior to the disaster have to be reviewed and all of the data, functions, and activities that were not performed have to be brought up to date and reinstituted. This interactive session should be used to guide the participants in developing their post-disaster plans, the decisions of operating performance in the event of a disaster, and the methods for recovering pre-disaster performance.

9.20 Data collection requirements (Information To Be Collected). In order to properly document any plan, the planners need information. The review of the types and extent of data required to document the plan is the subject of this interactive session. The details of data collection are found in previous sections of the handbook.

9.21 Developing an information and security policy manual. The protection of information assets in any organization is vital to the success and completion of that organization's mission. Information management is key to the protection and integrity of confidential and proprietary corporate information and data. The presenter should develop a model for such a manual, and use this model as a guide for this interactive session. This manual will reflect management's position relating to the security and protection of information assets and respond to the legal requirements of the organization.

9.22 User and business unit disaster/recovery planning. Corporations are recognizing the importance of developing disaster/recovery business resumption plans outside of the data processing environment, and coordinating such plans with the master data processing activities. Users participating in this program will obtain an excellent perspective and perception regarding the requirements of contingency planning. This interactive session should highlight the importance of users and business units developing and implementing plans within their activities.

9.23 Establishing/manning/occupying/operating a disaster/recovery control center. In the immediate aftermath of a disaster, it is assumed that the information processing center and all other facilities of the primary site will be inaccessible. The data processing function will move to a backup site, and other units of the business activity (if planned for) will move to alternate sites to continue to perform business services. The Disaster/Recovery Management Team and other senior management of the organization have to be provided with a facility from which they can manage the disaster/recovery activities, as well as control other aspects of business performance. The use of a disaster/recovery control center, developed prior to any adverse incident, is necessary to perform specialized activities. The control center provides the office space and pre-installed communication capabilities for the management of the disaster/recovery operation. This interactive session should establish the criteria and the occupancy as well as the selection of a site for the control center.

9.24 Continuity of corporate management in the event of a major catastrophe. In the development of disaster/recovery plans, most organizations do not include the importance of the continuity of corporate management, or the loss of one or more senior managers at the time of a disaster. A senior executive could be critically injured, or be unavailable at the time of an adverse incident. It is vital to the success of any business resumption plan that the line of authority be designated as part of the planning process, documented in the minutes of the corporation, and approved by the board of directors. The responsibility of this activity should be designated to the Disaster/Recovery Plan Administrator. The methods and procedures for the continuation of management should be reviewed in advance with senior management for the corporate level, and presented at this session. Each of the participants should develop the same process for their particular organization.

PART 6

DATA COLLECTION REVIEW

1.1 DATA COLLECTION

1.1.1 Information Collection

A reasonable systematic approach to disaster/recovery planning and documentation of the plan demands adherence to a carefully conceived structure. This structure is needed to:

- Ensure that every area is addressed.
- Permit easy reference to each of the sections. Each section is so structured and detailed that anyone can respond to the action plans and team assignments in the event of any disaster.
- Facilitate revisions as conditions, which would affect the whole or part of the plan, change.
- Utilize the structure to establish the task forces/action teams.
- Establish procedures to test the plan (or sections of the plan).

In order to accomplish all these tasks, you need *information*. Every aspect of the operation must be addressed to ensure that those who must effectively respond to a disaster by implementing the plan can do so. Remember, this is a *business resumption plan*, not a data processing recovery plan.

The request for information must not be limited to the data processing operation. As we have continuously emphasized, the plan is a business resumption plan for the entire organization; thus, the rationale for the various task forces and action teams. The first commitment to the plan will be exhibited by the cooperation and dedication to this important phase of the project by every function of the organization. The information on data collection from some areas may be as simple as a response to a memo; others may require interviews; still others may have to be constantly "dogged" to obtain information and details "dug" out of other existing plans and policies. Senior management's *support* and *mandates* will go a long way in accelerating this phase. Therefore, the recommendation is for at least one senior management representative (the Disaster/Recovery Administrator) to be involved from the initiation of the program.

The gathering of as much existing documentation as outlined on the following charts, checklists, and questionnaires is vital to the entire structure of the program. If a systematic approach is taken (the proper studies and detailed analyses are prepared based on the material indicated here), you'll find that little raw (new) material has to be developed. Much of what you will find difficult to obtain pertains to individual judgments, such as the determination of priorities and critical applications.

The lists of requests for information should be developed as quickly as possible, individually tailored to your specific operation and activity, and distributed to the proper supervisory personnel or task force leader. Every effort must be expended to obtain the information from the various departments as quickly as possible. In most instances, we have found that at the first or second meeting of the planning group(s), or interviews with specific department supervisors or manager, the response to the first request resulted in little or no activity. Therefore, immediate personal follow-up by the Disaster/Recovery Coordinator is mandatory. This will at least generate some activity and keep a high profile for the project prior to any meeting.

The requirements for data collection will vary greatly with each organization. The *request for information* must be tailored based on the complexity and functions of the company. Most of the information requested is readily available, but will be located throughout the company. The information and decisions to be made during the data collection activity will have a significant impact on the effectiveness and performance of the finished plan.

User departments play a key role in the successful implementation of a disaster/recovery plan. They must accept a full role and commitment in the preplanning, planning, and implementation phases of the program.

1.1.2 Personnel Assignments and Participation

Personnel commitments to specific responsibilities are necessary to achieve the objectives of obtaining the required documentation and information for the actual disaster/recovery plan preparation. It is not a "one-person" job; it requires the cooperation of all organizational and user groups that use or may be affected by the data processing operation support services.

The first step is to appoint a *data collection task force*

coordinator. This person may, or may not, be the final individual selected to be responsible for the disaster/recovery plan once it is implemented (that is, the Disaster/Recovery Coordinator), but it would be advisable if he or she were. Under the data collection task force assignments, the group will be responsible for collecting, indexing, and verifying the completeness of the material.

1.1.3 Task Forces and Disaster/Recovery Action Teams

We have listed two sets of teams that are required to initiate the plan: the disaster/recovery planning data collection teams, and the disaster/recovery action teams (disaster initiation action teams and recovery action teams). Figure 3-1 indicates 20 preplanning functions (teams), seven emergency response functions (teams), and 27 action plan functions (teams).

Efforts of team assignments can be combined based on available personnel. Here again we suggest that the initial assignments, whenever possible, from the data collection phase be carried over when selection and assignments are made for the disaster/recovery initiation action teams and recovery action teams.

The initiation action teams are the most critical. These are the groups that will evaluate the situation and determine the actions to be taken (level of response based on magnitude of adverse event) and determine the declaration of a disaster mode.

1.1.4 Senior Management Team

Unless management is willing to waive completely its decision-making responsibilities at the time of a disaster, there must be a senior management group, outside the data processing activity, that will perform or be responsible for the following at the time of a disaster:

- Making the final decision based on the recommendations of the initiation team to go into an emergency disaster mode of operation.
- Ensuring that organizational policy is strictly adhered to under all conditions.
- Ensuring the safety of all employees.
- Ensuring that the company's legal rights are protected.
- Providing the necessary authorizations for expenses that are to be incurred.
- Providing initial funding (having the funds available in the event of a disaster).
- Ensuring that all the company's vital records are protected.
- Convening the necessary executive staff (and board of directors) to maintain the continuity of the company's operations.
- Providing for the necessary credit required under emergency conditions.
- Ensuring that none of the actions taken will legally affect the company or encourage litigation.
- Protecting the company from fraud, embezzlement, or exposure of sensitive data.

1.1.5 The Workshop Task Force Coordinator

In addition to the tasks outlined in the preceding section, the task force coordinator's (and administrator's) responsibilities during the data collection phase should cover:

- Assigning all task forces, and authorizing them to perform their assignments.
- Chairing all the data collection phase meetings to ensure that each group is performing in accordance with its specific assignments.
- Following up on each task force leader to see that the information required is being collected in the manner agreed upon.
- Developing all the personnel lists of those involved in the data collection phase (and ultimately in the disaster/recovery program). This list should contain the addresses of team leaders and alternates, telephone numbers, vacation schedule, and so on.
- Seeing that all deadlines are kept and, if necessary, assigning those who have completed assignments early to assist other groups.
- Selecting those individuals who will participate in the plan preparation.
- Keeping senior management aware of the activities and the level of completion of each task and assignment.
- Establishing the format for disaster/recovery documentation and data.

The data collection task force coordinator should be permanently assigned as the disaster/recovery plan coordinator. The detailed project plan will require an individual who is familiar with each aspect of the data processing operation and its operational support services, as well as the overall operation and functioning of the organization. Greater expertise will be gained during the data collection phase: what is learned during these activities should not have to be communicated nor should another person be trained to perform the tasks of the operating plan. As a review, we have supplied a list of the phases of the program, stating when they will be accomplished and the phases to be reported to management:

- *Definition phase*. The objectives and assumptions of the program will be developed during the preparation of the preplanning seminar and the data collection phase to establish the parameters for the actual plan actions. The definition phase is also the basis for reporting to management the objectives, purposes, and necessities for the plan as well as the scope and extent of the plan.
- *Data collection phase*. Management should receive periodic reports on the status of this phase to ensure that all

the elements being collected meet the policy objectives of the company in developing and implementing the plan.
- *Design development phase.* The level of completion and effectiveness of the plan will depend on the ability of the task forces to obtain the data requested.
- *Plan preparation.* This activity covers the final plan creation. In most cases, the plan preparation is performed routinely and successively within the data processing and organizational environment. It is a phased-in process. If the plan is prepared in an organized, structured manner, the individual action teams should be able to develop their own sections of the plan. Upon completion, the combined efforts of the team efforts will provide the overall disaster/recovery plan, requiring the plan coordinator to organize it into a single, viable document.
- *Installing the plan (at your facility) and training personnel.* Once the plan has been documented and approved by management, all data center and user personnel must be assigned final responsibilities and trained.
- *Testing (monitoring) the plan.* In order to achieve the timely restoration of data processing support operations, the disaster/recovery plan must be tested. Any deviations or nonperformance aspects of the plan must be modified.
- *Maintenance of the plan.* Any changes in personnel, location, equipment configuration, applications, users, priorities, and so on require immediate upgrading or modification of the plan.

1.1.6 The Disaster/Recovery Information-Gathering Task Force

Recognize that prior to obtaining the knowledge of the methods and procedures for developing the plan, it is impossible to mobilize the working teams to any great extent. This section provides the initial task force requirements to fulfill the mission of data (information) gathering. It describes the organizational groups that may be affected or that may be required to provide answers to the questionnaires and special work forms contained in this manual. It also outlines some of the special assignment and methods of collecting the data expeditiously.

This manual, in effect, serves as the "plan-to-plan" and will accelerate the data- and information-collection program, as well as provide the maximum information needed to create the plan.

1.1.7 Management's Participation

Disaster/recovery planning coordination deals with the realities of the corporate environment and involves people, politics, and special interest groups. Therefore, without direct mandates and guidelines and support by management at the initiation of the data- and information-collection state, it will be impossible to develop the plan in a reasonable, cost-efficient, productive manner. Corporate policies must be developed, published, and disseminated to ensure that the program will be accomplished expeditiously.

1.1.8 Questionnaires and Forms

In order to accumulate the information required to set up the plan, it is important to use the checklists, guidelines, and forms provided in this manual. These checklists have been derived from numerous data-collection and facility-review programs performed by professionals when establishing their own plans. Combined with the policy, standards, and procedures established for your specific data center, this information will be invaluable in accelerating the preparation and implementation of the disaster/recovery plan.

It is important to note that the checklists, questionnaires, guidelines, and forms provided are generic in nature. Some of the material and information requested will not be applicable to your specific computer complex. There is also the possibility that some of the information needed by *you* for your activity may not be covered in the material that follows. Obviously, there is no substitute for experience and knowledge and the use of appropriate procedures within your organization.

The checklists, guidelines, and forms are an excellent source for determining how well your computer function is being managed and how well it performs in relation to operational effectiveness, technical efficiency, service of user needs, and the quality of its safeguards and controls.

1.1.9 Questionnaires, Checklists, Guidelines, and Forms

In "auditing" the data processing operations and activity for the purposes of developing a plan to mitigate and/or recover from an adverse occurrence, it is necessary to have a structured criteria. The time that can be expended using personnel in a haphazard information-collection program can be expensive and, if not properly directed, will result in a set of disorganized and unrelated information, totally inappropriate for the development of a sound program.

The questionnaires, checklists, guidelines, and forms provided in this manual deal with specific functions. In many cases, the same questions appear in one or more activities to be reviewed and require the duplication of information to ensure the continuity of the details in a specific area. The response to the same question in each section is important, since it avoids time-consuming "hunting" when addressing specific functions of the plan.

Further, responding to those questions that might not specifically pertain immediately to your facility will initiate additional questions that should be responded to, or give you the opportunity to devise questions more specific to your activity or those of the users. What we are attempting to do is pinpoint areas of concern, generate preaction/predecision to the plan development seminar/workshop, and further ensure that steps are taken to correct identified "security" conditions.

Sound computer practices respond to *if, how, when, with, what,* and *by whom* in order to determine what has to be accomplished to achieve the results and objectives of management's policy and the data center's security. Remember, responses have to be objective, concise, and reflective of management's policy and operating standards which apply to your installation; they should not be what you would like the conditions to be, but what is practical.

If the computer complex is presently operating without a policy or established management guidelines, these questionnaires, checklists, guidelines, and forms will serve two purposes:

- Provide a means for establishing a policy and procedures manual based on the requirements, strengths, and weaknesses of the operation.
- Provide the data needed to develop the disaster/recovery plan.

1.1.10 Limitations Concerning Questionnaires, Checklists, Guidelines, and Forms

Redundancy emphasizes importance, as we have previously noted. The broad scope of questionnaires, checklists, guidelines, and forms presented in this handbook may not be applicable in many areas with regard to your facility. We have used the broadest approach possible to ensure that every level relating to the size and complexity of the operation is covered and properly reviewed. You may have to make specific adjustments in order to have the questions apply to your situation. Sound judgment based on knowledge will provide the desired results. Do not ignore the benefit of applying baseline security techniques.

The quality and level of knowledge of the activity by the personnel assigned to obtain the information is extremely important. Assigning this task without a complete indoctrination by the Disaster/Recovery Administrator and Coordinator as well as senior data processing management is a waste of time, especially when these people will have to confront users and obtain highly confidential, proprietary, and extremely sensitive information.

1.1.11 Data Processing Planning Coordination: Selecting Personnel; Developing Task Forces and Plant Teams

In general, the pre-seminar and data collection (information-collection) phase will encompass broad groups of activities and require communication with virtually every function of the organization. In order to accomplish this task, it is the responsibility of senior management to first assign a data processing disaster/recovery planning administrator and co-ordinator, and then inform the entire organization of its commitment to the program and the cooperation it expects from each operational division or unit of the organization. In making this organizational commitment, expression of its support must encompass the management support and staff participation expectations.

1.1.11.1 Management Support

Management will need to provide the funding and personnel resources to develop, implement, maintain, and test the plan. Since funding decisions may be premature, lacking the actual plan requirements, it is sufficient for management to state that it intends to appropriate funds.

1.1.11.2 Staff Work

The Disaster/Recovery Administrator will select the initial information collection task force from the data processing department and all of the user departments. The task force will then:

- Provide a list of forms and distribute them as per the various team assignments and tentative appointments.
- Notify all related departments (purchasing, security management, risk management, OSHA administration, and so on) of the requirements for data collection and the use and application of the material.
- Establish deadlines for completing all materials, questionnaires, guidelines, checklists, and forms. Each document is to have a deadline date indicated.
- Review the materials as they are returned.
- Correlate final materials, questionnaires, and so on in the same order as they were requested in the preworkshop manual.
- Review the materials with appropriate management personnel.
- Select teams (groups) that will share the responsibility throughout the project. [Suggested participants: data processing management, internal auditing, security, OSHA or risk management administration, and user(s).]

1.1.12 Task Force Assignments

In order to obtain the information necessary, an immediate planning and assignment meeting is required. We have indicated some of the functional business units that are required to respond to the various questionnaires during the pre-seminar and data-collection phase. In most cases, it will be impossible to get the number of people listed, so additional assignments will have to be made as the work progresses.

1.1.12.1 Recommended Courses of Action

It is suggested, owing to the normal daily burden of work, that alternative methods of information collection be implemented under the coordinator and within each department. These alternatives have proven quite successful in many companies; without affecting normal operations, the coordinator can achieve the results necessary. Implementing these alternatives are reviewed in the Program Profile (Part 7). What you are attempting to do is create an immediate impact and generate the program without giving anyone a chance to back-off or procrastinate.

1.1.12.2 Task Force Requirements

The requirements of the task force cover a broad range of necessary personnel assignments. Depending on the size and complexity of the data processing operation, and the size of the organization, duplication of responsibilities will occur.

The initial task forces, with slight modification, should become the recovery teams; the assignments made for the "plan-to-plan" program are the basis for the follow-through in implementing the plan once it has been developed.

1.1.13 The Workshop Disaster/Recovery Administrator and Coordinator

The task force administrator and coordinator should be the primary functionaries at the seminar and workshop; these individuals can create the visibility and continuity for the entire disaster/recovery program. The choice for the workshop task force administrator should be someone from management who has a full understanding of the duties and responsibilities inherent with the position. The plan coordinator is a full-time responsibility.

1.1.14 The Task Force Disaster/Recovery Coordinator

The following is a list of some of the duties and responsibilities of the task force "d/r" coordinator. These responsibilities should be communicated to senior management in order to obtain the authorization required to perform and organize the various activities required to design, implement, maintain, and test the plan.

The task force "d/r" coordinator is responsible for:

- Establishing the necessary information-gathering (data-collection) teams and directing their efforts.
- Setting up the procedures under which these teams are to work.
- Obtaining all the information required for the data collection manual, and mobilizing all the task forces to collect this information.
- Correlating all the information in accordance with the workshop preplanning data collection aspect of this program.
- Maintaining all the files and information.
- Working with internal and external auditors to determine the security and recovery requirements of the company.
- Reporting to management the status of the data-collection activity as well as the development of the disaster/recovery task forces and teams.
- Maintaining the completed plan, updating the information, and keeping the plan current with the changing or expanding data processing operation.
- Keeping current all documentation concerning the plan.
- Integrating other activities that relate to disaster/recovery procedures, as well as maintaining the continuity of operation and integrity of the data.
- Coordinating activities with other groups outside the data processing activity, as well as other companies in developing, implementing, and maintaining the plans.
- Disseminating (on a need-to-know basis) information on the plan, conducting tests, holding meetings, and keeping those concerned apprised of new problems and new techniques in disaster/recovery planning.

As disaster/recovery planning is realistically a management responsibility, dealing with management issues and policies, the coordinator should be part of the management staff, rather than a line manager. The ability to work with and understand the needs of auditing is an important qualification. The greater responsibility for coordination will be involved during the data-collection phase. Therefore, the individual must have the authority to:

- Call and hold meetings to obtain and disseminate information to all operating (user) departments.
- Review each task force member's material to ensure completeness.
- Obtain all the information from the various departments and users relating to personnel data.
- Gather all information in a timely manner.

1.1.15 Plan Economics and Personnel

The primary objective of the data-collection phase of activity is to provide all or most of the information required to establish the disaster/recovery plan. In addition to the specific details that must be accumulated, it is important that the coordinator create the schedule of time and cost estimates for the actual preparation and implementation of the plan. If this is not feasible at this time, or there is little understanding of the total economic aspects of the program, it is better left undone for the moment. Poorly prepared budgets will reduce the overall impact of the project.

If management has already committed itself to development of the plan, then there is recognition of the need for a plan, and cost will be of little consequence in achieving the objectives of the program. Time commitments and business/auditing pressures may have a greater bearing than financial considerations.

Experience has shown that if there is a full-time coordinator assigned to the project and supported by management, it is not difficult to obtain the cooperation of other departments or to get other staff members assigned to the project. It will usually take fewer people to accomplish the data-gathering task than it will to actually implement, maintain, and test the plan.

2.1 DATA COLLECTION: INFORMATION REQUESTED (DEFINITIONS)

2.1.1 Outline of Information Requested

2.1.1.1 Data Processing Facility Layouts

Building, data center, floor plans.

- *Layouts (all buildings).* This includes all building schematic layouts, indicating where data processing facilities are located. (See Figure 20-1.) The layout should cover all the functional areas of the same building where user departments are located, and all other locations where data processing supports other divisions or other organizations, or where other independent data processing operations occur.
- *Layouts (data center).* The layout of the actual data processing center would include floor plans; hardware and peripheral equipment; power and communications input; emergency controls; environmental controls; and fire protection, detection, and extinguishing devices. (See Figure 20-2.)
- *Details of security provisions, systems, and procedures.* This details all external and internal security precautions taken. Prepare schematics, locations, and security of input power protection, communications protection, environmental equipment protection (external and internal), UPS (if any), motor generators (if any), and power filters (if any).
- *Area description.* In schematic form, locate your facility in relation to other buildings, industries, and so on that surround the data center. Note especially those buildings and/or industries that could cause any severe local condition that would affect or interrupt your data processing operation. This would also include evaluation for flood areas. (See Figure 20-3.)

2.1.2 Organizational Charts

Supply the following information:

- A detailed corporate organization chart that would include every entity and function of the company.
- A detailed management information services chart indicating the relationship of the data center to all user groups.
- A detailed data center management and operations chart.
- For all of the above, and for all persons involved in the entire disaster/recovery activity, lists indicating all supervisory personnel, home addresses, business locations, home phone numbers, and business phone numbers and extensions.

2.1.3 Risk Assessment

It is essential to recognize the potential consequences of undesirable events against which the facility needs protection and the need to develop a disaster/recovery plan to minimize the damage resulting from losses or damages to the resources. Such recognition is dependent upon understanding the threats, hazards, and vulnerabilities to which the data processing center and/or users are exposed. The data processing operation and facility is an assemblage of many resources. Some particular subset of these is needed by the facility (organization) to provide data processing support. Resources include people, programs, data, data processing hardware (data center and users), communications facilities, power, environmental control, the physical facility (building and specific data processing location) and access to it, supplies, and even special forms and documents.

All these elements are not equally important to maintain the continuity of operation and the integrity of the data. And, each of the elements is not equally susceptible to harm. Therefore, the selection of safeguards to protect the facility, and the elements of the disaster/recovery plan, must be accomplished with informed awareness of which critical business functions are supported by each computer resource. In short, the development of a cost-effective, efficient, and workable disaster/recovery program for a data processing facility is heavily dependent on:

- An awareness of the facility's relative dependence on each of its component parts.
- Knowledge, at least in an overall way, of the probability that some undesirable event will occur.
- An evaluation of the probable results of undesirable or unexpected events so that one can minimize or eliminate the possibility of an undesirable event happening, and the effects on the function of the operation if one does occur.

Knowing (a "what-if" scenario) the consequences of not being able to perform each system function for a specific (or unknown) time interval is essential to the development, implementation, and use of effective disaster/recovery plans, which can be adequately designed and sufficiently responsive to the supported organizations.

For the program to be successful, it is essential to be able to determine which undesirable occurrences would affect what operations, the impact on the normal business activity, and the work load of the facility's activity. Usually, it is only a small percentage of the work load that would cause the greatest intolerable disruption. Guessing into which category each activity would fall (or should fall) is almost impossible for data processing. This is the purpose of "total" investment of *all* user departments.

A properly conducted investigation and risk analysis should yield this data, which then can be used to properly construct the disaster/recovery plan elements based on actual, quantitatively expressed needs of the supported organizations for data processing services.

2.1.4 Hardware/Auxiliary Equipment (Configurations)

Each facility is unique in its operation and dynamic in its function; therefore, the configuration of hardware and auxiliary equipment that has been designed to meet the operation's specific needs does not remain static. The diagram should reflect the latest configuration available. Realistically, there are no two data facilities that are identical. Unless such a redundant facility has been installed intentionally as a backup facility, each configuration must be identified and a plan must be developed for each. The necessity to identify *every* piece of equipment and associated auxiliary equipment, its system utilization and application dedication, is vital to a successful disaster/recovery program.

2.1.5 Layout of the Facility

Prepare a layout of the complete data center. This diagram should include the locations of all equipment; power inputs; communications input; media vault; auxiliary and peripheral equipment; entries and exits; power switches; fire extinguishers; air-conditioning equipment; Halon tanks and outlets; fire, water, and smoke detection devices; and so on. The diagram should be as close to scale as possible. By using the raised floor panels as a grid, developing and drawing to scale should be fairly easy. (See Figure 20-2.)

Part of the drawing, or an overlay to the drawing, should indicate sprinklers, water feed lines, any water mains, ceiling and floor air outlets and returns, and so on. It should also indicate the location of all special alarm detectors: water, smoke, and heat. If the facility is part of a larger building or located in a multitenant building, you should obtain an engineer's drawing (floor plan) of all surrounding areas, indicating access, emergency exits, and so on, if the location of the data center could be directly affected by any incidents (fire or intrusion, for example) occurring in these adjacent areas. A similar layout should be obtained from user departments indicating equipment layout.

2.1.6 List of Hardware/Auxiliary Equipment and Contacts

You will need a complete list of vendors, including hardware vendors. This may seem like a duplication of effort, but having a specific list, by priority requirements, will save a great deal of time in the event of a catastrophe. The list should include everyone (hardware vendors, service organizations, forms suppliers, supply vendors, and so on)—any organization that provides equipment, supplies, or services. It should also include any consultant who would be instrumental in aiding in recovery operations.

2.1.6.1 Hardware (All Primary Equipment Vendors, Auxiliary Equipment, Etc.)

List regular hour telephone numbers as well as emergency numbers (if available) for local representatives, field service personnel, or main office contacts. At the time of an emergency, it is imperative that you reach anyone who can act in mobilizing the vendor for immediate assistance.

2.1.6.2 Software (By Critical Applications/ Vendors or Backup Facilities)

Your software list should be formed utilizing the priority recovery selected and detailed within the plan. There is a possibility that the recovery priorities will vary according to the day, month, quarter, or annual basis; therefore, the list should be broken down accordingly. This requires adjusting priorities. Purchased software poses some unique recovery problems if enhanced versions are not properly documented, duplicated, and stored off-site. The vendor should have a copy of the latest version of the software available for immediate delivery to a backup site or the primary facility if it is functional. If all software is developed in-house, other recovery and documentation conditions exist.

Let's assume that all software is backed up (that is, the latest version with appropriate documentation) off-site, and that recovery can be made from this facility 24 hours a day, 365 days a year. In the event of total destruction of the facility or building, because damage cannot be accessed for a considerable length of time and access may be denied to retrieve working tapes, an off-site storage program is vital to the recovery process.

Do not forget about specific production requirements that the data center must perform as part of its daily, weekly, monthly, quarterly, or annual activity—which is not part of its user-application-oriented production.

2.1.6.3 Forms (Inventory Requirements, Emergency Response Replacement)

Each specific application has related forms. In most cases, such forms are kept on the premises and would be unavailable to recovery operations at the time of a disaster. Such items as preprinted and numbered checks, invoices, statements, purchase orders, confirmations, envelopes, and so on should be backed up with a sufficient supply, off-premises, at a secured location. The supply quantity would be predicted on the amount used for specific operations and the length of time necessary to receive a replacement. At least one month's supply must be available.

A complete list, by specific form number (including blank paper, size and type) should be made. This should also include cost of special forms and an alternate supplier. A sample of each form, and a description of its use (specific application) should be included in the appendix of the plan. The names of the suppliers of these forms, their emergency phone numbers, and emergency contacts should be provided. Discussions should be held with vendors concerning emergencies and how they can assist in providing inventory, emergency deliveries, and replacements. (Numbered forms should always be logged to see what was the last serial number used. A record should be maintained on-site as well as off-site.)

2.1.6.4 Supplies

From paper clips and staples to tape reels and disk packs— every item must be listed with its source, contact, and vendor's phone number. Most commercial backup facilities have a limited amount of supplies that can be used until a regular supply can be obtained. *Do not depend on this unless previously agreed upon.* An off-site inventory maintained in a proper environment (i.e., magnetic media) may seem insignificant, but these insipid little items may quickly frustrate recovery backup efforts.

If there is a cooperative agreement between company divisions to each other's center, or if you intend to use (or are using) a service bureau, these items could be critical.

List all vendors by supply categories and the quantities needed for minimum and long-term activity.

2.1.6.5 Office Equipment

If a commercial backup facility is properly contracted, items such as file cabinets, desks, and chairs would be part of the initial arrangement but available in limited supply. Where other situations exist, it may be necessary to purchase and warehouse such office equipment, especially if a cold or shell site is contemplated. Arrangements can usually be made in most areas to rent equipment on an emergency basis. These arrangements must be made and contracted for in advance. Quantities and types of office equipment must be decided upon, contracts arranged for, and emergency deliveries guaranteed by the vendor.

2.1.6.6 Services

Most facilities require special services (consulting, repair, maintenance, and so on) to maintain all the hardware and peripheral devices, electrical supplies, UPS systems, air-conditioning systems, security devices, and building maintenance personnel (plumbers, electricians, and so on). Consultants who are involved in any aspect of operations, application, or systems development; construction; or security will be needed during an emergency.

Each service should be listed by category. At least two organizations should be part of the list for emergency activity.

2.1.6.7 Office Recovery Sites

A survey should be made of the area (a real estate broker would be of value) to determine where office support (executive, administrative, and clerical) personnel could be housed. Possibilities would be other company facilities, empty office buildings, hotels with available rooms (or banquet facilities).

In doing this survey, environment, ease of access, security, communications, type of area security at the facility, and length of probable occupancy are important considerations. An empty office building, or one with some space, may be available only for a short period of time. Therefore, a continuous activity for space provision is necessary, as the availability of office facilities changes daily. If possible, some form of agreement should be made between the company and the owners of the property that could serve as a temporary location.

2.1.6.8 Contracts and Agreements

This section should list, but not contain, the agreement made with any vendor, supplier, or consultant. Begin with a list (or lists) of appropriate backup sites (hot or shell) service bureau, other divisions of the company, cooperative agreements between two companies, and so on. These agreements are "vital" records of the company and should be protected in a secure location.

Also include a list of the following:

- All agreements with hardware and auxiliary equipment vendor; include the item, date of agreement, and a short summary of the agreement.
- Backup facility agreements.
- Supplier agreements.
- Any agreements with other firms or services that can be of assistance in the time of an emergency.

Note: All agreements should be reviewed by the company's legal department.

2.1.6.9 Emergency Services

List all services (describe the services where applicable), contact names, emergency numbers, special contacts (if any), and alternate companies or organizations where available. Here is a partial list:

- Police (local and state)
- FBI (local office)
- Fire department
- Security guards (central station)
- Security systems (central station)
- Water company
- Power company
- Building engineer
- Generator (electric backup)
- UPS backup
- Air-conditioning service
- Sprinkler service
- Fire extinguishers and Halon suppliers

2.1.6.10 Off-site Storage

The following criteria establishes the necessity for protecting data, software, and documentation to ensure the recovery of data integrity:

- Accuracy of data
- Completeness of data
- Latest versions of software
- Legal or authorized data
- Data in accordance with accepted accounting procedures
- Data that is in compliance with management's standards, policies, and practices

- Data that meets legal requirements (activity and retention)
- Documentation that supports construction of data
- Accurate operating and run instructions
- Latest versions of all software, systems analyses, and application systems
- Location and handling of source documents

The criteria for maintaining an off-site program must include:

- Availability at all times (precludes use of bank vaults and secured buildings after hours)
- Environmental conditions
- Professional management
- Historical business activity
- Security
- Transportation under environmental and security conditions
- Handling of magnetic media or other sensitive data by a third party (not an employee of the company)
- Type of vault
- Location with respect to the data center and backup site
- Emergency procedures
- Proper documentation of media movement
- Maximum access security
- Certification by a federal agency

Further details will be covered at the workshop concerning media storage, off-site security, recovery procedures, and emergency cooperation.

2.1.7 Data Center Operating Procedures and Policies

It is important that the disaster/recovery plan contain a summary of the standard operating procedures of the data center's operations. In the event of any interruption, or major disaster, these procedures will provide the specific action plans under which the organization and personnel will perform. During such emergencies it is difficult to maintain normal operating procedures. However, in order to ensure a continuity of activity, critical application recovery, and data security and integrity, it is vital to follow standard operating procedures. It may be necessary to develop an entirely new set of operating procedures based on recovery requirements and the type and location of backup site(s).

This section should include a complete list of all operating procedures, the specific documentation descriptions, the custodians of such manuals and their location, and special user requirements. Backup copies of all operating procedures should be available from the off-site vault in the event of an emergency.

Note: The specific procedures for recovery of the most critical application, as defined in your listing of application criticality and selected recovery strategies, should be included in the disaster/recovery plan. Distribution of operating procedures should be limited to a "need-to-know" basis.

2.1.8 User Operating Procedures

There is a possibility that in the event of a disaster more than one data facility would be affected. User operating procedure manuals and application documentation must be included in this plan as well as the development of miniuser plans. Security of these procedures, like those of the data center operation, is vital. Information as to the location of complete user manuals should be noted in the plan. Surveys have indicated that there is a lack of proper user documentation and procedures. Remedial measures should be taken to correct this condition.

2.1.9 Security Systems

The following procedures should be followed for documenting all security systems, including alarms, guard services, detection devices, warning devices, emergency controls, emergency communication systems, annunciators, and surveillance cameras.

- List all systems by vendor, contact, and regular and emergency phone numbers.
- Prepare a layout of the location of all devices (for example, emergency power switches, emergency controls, Halon actuating and abort buttons, fire extinguishers, sprinklers, and so on). This layout can be an overlay on the facility layout.
- Prepare a layout and list of all emergency communications equipment; list vendor, emergency numbers (verify standby communications lines to ensure their "live" operation).
- Include a copy of, or a description of, guard responsibilities.

2.1.10 Communications

On-line and real-time computer communications are critical to the success of any data processing facility, and imperative during an emergency. Three elements must be considered as part of the disaster/recovery plan:

- The continuance of normal communications
- Communications security
- Communications continuity in the event of a disaster

2.1.11 Normal Communications

The following procedures are appropriate when developing the disaster/recovery plan:

- Outline in detail the extent and complexity of the system's communication network.
- Describe the communications requirements to fulfill user activity.

- List all contacts and emergency numbers for the telephone company and all vendor equipment. Usually, plans and communication layouts are not available—attempts should be made to obtain such drawings for the data center since it will save hundreds of hours in installation time in the event replacement is required.
- Indicate on the layout the routing of all communication lines including all terminations and pass throughs.
- Develop alternate communication plans.
- Outline (or obtain from the commercial backup facility) a list of all communications lines reserved for you.
- List the minimum number of terminals, modems, controllers, and so on needed to maximize critical application recovery.
- List all dial-up lines and dedicated lines for the facility (this is *top secret*).
- List restoration time requirements for recovery of the entire system.
- Consult with your local area phone company.

If the communication network design is not carefully planned for the primary and backup recovery site, conditions may exist in which existing software will not function because it is incompatible with the backup configuration.

This aspect of the program is probably one of the most critical, since recovery of its activities must be rapid. *The greatest problem is the availability of the proper software that will conform to backup equipment and change of mode of operation.* If this is the case, then it is the responsibility of data center management and the communications team to identify the problem, correct the condition, and provide proper software design. Such versions must be clearly marked and secured. Testing of communications software at any backup site is critical to the continuity and recovery of the data center.

In a major local or areawide disaster, there is usually an excessive burden on telephone restoration activities. As part of the communication investigation, management must identify the historical aspects of phone failure and the capabilities of specific central stations to switch leased lines from initial termination to alternate sites. The methods and procedures established by the phone company should be included in the disaster/recovery plan.

Note: Every change of normal condition exposes the communication network to illegal intrusion and/or breach of security measures. Careful analysis of the problem is warranted, and procedures, once developed, should not be indiscriminately disclosed.

Systems diagrams and methods of switching to alternate sites should be illustrated and verbally described. Where a commercial backup site is used, communications experts can provide the expertise to develop the systems and software to ensure total compatibility. The redundancy of lines and associated equipment can be expensive; therefore, critical dependencies and recovery requirements should determine the value of this approach to communications backup.

2.1.12 Emergency Action Plan Responsibilities

The emergency response action plan is probably the most important aspect of the entire program. The most critical part of any emergency plan is *people*. The second most important aspect is *decision*.

In the event of a fire, the decision must be made to either evacuate or fight the fire. This decision is separate from any other procedural activity designed to combat an emergency. Therefore, we must understand that training plays an important part in the successful avoidance of a major catastrophe.

2.1.13 Emergency Response Plan

Your emergency response plan (see Part 9) should include:

- Identification of the location for which the plan is designed.
- A clear, big-type index covering the location within the plan to find specific information
- A clearly printed (set in large type) emergency call list, broken down into categories:

  - Security
  - Plant engineering
  - Personnel
  - Emergency control center
  - Responsibilities (names of persons responsible for specific functions, their extensions, their home phone numbers, and their personal message annunciator numbers)
  - Fire department
  - Law enforcement agencies
  - Medical services (Poison information center, hospital, medical emergency, and house nurse or doctor)
  - Utilities/services

    —Gas company
    —Electric utility
    —Water company
    —Telephone company (and after business hours phone number)
    —EPA (federal and state)
    —Sanitation district
    —OSHA office (internal administrator and federal agency office in area)
    —Insurance company
    —Civil defense office
    —Transportation commission
    —Highway department

- A diagram (in the form of an organizational chart) indicating specific responsibilities by departments
- Clearly printed emergency responsibilities. See Part 9.
- Emergency procedures, including:

  —Index listing all threats and hazards and references
  —Diagrams showing:

- Location of fire extinguishers
- Evacuation routes (use building diagrams)
- Plant shutdown procedures
- Emergency utility shutoffs

These procedures should be consistent with the manual prepared for OSHA administration and conform to the requirements of OSHA regulations, local building codes, and fire regulations.

2.1.14 Pyramid Calling Tree

All through this manual, we have discussed the requirement of having *lists* of all the key people who will be involved in the event of a disaster. Lists are to be prepared for facility personnel by specific responsibility. For example:

> The communication coordinator is called by the disaster/recovery team coordinator. The next sequence should proceed as follows:
> Telephone company emergency service
> Telephone company emergency executive contact
> Alternate communications coordinator
> Communications equipment vendor contact

One sheet should be specifically designed for each of the recovery functions, each of the specific suppliers, each of the equipment vendors, and so on. Once the recovery teams are organized, the setup of these pyramid calling trees is fairly simple.

The disaster/recovery manual should contain the master pyramid list and each of the individual calling lists. Each team coordinator should have in his or her possession at all times a list of the people he or she has to call. The disaster/recovery team coordinator list must contain information about the team coordinator and alternates in the event any one person is not available.

The development of the list, if personnel will release the information in total, is easy. If not, it will be necessary for each team coordinator to personally obtain the information. A successful method of obtaining this information is for the team coordinator to prepare a memorandum from the company indicating the purpose of the information and advising employees that this information will be held in strict confidence.

2.1.15 Application System Recovery— Software Data and Documentation

List in this section all the applications deemed critical that require recovery. The preamble to your request for establishing the critical application recovery priority could be as follows:

> The applications listed by you (the user) must be considered a critical priority as established by company policy. When listing the requirement for recovery of the specific applications, please qualify by indicating on-line, real-time or run frequency of the report. Final determination will be made by management, who will establish the critical priorities and mandate their recovery. If there is a future requirement, or if it is determined that under emergency conditions more applications can be run, you will be notified upon completion of this priority listing.

The plan should contain the listing and specific details and specific authorization to build those batch interfaces onto the on-line systems necessary to support the company's on-line (or batch) applications.

The listing should contain the following:

- Application system name and group of programs or software source
- Number of reels or disk packs involved
- Identification by dataset name or number of each reel and/or disk pack
- Frequency of report (response time for on-line)
- Minimum time for recovery
- A separate listing for on-line or batch
- Name of systems analyst for specific system
- User contact
- Equipment (processing) requirement (hardware, memory, and operating system to be used)
- Minimum terminal requirement
- Recent modification/new development (current version)
- Location of backup programs and data
- Documentation backup (list volumes and location)
- User or management, special conditions, forms, requirements, and so on
- Audit verification and sign-off

With each application, the following off-site rotation outline should be provided:

- Description of material available at in-house vault (description of material at the off-site vault relating to the specific application)
- Normal cycling of data to determine what period of time has to be reconstructed (if possible)
- Last update of application programs
- Generation (grandfather, father, or son)
- Location of possible third generation
- Archival sources for past detail construction
- Documentation, operating procedures, specifications, and so on
- Security controls (entry or access codes, encryption information, and so on—limited distribution)

It would also be advisable to have available a complete and up-to-date inventory of all data and programs, stored in the in-house vault and off-site, to ensure that:

- They are both compatible.
- In the event of destruction of either location, you are assured of one complete set.
- An audit trail can be maintained.
- One copy of each data update is secured in a third location.

2.1.15.1 Application System Documentation

Your application system documentation should include:

- A complete list of all application systems including:
  - User and data processing contact responsibility
  - Processing requirements and scheduling

  - Terminals required
  - Level of personnel for operation
  - List of current personnel

  - Applicable programs and data files
  - Documentation available:

  - The person responsible for the documentation
  - Location
  - Backup (location)
  - Control of documentation

- Up-to-date list of developments (made or in progress), program maintenance, and test work completed

  - Personnel assigned to each activity
  - Hardware and supplies requirements

- List of purchased or leased systems (complete list of vendors and contacts)
- List of all operating personnel, their addresses, and their telephone numbers

2.1.15.2 Technical Support Documentation

List of all system software. Prepare a checklist of each application documentation and tapes (or disk packs) necessary to support the application.

Note: A copy of the priority listing should be included in the documentation at the backup site to alert the facility what has to be loaded at the initiation of emergency procedures.

2.1.15.3 Communications

List all teleprocessing alternatives and the necessary revisions to existing configuration of the data entry network.

2.1.15.4 Backup Facilities

List all alternative run capabilities and the applications operations to be performed. Include where applicable:

- Alternative company site
- Time-sharing service
- Service bureau
- Use of other internal computers
- Cooperative facility site
- Vendor cooperative site
- Commercial backup site (hot site)
- Shell site
- Other

2.1.15.5 Other Computer Activities

List the following:

- Alternatives for word processing operations
- All programs, applications, and so on
- All tapes, diskettes, and so on required to support activity
- All run specifications
- The location of backup data and programs

2.1.15.6 Database Management Support

List all disaster coordination activities developed (or proposed) with the database group.

2.1.15.7 Vendors, Contracts, Contacts

List all vendors, addresses, phone numbers, and normal and emergency contacts.

2.1.15.7.1 List of All Personnel

List addresses and telephone numbers, including phone numbers for call annunciators, for all personnel.

2.1.15.7.2 List of All Support Services

List emergency contacts and phone numbers for all support services.

2.1.15.8 Documentation (Database Administration)

List all of the following:

- Security measures to protect database (on-line as well as library) activities
- All personnel involved in database administration and activities
- Procedure manuals that outline organizational policy and procedures relating to database administration and security
- All general documentation concerning database(s):

  - Systems/application on database and new systems proposed and in progress for the following year
  - Size and projected size
  - All test databases
  - All production databases

Note: Test and production databases should be kept separate.

- All vendor-supplied programs and database information together with description and contacts
- All vendors who have agreed in writing to support recovery by maintaining updates, modifications, and so on and will supply duplicate programs

2.1.15.9 Documentation (Office Support Services)

Documentation for organizational policy and procedures should include the following emergencies:

- Fire explosion
- Chemical spill
- Earthquake (if applicable)
- Tornado
- Flood
- Civil disturbance
- Bomb threat
- Evacuation routes
- Shutdown procedures (building, computer)

The organizational policy and procedures should also cover *security* prior to, during, and after the emergency event. It should also cover security in the event the area is under the Federal Emergency Management Agency (FEMA), or subject to special evacuation conditions resulting from a nuclear power plant or other hazard. A copy of the evacuation preparedness plan should be made part of this plan.

2.1.15.10 Emergency Communications: Postal and Special Services Plan

List the following:

- Post offices servicing the facility
- Local messenger services
- United Parcel Service (UPS)
- Federal Express, Purolator and/or any other specialty carriers
- Telephone company
- Private telephone company (anyone that can assist in recovery of communications)
- Trucking firms servicing company
- Western Union

At the bottom of the page which lists all these services, with contacts, should be a script that can be used to ensure that proper information is communicated. Sample:

> This is the _____ company calling. Our existing building at _____ is no longer in operation. Please deliver all (mail, packages, etc.) to _____.

A series of postcards *preprinted* with similar information should be inserted into the manual along with a form issued by the post office for a change of address.

2.1.15.11 Vital Records Reclamation Team Action

The ability to recover as much of the working hard copy and magnetic media to aid in maintaining data integrity and application recovery is vital to successful continuance of the operation.

In order to ensure that such recovery of vital records is possible, during normal activities, work should be kept in batches in folders with proper identification labels. Attempting to retrieve single pieces of paper is impossible.

- *Vital records.* Vital corporate records should be maintained in original copies either on-premises in fire-resistant safes, or in a safe-deposit facility off-premises. A duplicate copy, preferable utilizing micrographics, should be stored off-premises.
- *Working documents.* Files in offices, in the event of a disaster (fire, flood, and so on), are not retrievable in most cases. Again, reducing these to micrographics on a scheduled basis is vital to the continuance of activity. Micrographic duplicates should be secured off-site. Daily work papers during process should be handled the same way.

This section should specifically outline all processes combining paper and document procedures and the locations (preferably on a drawing) where such activities are performed. Knowing location can aid recovery teams in quick retrieval if the documents are still whole. In the event that a disaster occurs, employee training should include details on how to protect and salvage material at the time of evacuation. New processes have been developed by Dow Chemical Company for the resurrection of water-damaged documents. When retrieving water-damaged documents after a fire, extreme care should be taken not to disturb or separate papers. Even under certain conditions, magnetic media may be salvaged if handled with care.

2.1.15.12 First Aid, Emergency Medical Services, Company Medical Facilities, Hospitals, and Emergency Shelters

Personnel safety is utmost in the event of a disaster. Other considerations include accounting for personnel if the building has to be evacuated and providing a place where medical services and first-aid people can assist injured employees. List all available locations where employees can regroup and be accounted for. Each supervisor should have a list of personnel to ensure that everyone has left the premises. Sending rescue workers to find unaccounted-for personnel who leave the area or do not follow regrouping instructions is dangerous and unnecessary. A plan (drawing) showing where and to whom employees should report after evacuation is important.

A vacation list should be sent to all supervisors and team members to make sure that those away from the facility do not have to be accounted for. The same condition goes for personnel who are traveling.

2.1.15.13 Mutual Assistance

Most municipalities will provide shelter locations and assistance during an emergency under mutual aid agreements. Arrangements for shelter, assembly points, and so on can be negotiated. In some cases, even empty city- or municipal-owned buildings (e.g., unused schools) can be used for short periods of time.

Other firms in the area may have excess (unoccupied) areas of buildings or warehouses that can be used under mutual agreement.

Contacts should be made with emergency preparedness and action groups in the community to determine what is available, and what arrangements have to be made.

Once these programs are established, the following details should be included in this manual:

- Procedure
- Accountability
- Assembly point
- Location
- Transportation
- Employee responsibilities and assignments

2.1.15.14 Insurance Reporting and Loss Recovery

List information about all insurance companies by the following categories.

- Information:
  - Carrier/agent
  - Loss coverage (description)
  - Effective date of insurance
  - Policy number
  - Exclusions (summary)
  - Emergency telephone numbers and contacts

- Categories:
  - Data processing equipment
  - Employee infidelity
  - Media reconstruction (data, software)
  - Extra expense
  - Business interruption
  - Errors and omissions
  - Loss of items in transit
  - Liability to customers as a result of the electronic funds transfer system (EFTS) activities (financial institutions)
  - Extended coverage, operating off premises
  - Third-party liability, disclosure
  - Outside computer services
  - Lease/rental coverage
  - Transportation coverage (employee transportation to alternate site)
  - Employee losses on company premises

2.1.15.15 Insurance Notification

In the event of a disaster, it is imperative that the insurance carriers be immediately notified to properly assess damages and recoverable losses. In addition, it is important that insurance mechanisms to recover costs (extra expense) and business interruption be done quickly to avoid excessive cash drain on the company. Decisions may be arrived at by adjusters who require negotiation (e.g., an insurance company has the right to determine whether equipment should be rebuilt instead of replaced). The concern of the company may be to recover operations and to have new equipment in place within days; rebuilding may take months. The faster these problems are resolved, the quicker the recovery teams can perform their tasks in rebuilding and supplying the data center. Delays in notifying the insurance company could seriously hamper the time and amount of dollars recovered from claims. Many claims to insurance companies have been substantially reduced because procedures were not followed (or in force) to properly protect equipment before or after an adverse condition. During the data-collection phase, it is advisable to have an insurance carrier loss prevention engineer review the facility and report on findings. If you have a "clean" facility, claims by adjusters can be negated by having, on record, positive documentation. A public adjuster may be necessary to assist in establishing a claim.

2.1.15.16 Recovery/Repair and Restoration

Once the details are completed with insurance people the recovery team can start to work. (*Note*: In many cases insurance carriers will give immediate approval for restoration in order to avoid further damage or deterioration of equipment and facilities.)

List all those who will be involved in the restoration of the home site:

- Building engineers
- Contractors (construction, electrical, plumbing, electronic, communications)
- Cleanup services
- Public garbage/private waste disposal firms
- Public utilities (water, sewerage, EPA, and so on)
- Dumpster leasing firms

Note: The restoration and reconstruction group is dependent upon the support of corporate financing in order to pay or obtain credit. All of this must be established prior to a disaster.

Further details are covered under the recovery team action plans at the workshop.

- *Office support*. List all needed supplies, furniture, forms, materials, calculators, terminals, and so on. List all vendors (see municipal and industry cooperative support as developed by the Federal Emergency Management Agency for local municipalities and civil defense organizations). Obtain agreements and list dates of agreements with organizations who have and will provide office recovery space.
- *Personnel*. List all personnel, and indicate vacation schedules next to each person's name, address, telephone number, and so on.

2.1.15.17 User Coordination/Cooperation

The ability to recover from any form of disaster that will cause an interruption of the data processing activity is complex and requires a unique team of dedicated people. In fact, the authors would love to avoid any dependency on any other group to achieve the results the plan is designed to produce. We're sorry to say, however, that this is impossible. The coordinated efforts of a coalition of data center personnel, user groups, management, and so on is essential to the plan's success. We have listed the many other groups that will be involved directly or indirectly with the design, development, implementation, and testing of the disaster/recovery plan.

- Auditing (internal and external)
- Building management
- Engineering (architect or designer of the facility)
- Finance
- Labor relations
- Legal
- Medical services (internal and/or external)
- Personnel
- Public relations
- Transportation

2.1.16 Plan Preparation Workshop

The Disaster/Recovery Administrator will determine the most appropriate time for setting up the workshop so that individual groups can develop their plans. The plan model has been established by the Table of Contents Review section. The Disaster/Recovery Administrator's responsibility is to guide the plan preparation team(s) through the writing phases of the plan in a workshop environment.

2.1.17 Testing the Plan

The Disaster/Recovery Administrator, data center management, and the various user groups will establish a phased-in process for testing segments and ultimately the entire plan under disaster-simulated conditions at the backup site. The test plans have to be defined in the plan.

2.1.18 Plan Test Monitoring

The Disaster/Recovery Administrator and data center management will establish test criteria as part of the plan development. The internal auditing department, or an impartial referee, should monitor each test, review the procedures, and make suggestions for modifying the plan based on the test results. Each test should be carefully debriefed to determine those areas of activity that have to be enhanced.

2.1.19 Personnel Training

When the completed plans have been approved, it is necessary to provide training for all participants, team leaders, alternates, team members, and employees with regard to how the plan will perform and affect their group. The training procedures are to be established by the Disaster/Recovery Administrator for each team leader, and it is the responsibility of the team leaders to train their groups. Training is one of the most important aspects of the planning process, since each individual must understand how his or her actions link the entire plan into one cohesive, structured program.

2.1.20 Maintenance of the Plan

Periodically, and more so on a specific schedule, the plan has to be updated to reflect existing conditions, change of personnel, review of critical applications, and so on. The Disaster/Recovery Administrator is responsible for surveying each of the groups involved in the plan and updating the changes as quickly as possible. Subsequently, change sheets reflecting modifications should be distributed to those team leaders who have entire plans or sections of the plan in their possession.

PART 7

Section 1
PROGRAM PROFILE: CORPORATE BUSINESS INTERRUPTION RECOVERY PLAN

OVERVIEW

Management has viewed disaster/recovery planning as a technical issue and thereby delegated the development of plans to technicians, thus the past data processing involvement in the planning process.

In this section, we take the broader view, and in order to assist those concerned with maintaining the continuity of overall business performance, we have provided the ammunition to direct towards management to ensure the widest possible application of the disaster/recovery plans.

This section includes material that can be readily incorporated into reports and memorandums to describe the project that is to be undertaken.

PERFORMANCE AND PROCEDURES PROFILE: ESTABLISHING A CORPORATE PREVENTION SECURITY AND DISASTER/RECOVERY PROGRAM

Executive Summary

The business and computer capabilities and technology that have been implemented within your corporation have provided the basis for immeasurable contributions to growth within the organization. The deliberate and systematic application of this technology and business knowledge has, in turn, contributed to the improvements, credentials, and confidence of the business entity by its customers. In addition, it has further advanced the ability of the organization with respect to productivity, economic growth, and organizational stability.

This program is based on the premise that the unmistakable influence of computers, and the dependency of all business operations on technology in contemporary trends and outcomes, will not abate. This dependency, together with other computer and noncomputer activities, brings to the forefront and identifies an array of issues that confront the organization for years to come. Therefore, it is important that present and future opportunities and constraints associated with technology and the use of technology dependency be substantially addressed by this organization.

The basic issues of computer operations continuity have been identified, and strategic plans have been set in place to ensure that the organization's missions involving computer technology can be resolved in the event an undesirable, unexpected adverse incident occurs. *The partial or whole resolution of these technical problems—the ability to recover from such an adverse incident that will cause an interruption of the information flow—addresses only part of the problem.* The issues of business continuity in the event that other activities of the institution are affected, transcend or cut across specific substantive activities and acknowledged business operation areas of application. The problems that affect the present and future opportunities and pose sufficient constraints on normal business functions must be identified, and plans and strategies must be put in place to ensure the continuity of activity of the business performance can be maintained.

In addition to the requirement of the business mandate and mission, management must also address the legal constraints and implications associated with the failure of one or more activities of the institution.

Addressing the Corporatewide Problem

Critical decisions regarding the adequacy of "security" safeguards within the sensitive applications and business unit operations of the organization must be made by authorized managers and based on reliable technical information. Computer fraud prevention, detection, deterrents, investigation, loss recovery, and risk management comprise only a fraction of the existing problem.

The introduction of computers into the daily operation of the organization has substantially obscured the issue that management (managers) are still responsible for protecting *all* the organization's vital assets and resources, including, but not limited to, its information systems. Although computers can drastically affect the "security" and viability of any organization, and more so in the financial services field, the entire organization is vulnerable to adverse conditions that will affect the normal flow of business activities. It is recognized that computers increase the complexity of systems; thus, such systems are more difficult to understand and protect. The interrelationship and interaction of centralized, dispersed and stand-alone systems, as well as many noncomputer functions and support services, pose sufficient risk to business activities so that they, too, must be viewed as a means for potential loss that will affect business performance.

Safeguards, although complex within the computer environment, are equally vital to all other business functions. The ability to recover from any adverse incident that will interrupt (for any period of time) these functions is as important as those addressed in computer performance.

To counter these problems, management needs techniques and methodology to assess and cost-effectively improve their prevention, security, and recovery posture.

Embodied in the program is the essential realization that the organization is in business to do much more than protect itself against computer failure. Ignoring what are known to be clear risks involving the entire organization and their ability to provide the basic services does not qualify as due diligence and reasonable care.

Therefore, it is necessary to approach the issues in an organized manner, which will basically include:

1. Recognition and understanding of the problem
2. Management policy, standards, and requirements
3. Task force (steering committee) establishment, directives, and mandates
4. Prevention, detection, and security (deterrents)
5. Disaster/recovery (contingency) planning
6. Loss recovery and legal considerations
7. Implementation (training, testing, and maintenance)
8. Inclusion in the minutes of the board of directors:
    a. The ability to perform with less than a quorum vote
    b. Disaster succession
    c. The ability to perform corporate activities outside the incorporated area (state) if such are the laws or inclusions in the present minutes
9. The establishment of the Disaster/Recovery Management Team as the hierarchy of the organization until the disaster has been normalized.

General Outline of Work of Project Initiation Performance

The best way to view the project on computer and corporate prevention, security, and disaster/recovery strategic planning is to consider it as a form of *quality control* for the "security" of the business, sensitive applications, systems operation, and functions (those with significant potential for loss) that will create a high business loss or impact substantially on the organization's ability to perform its services. The critical decisions regarding the adequacy of "security" safeguards must be made by a coalition of authorized managers and company components based on reliable administrative, business, and technical information. It is the purpose of this project, as we foresee it, *to define the roles of each participant organization based on its specific mission objectives*, critical success factors, and critical activities while forming a cohesive, organizational and viable steering committee for the purpose of certification, accreditation, and user "security" requirements. These criteria are required for the administrative, business, and technical evaluation, utilizing certification and accreditation. They are also used as management's audit tool for the approval of each planning phase.

The primary function, and the role of the multiparticipant steering committee, is to act as a vehicle and provide methodology for information resource "security" enforcement. In addition, it provides the guidelines for "security" certification and accreditation, computer interaction and interfacing guidance, as well as expands the aspect of certification and accreditation activity into a general business information assurance program to ensure that all functions of the organization perform within prescribed "security" requirements as established by management's policy, standards, and procedures.

These program concepts and the responsibility of the participating units provide the methodology to:

1. Establish the parameters and define the organizational structure for a coalition of organizational functions and participants into a cohesive steering committee.
2. Establish a "security" program.
3. Establish policies, charter, standards, procedures, operations, and activities.
4. Perform a certification and accreditation methodology for all participant organizational groups.

The following sections summarize the approach that management will be required to take prior to preparing for this activity.

Preliminary Activities

Phase 1: Establishing the Program

There are six major issues that need to be addressed.

1. Policies and procedures
    a. Management information program directive
    b. Program manual/standards and policies
2. Roles and responsibilities. (*Note:* Each participant operational unit may have different titles for specific functions; here we attempt to establish a uniform approach utilizing common terminology. All roles enumerated are functional.) Each participant unit from each functional group will have the following members:
    a. Senior executive officer
    b. Project manager

c. Operations-applications/systems manager
   d. Security evaluator
3. Definitions of entities involved in the program
4. Concerns regarding the organizational structure
5. Scheduling
6. Staffing, training, and support
7. Steering committee
   a. Senior executive (chairperson)
   b. Alternate chairperson
   c. Participant coordinator/administrator (project director)
   d. Associate project director
   e. Senior members of participant functional unit of the organizations

Phase 2: Performance Activities

In Phase 2, we will establish the policies, standards, procedures, and guidelines for the certification/accreditation of a corporate steering committee to evaluate plans, approve security provisions, monitor tests, and verify plan maintenance.

1. Certification
   a. Planning (policies, procedures, and methodology)
   b. Data collection (information on data collection supplied)
   c. Basic evaluation
      (1) Security requirements evaluation methodology
      (2) Security function evaluation
      (3) Control implementation determination
      (4) Methodology review
2. Detailed evaluation
   a. Functional operation
   b. Performance
   c. Penetration resistance
3. Reports and finding methodology
4. Accreditation processes
   a. Prevention
   b. Security
   c. Disaster/recovery planning
   d. Backup evaluation
   e. Emergency response (procedures)
   f. Mission control
   g. Primary recovery processes
5. Recertification and reaccreditation methodology (dealing with environmental change)
6. Evaluation techniques for certification and accreditation
   a. Risk assessment
   b. Validation, verification, and testing
   c. Security safeguard evaluation
   d. Role of EDP and internal audit

Note: The establishment of security-disaster/recovery steering committee is vital to the overall success of the completed plan. If the steering committee has not been formed, participants should outline the structure for the formation of the committee based on the preceding outlined activity.

Initiating a Corporatewide Disaster/Recovery Plan

Project Purpose

The organization, over the years, has substantially addressed the subject and planning requirements for contingency planning as it relates to data processing operations. This program has been supported by management to a limited degree, and the success or failure of this program can be directly attributed to that level of support.

Functionally, the contingency disaster/recovery plan for data processing is a stand-alone function. Although it addresses the requirements for supporting critical applications that will ensure maintained continuity of data processing operations of the organization in the event a catastrophic event should occur, the problem of failure on the part of one or more functional units of the organization has only been superficially addressed. The emphasis placed on data processing technical dependency has necessitated this approach; however, a realistic need to address corporate, regional, and branch office contingency planning exists. There also exists a need for the overall consideration of developing such plans for every functional unit in the organization. These include, but are not limited to:

1. All computer-dependent operations (central computer support)
2. All stand-alone computer-dependent operations
3. All noncomputer operations
4. All regional and branch office activities
5. All international activities

The approach and methodology discussed here is perhaps one of the few places this concept is presented. It is our objective to create as stringent a priority on this type of planning, as that addressed in data processing. Without total organizational participation in the actual writing of the plan, there is no plan. The final product will be required to interface and interact with existing organizational policies, standards and procedures, and the information resources and processing capabilities of the company.

Overview

This overview is intended to provide senior executive management, data processing policy managers, information resource managers, data processing technical managers, data processing security and contingency planning managers, the user community, the internal and external auditing functions, as well as all functional units of the companies' operations management with a summary and guide to the process the company will have to address in order to achieve a comprehensive program for the development of a business interruption recovery plan (contingency disaster/recovery plan) for data processing and all other executive and functional units and service divisions of the company.

Expanded Scope of the Program

Security risks threaten every aspect and the very existence of an organization. Over the past years, a great deal of concern has been directed toward developing, implementing, and testing plans for the recovery of data processing operations. The deep concerns that have directed security and disaster recovery extensively towards the computer operations have obscured the issue of a disaster severely affecting part or all of the user community and noncomputer activities. Managers are responsible for protecting the organization's vital resources and have an obligation to take prudent steps to also protect its information systems.

First, while computers have radically changed the methods of doing business, computers themselves are dependent on the organization as a whole in order to provide their unique support service, and vice versa. Second, computers have increased the complexity of operations and systems while allowing us to expand these operations. However, they are still dependent on the functional units of the organization for their activity. This complexity of activity has led to an even greater dependency on noncomputer and computer-based services and safeguards for both of these functions; it has also made their ability to maintain continuity more complex. Some examples of changes in organizations that have to be addressed relating to computers, noncomputer- and computer-based operations, and the "security" problems raised by these changes are as follows:

1. Data assets are more centralized, whereas computer support functions are dispersed in branches, users, and in other cases mini-data centers and information processing systems not integrated with central systems.
2. Decisions are based on information provided by data processing activities (central or dispersed), and these functions cease if operations are interrupted at either the originating site or wherever the data processing operation(s) may be.
3. Functional units of the organization, regardless of their dependency on computer operations, must be capable of performing their services.
4. If noncomputer, administrative, and general business activities required to support the organization fail owing to a disaster, the same concerns that have motivated plans to "back up" and provide disaster/recovery plans for computer activities must be developed for these other business functions.
5. Organizations' management structures and business methodologies are radically shifting to accommodate computer technology. This has often resulted in fuzzy allocation of management responsibility and resources, and has not considered the efforts and resources that are necessary to maintain the overall organizational structure.

To counter these and other problems that exist beyond the "computer," managers will establish techniques and strategies during the workshop and implement cost-effective programs to improve their corporate "security" and contingency disaster/recovery planning posture as an addition to, not in place of, data processing disaster/recovery planning.

Management's Considerations in Corporate Disaster/Recovery (Contingency) Planning

Management for every organization must address what the consequences of losing their operating environment and resources could be and consider their exposure. The four primary areas of exposure that management should review are:

- Financial loss
- Legal implications and responsibilities
- Business activity interruption
- Loss of business and/or opportunities

It is important that a plan be developed and the facility organized in such a manner that any catastrophic event will have the least impact on the continuity of business operations.

Each functional unit of the organization should be involved in the interactive session, "Maximum Allowable Downtime/Cost of Downtime." Identifying, in stages (time), how an individual business function will degrade, will certainly highlight the function's computer dependency.

The Approach to Management Considerations in Prevention and Security Planning

The primary objective of any contingency plan is to identify the area of risk and the levels of prevention, security, and operational policy, standards, and procedures that will reduce (or limit) the chances and effects of a disaster (catastrophic event) beforehand. Therefore, the plan preparation team(s), at the workshop, will tailor the "plan" to their particular organizational unit and organizational structure.

The plan should address those aspects of prevention and security that relate specifically to the organizational units for whom each individual plan, as well as the master plan, is being developed.

Components of Business Interruption Recovery Plans

The subject matter and components to be addressed within the scope of the plan are:

- Planning scope
  — A realistic evaluation of need
  — Legal obligations (regulatory and civil)
  — Cash flow maintenance
  — Customer service
  — Competitive advantages
  — New product and/or services introduction
  — Administrative activities

- Logistics and operations control
- Ongoing project control
- Branch, agency, or division communications
- Personnel relations
- Purchasing functions and vendor relationships
- Shareholder and public relations
- Organizational credibility
• Plan methodology
- Prevention
- Security
- Assumptions and considerations
- Critical operations and/or services
- Scope
- Recovery requirements
- Required resources (critical resources)
- Strategies
- Detailing recovery procedures
- Emergency response plan
- Backup plan
- Staffing and responsibilities
- Maintenance of plan
- Testing of plan
• Plan components review
- Funding and maintenance of the plan (management commitment)
- Personnel assignments
- Interfacing of a division, branch, and so on with the plan
- Other organizational department activities
- Prioritizing activities
- Interfacing of activities with data processing
• Internal audit requirements
- Verification, certification, and accreditation of security practices
- Establishment of vital records program
- Planning and allocation of resources
- Monitoring of tests
- Verification of plan maintenance
• Development of the plan (project definition)
- Definition phase (objectives and assumptions)
- Functional requirements phase (data gathering and decisions)
- Testing and development phase (evaluation of alternatives)
- Implementation phase (creating the plan)
- Testing and monitoring phase (postdevelopment review)
- Maintenance phase (updating the plan)

Personnel Participation: Team Assignments and Responsibilities

An organization's commitment to develop functional unit plans based on mandatory policy requirements for achieving the level of backup and recovery required to sustain the critical operations of a business, as well as other critical services, is imperative. The following is a list of some of the affected departments:

• Security
• Medical
• Public relations
• Transportation
• Buildings (building maintenance)
• Financial
• Legal
• Auditing
• Personnel
• Engineering
• All functional (operational) divisions of the organization
• Interfacing regional and worldwide activities

In-depth discussions should be held to define each group's responsibilities, interaction, and interface to determine backup and recovery operations.

Functional Requirements Review (Fact Gathering and Decisions)

One of the first responsibilities of the corporate disaster/recovery team is to gather from the various executive, administrative, and operational groups as much information as possible. This would include the activities of each department, where the operation is computer dependent, and how the operation interfaces with the data processing department. The data-collection (fact-gathering) activities are covered in other sections of this book. The following are a few of those already mentioned:

• Operations documentation
• Methods and procedures information
• Determination of critical operations and resource support requirements
• Personnel requirements
• Technical support documentation (data processing support requirements; mainframe or stand-alone systems)
• Administration Documentation
• Office services documentation
• Space, furniture, and supplies

Organizational Activity for Initial Response

The following list can be used to review the actions and ongoing responsibilities of the Disaster/Recovery Management Team:

• Team definition, charter, and actions
• Disaster evaluation
• Declaration of a disaster
• Mobilization of teams
• Control center initiation
• Ongoing disaster/backup/recovery responsibilities
• Operations monitoring

- Neutralizing disaster status
- Review (debriefing) of disaster/backup/recovery activities

PLAN PREPARATION WORKSHOP
Functional Plan Components

The prompt and timely recovery of an organization's corporate and functional operations from a loss of capability depends on the availability of a broad spectrum of resources. The procedures necessary to restore operations are defined as the "plan."

The functional plan is composed of four basic plan components:

- Continuity of management at the time of a disaster
- Emergency response plan
- Backup operations procedures (action plans)
- Recovery action procedures
- Normalization of disaster procedures (return to the home or a new site)

Functions, Operations, and Services Criticality: Requirements

Not all functions, operations, and services are critical to an organization's survival at the time of an unexpected, undesirable, or adverse incident. The steering committee will establish the methodology for identifying critical applications and prioritizing the work performance vital to the operation of the company. This selection methodology should be based on critical applications, the period in which disaster occurred, and the time mandated for recovery of critical applications. This will include, but is not limited to, a definition of criticality based on:

- Function, operation, and service
- Computer- or noncomputer-dependent activity
- Executive, administrative, production, and financial activities
- High customer profile (public relations)
- Legal implications (regulatory, civil, and so on)
- Unacceptable period of loss of availability
- Minimum and maximum facilities and resources requirements
- New development project
- Deadline requirements to fulfill public announcements

Records Management

Executive, administrative, operational, and special function activities have "vital" records within the scope of their operation. The ability to manage, protect, preserve, and recover these records (computer—central or stand-alone—noncomputer, source documents, and so on) is extremely important. In fact, any record that establishes or maintains the identity, financial status, or business structure can and should be considered a *vital* record. In a pure definition, a vital record is any document or group of documents necessary to ensure the survival of the business. Records management review will include, but is not limited to:

- Classification of documents
- Retention requirements (based on management policy, legal requirements, business functions, and so on)
- Source document control
- Backup procedures for documents
- Off-site storage systems and procedures

Facilities, Office Equipment, Supplies, Forms, and Communications

Each office environment requires a broad range of materials, equipment, and communications systems (from voice to data processing lines) in order to perform daily tasks. Every aspect of the specific function will be analyzed in order to ensure that these requirements will be addressed in the plan development phase. In the context of equipment, materials, and supplies, the program will dwell heavily on the alternatives and strategies to provide adequate *space* to perform required tasks. Communications require special consideration; and backup and/or disaster initiation strategies will be discussed and developed.

Other Resource Requirements

The following is a list of other resource requirements:

- Personnel and personnel policy
- Supplies and forms requirements (special conditions)
- Transportation
- Power and environmental systems requirements
- Documentation, policies, procedures, standards, and so on

Other Considerations

Also to be considered are recovery and backup strategies:

- Acceptable backup and recovery strategies
- Service degradation
- Internal backup recovery capabilities
- External backup and recovery strategies
- Personnel training

Cost and Expense Considerations of Disaster/Recovery Strategies

Review the methods to summarize the cost of measures for disaster/backup/recovery strategies.

Risk Assessment: Evaluating Probability of Occurrence

Establish a definition of terminology as it relates to the organization. Develop loss potential scenarios and probability of occurrence evaluations.

Other Plan Components

Other components of the disaster/backup/recovery strategies include:

- Insurance evaluation methodology
- Certification and accreditation
- Plan testing
- Plan maintenance

ONE-DAY BUSINESS INTERRUPTION RECOVERY PLAN WORKSHOP
Phase 1: Training Methodology
Participants

Main Office

Participants should include (but should not be limited to) Main office: Executive staff representation, departmental/user managers, group or division systems development managers, representative from the auditing (internal) department or a representative of financial services, internal security officer, personnel department representative, manager training, medical director, manager of human resources, and any other group manager and/or alternate of special operations located within the main office activity that will be affected by the loss of any facilities. Those data processing personnel directly involved in the disaster/recovery planning operations should also be participants.

*District or Regional Offices/
Special Services Organizations*

Representatives from special operations, facilities, security and (as above) not located at the main office.

Branch Offices

Managers, assistant managers, or any representative who would be involved in the recovery operation of any branch office.

Outline of Presentations and Interactive Discussions
Presession Data Collection (Advance Questionnaire)—Management Information

1. Profiles of individuals participating in the workshop
2. Job title and organizational, division, subsidiary, or regional, district, or local branch responsibility
3. Description of general business functions of the department, division, subsidiary, or regional, district, or local branch
4. Description of computer utilization (list of applications and/or direct/on-line activities)
5. Departmental overview on critical applications and services, length of time of acceptable outage, and estimated cost of outage (per day, week, month)

Other aspects of the preplanning data collection phase cover identifying threats, hazards, and vulnerabilities; facilities evaluation; feasibility studies; risk analysis; emergency procedures; personnel policy; security; management considerations; standards, procedures, and policies; and level of existing plans.

6. Management leadership
7. Assignment of responsibilities
8. Initiation of controls
9. Employee, manager, and supervisor training
10. Incident record keeping and management reporting
11. Employee awareness programs, public relations, and so on
12. Records management procedures
    a. Identification of vital records
    b. Classification of records
    c. Off-site storage

Operations Information

1. Organizational (department, branch, subsidiary, and so on) structure
2. Departmental (internal) description of the area of responsibility
3. Critical status relative to organizational activity, revenue generation, and interface activities with other departments
4. General description:
   a. Personnel (amount)
   b. Personnel in critical positions
   c. Space requirements
   d. Equipment requirements
   e. Communications requirements (computer and non-computer operations)

A One-Day Preplanning Workshop for Company Disaster/Recovery Planning

(This program will precede the four-day plan preparation workshop in order to bring all participants up to date on the plan preparation and requirements methodology. It will be aimed primarily at those participants who have not participated in the three-day preplanning seminar.)

The introduction of the concept of business interruption recovery planning beyond the scope of data processing operations is relatively new and innovative to many organizations. Although there has been the implementation of some elements such as emergency procedures, most companies have not recognized the importance of preplanning in order to maintain the continuity of operations and the integrity of information and service flow. This type of planning has its base in regulatory requirements of various federal and regulatory laws, rules, and regulations.

Beyond the requirements and effects of regulatory requirements, the necessity for the company to sustain normal business functions in order to support the financial base, customer commitments, and shareholder confidence are the basic tenets of cohesive organizational activity. The loss of any key functional unit, for whatever period of time, can and will exert hardship not only on the affected area, but ultimately throughout the entire structure. Although many organizations perceive that an interruption affecting stand-alone activity will not have an impact on other activities, they soon find that minor interactions between functions can, over a period of time, create a domino degradation of all corporate services and activity.

In most organizations today, the focal point of many functional activities has, within a functional unit capability, to be supported at the data processing operation. This focal point becomes extremely sensitive, and when external (user) activities cease to provide a continuity of expected input, the interaction and data-gathering process will cease to function under its preplanned program.

Therefore, the overall scope of the one-day workshop is to highlight, emphasize, identify, and protect those company activities that will, in the event of an adverse, unexpected, or undesirable incident, provide the methodology, discipline, structure, procedures, and action plans to back up and recover those functions that have been affected. It will also prepare participants for the following four-day plan preparation workshop.

Developing a business interruption recovery plan requires the total support of management, an assumption of responsibilities by *every* organizational unit (computer or noncomputer), and a comprehensive predecision document.

The completion of the preplanning documentation will assist the instructor/consultant to direct the efforts of the data center and users, and assist in the formalization of disaster/recovery plan workshop materials specifically for the organization for which this program is being held . . . in-house.

Administrative, Functional, and Operating Units

The following administrative, functional, and operating requirements will be addressed during the workshop. (They are not necessarily listed in order of importance.)

- Corporate directors
- Executive management
- Finance and accounting
- Human resources
- Personnel
- Data processing (centralized, distributed, or PC/mini stand-alone activities)
- Manufacturing
- Purchasing (procurement)
- Administration
- Operations
- Marketing (research and operations)
- Sales and sales management
- Research and development (products, services, computer applications, and so on)
- Engineering (by specific discipline)
- Worldwide, national, regional, district, and local offices
- Contract administration
- Logistics and distribution
- Public relations (corporate, product, or service)
- Security administration
- Service departments
- Field support service
- Customer education
- Banking and financial services (specialty conditions)
- Maintenance services

This list is by no means complete. By providing information on the preplanning questionnaire, each organization has the opportunity to identify the functional and support operations of the company. This accomplishes the first approach in identifying the interrelationships, dependencies, and interactivities that are necessary to support the organization at the time of any crisis.

Emergency, Backup, and Recovery Planning

Definition

Disaster/recovery planning encompasses four distinct activities:

- Continuity of management
- Emergency plans
- Backup plans
- Recovery plans

Scope

Identify and determine, relative to the scope of the plan, interfaces, interactions, and interrelationships among functional units of the organization.

Assumption

1. Threats, hazards, vulnerabilities, risk, and exposure
2. Disaster scenarios
3. Critical recovery strategies

Emergency Plan

The emergency plan will address issues such as:

1. Strategies
2. Evacuation
3. Shelter
4. Containment
5. Supression
6. Testing
7. Maintenance

Backup Plan

The backup plan provides a structured mechanism for utilizing personnel skills in an orderly fashion, thus ensuring the continuity of business performance.

1. Activities (computer)
2. Activities (noncomputer)
3. Identification of critical activities
4. Identification of discretionary jobs
5. Identification of required environments
6. Identification of critical resources
7. Backup strategies
   a. Portability
   b. Manual procedures
   c. Other installations (offices and common activities at other locations)
   d. Temporary services
   e. Customer support (compatible activities)
   f. Distributed operations
   g. New facilities
   h. Homework
   i. Communications
   j. Shell (empty space)
   k. Office rental services
   l. Mail, courier, or messenger services
   m. Records management and recovery
   n. Personnel
   o. Multiple alternatives
   p. Data processing support

Careful attention to detail must be made when consideration is given to a worst-case environment and when both business support services and data processing are affected by an adverse incident. These interrelationships and backup considerations for the establishment of both environments are key issues to backup and recovery.

Recovery Plan

The recovery plan will address the following topics:

1. Decision factor considerations (time sensitive)
   a. Temporary restoration of critical mission capability
   b. Permanent restoration of critical mission capability
2. Personnel considerations (location versus loss of personnel)
3. Physical plant plan (Centralized versus decentralized)
   a. Space requirements
   b. Space availability (suitability to meet organizational requirements)
   c. Public image (service versus appearance)
4. Equipment and supplies
5. Communications
6. Transportation (emergency versus permanent)
7. Vendor cooperation
8. Administrative support
9. Data processing plan interface and support

Phase 2: Data Collection

Upon completion of the preplanning seminar, the participating group(s) will be required to assemble a number of documents and special information about their specific operating environment, and identify and prioritize operations (jobs), as well as determine on a corporate basis specific, critical performance/service requirements. The material requested will be used and integrated into the business interruption recovery plan.

Phase 3: Business Interruption Recovery Plan Preparation Workshop

The plan development phase is designed to integrate and interface the disaster/recovery plan of the data processing operation with the business interruption recovery plan program. It is the intention of the instructors/consultants to review the company's existing plans and make recommendations concerning the integrated program. Owing to the complexity of the *two* programs (plans), disaster recovery for the data processing operation and contingency planning (business interruption recovery plan) for corporate activities, it is suggested that the disaster/recovery plan for the data center be converted into "book" format so that the business interruption recovery plan will function in a totally coordinated manner.

The following plans, systems, procedures, and so on will comprise the four-day workshop.

FOUR-DAY PLAN PREPARATION WORKSHOP

Business Interruption Recovery Plan

The following outline covers the major categories within the data center master plan. (Using master plan as a guide each participant group will develop its own unit plan.)

1. Purpose
2. Scope
3. Assumptions
4. Strategies
5. Prevention
6. Security
7. Emergency response and procedures
8. Action plans
9. Testing
10. Training
11. Maintenance
12. Addenda

Plans

This document provides the guidelines, strategies, and action plans to be executed when an adverse incident has interrupted the normal flow of business performance.

- Data center master disaster/recovery plan (one document)
- Corporate business interruption recovery plan (one document for corporate headquarters; separate plans to be developed and interfaced with both plans for each separate function in corporate headquarters, divisions, regional offices, branches, and so on)
- Data center users guide (establishing standards and procedures for users; describing the methodology of user interrelations with the data center)

Other Standards, Procedures, and Plans

- Emergency response plans
- Data center standards and procedures manual
- User standards and procedures manual

Plan Preparation Methodology

Each corporate group participating in the program would prepare its plans based on a *master plan* model. The master plan has been developed to cover every aspect of prevention, security, and contingency disaster/recovery planning. The functional unit team will modify, massage, or rewrite each paragraph in the model so that the master plan represents its specific function and operation, and interfaces with the data center master plan (where applicable).

Upon completion of the program, each group will have completed an *organizational subplan* that will cover its specific function as well as address the problems that may exist if one, two, or more corporate, administrative, or service operations are adversely affected by any form of disaster (for any period of interruption). Although the plan addresses disaster backup and recovery from a worst-case condition, smaller, shorter-term interruptions may be dealt with by utilizing subsets of the master plan components, action plans, and strategies.

Section 2
COMPUTER AND NONCOMPUTER DISASTER/ RECOVERY PLANNING QUESTIONNAIRE

OVERVIEW

When the computer goes down, how can the user community maintain the continuity of business performance? Many users (business units) have little or no interest, and more important, have little or no desire to be involved in disaster/recovery planning. As they fill out the following questionnaire, however, they will see how important disaster/recovery planning and stand-by provisions are.

Step by step we have provided a road map that will guide the user through the maze of disaster requirements. If the disaster/recovery administrator is looking for allies, have the users fill out this questionnaire.

Following this questionnaire, the user will need a model plan to follow. The model presented in this section deals with a bank branch or branch office of a financial services organization. This plan covers noncomputer operations to be interfaced with the data processing plan or used as a stand-alone plan in the event an unexpected incident immobilizes a specific business function of the company.

Our company is presently engaged in the development and implementation of prevention/security disaster/recovery plans for the data processing operation. In order to achieve maximum recovery, as well as provide sufficient support services, it is *vital* that each function (computer and noncomputer) provide information relative to its capability and requirements necessary to properly support the continuity of operations and performance of the company. Each user or functional department must address the problems inherent in sustaining critical services at the time of a disaster. This entire activity is totally supported by management (management should prepare a letter supporting the entire process of disaster/recovery planning and endorse the process for development and maintenance as well as providing authority to the group).

The enclosed questionnaire addresses many of the important functions and information that are imperative in the development of a user plan. It also provides the guidelines and activities necessary to evaluate whether critical activities, regardless of the extent of the disaster's impact on the company, can be sustained.

Although details and responses may be documented by subordinates in your department, it is important that the entire survey be reviewed and approved by a senior member of the operation, department, and so on. Information submitted will not be included in the data center or corporate disaster/recovery plans unless such authority is provided by senior management.

USER DEPARTMENT COMPUTER AND NONCOMPUTER DISASTER/RECOVERY PLANNING QUESTIONNAIRE

- Please complete one questionnaire per [user, department, functional operation, bureau, agency, manufacturing system].
- Please return to:

  Name _____

  Company _____

  Department _____

  Address _____

  City _____ State _____ ZIP _____

- This form must be completed and returned on or before _____ [date].

- The Disaster/Recovery Management Team is prepared to assist you in properly completing this form (e.g., in advising current backup arrangements for you, your department, and so on). Please contact the Disaster/Recovery Planning Administrator or Coordinator members below:

  Name _____ Ext. _____

  Name _____ Ext. _____

  Name _____ Ext. _____

- All information provided will be considered highly confidential, and will be handled with top-priority security. The information will be used for disaster/recovery planning purposes only.

- This document was prepared by:

  Name _____ Title _____

  Name _____ Title _____

  Name _____ Title _____

- Date prepared: _____

- Authorized by:

  Name _____ Title _____

  Date _____

- Reviewed by:

  Name _____ Title _____

  Date _____

- Approved by:

  Name _____ Title _____

  Date _____

- Effective date: _____

- To be reviewed

  ☐ Monthly   ☐ Quarterly   ☐ Semiannually   ☐ Annually

- Senior management approval:

  _____ Date _____

I. PERSONNEL OFFICE SPACE AND SPECIAL CONSIDERATIONS

A. Please provide a floor plan of your present facility. Show:

1. Number of people in department (by area)
2. Number of fully enclosed offices and location of each
3. Number of offices made up of partitioned units
4. Special workstations (office pods)
5. Storage facilities for:
   (a) Files (number of file cabinets and general list of contents)
   (b) Supply areas
   (c) Security storage (if applicable)
   (d) Special communication devices (if applicable; include intercoms if not part of telephone system)

B. What is the minimum office space necessary in the event of a disaster? _____

C. Suggestions on establishing an employee reporting center in the event of a disaster (manning, communications, suggested location). _____

D. What is the minimum number of personnel required to reestablish operations? (Consider various times of the week, month, quarter, end-of-year.) _____

E. Evaluate and list time frames necessary to increase the number of personnel on an ongoing basis. _____

F. How long will it take you to reestablish minimum support operations? _____

G. Could you continue to operate at an alternate facility for a period of:
   ☐ One week   ☐ Two weeks   ☐ One month   ☐ Three months
   ☐ Six months   ☐ One year   ☐ Permanently
   ☐ (Other) _____

H. What are your requirements for interaction with other departments? _____

I. What do you estimate it would require in financial resources to reestablish your entire activity in the event your area was totally demolished? _____

J. What other departments' assistance would you require to aid your group in reestablishing the functions of your department? _____

K. If they experienced a disaster, could you assist them (or any other department) in providing space, facilities, equipment, communications? _____

L. Relative to the preceding question, what areas of activities would you scale down in order to support other departments' operations? _____

II. **OFFICE AND PERSONNEL REQUIREMENTS**

A. What is the area (square feet or square meter) that your unit occupies? (See Section I.)

B. Have you recently taken an inventory of the following items within your department? (Please indicate in the spaces below each question the quantity, type, availability, and location of each device or item.)

1. Telephones (location, extension numbers, number of lines per phone, user's name as well as other telephone accessories). _____

2. Facsimile equipment (model, company, availability, dedicated lines, supplies). _____

3. Teletype equipment (include telex number). _____

4. Copy equipment. _____

5. Typewriters (manual, electric, electronic). _____

6. Terminals. (Identify the person and activity for each of the following. Also list the general activity for each unit.)

   (a) On-line _____
   (b) PC or mini cluster _____
   (c) PCs (stand-alone) _____
   (d) Mini- or micro-stand-alone _____

(e) Portable PCs _____

(f) Word processing stations _____

7. Modems (type, style, vendor, quantity). _____
_____
_____

8. Printer(s) (type, quantity, vendor). _____
_____
_____

9. Desks. _____
_____
_____

10. Computer furniture. _____
_____
_____

11. Supplies needed for at least one month's operation for each departmental unit or individual, whichever is the most appropriate (pens, pencils, desktop supplies, and so on).
_____
_____
_____

12. Special forms for computer and noncomputer operations. (Consideration should be given to using manual procedures in the event of a disaster, until operations can proceed normally.) _____
_____
_____

13. If you resort to manual operations, are there personnel within your department to perform these activities and train others? _____
_____
_____

14. Have you reviewed manual procedures with the internal auditing department? _____

III. IDENTIFICATION OF CRITICAL SUPPORT SERVICES, SYSTEMS, AND FUNCTIONS FOR COMPUTER AND NONCOMPUTER ACTIVITIES

Please complete one questionnaire per system, function, department, operation, facility, and so on.

A. Department name: _____

B. What is the system, function, operation, support service?
_____
_____
_____
_____

C. How can it be identified (job name, function, charge code, department code, and so on)? _____
_____

D. Who should the Disaster/Recovery Management Team contact for inquiries?

Name _____ Title _____

Phone number _____ Ext. _____

E. Provide a brief description of the function or system.

_____
_____
_____
_____
_____
_____
_____
_____
_____
_____

IV. EFFECT OF LOSS

Please determine the maximum downtime you could tolerate. (See question B.)

- ☐ Processing facility
- ☐ General activity support service

(Use a separate sheet.)

A. What would be the effective loss of the [check one or more]?

1. ☐ Central processing facility (resource)?
2. ☐ Support service?
3. ☐ Telecommunications network?

on ☐ Your operation?
- ☐ On other departments?
- ☐ On support services?
- ☐ Company performance?

(Use a separate sheet if necessary.)

_____
_____
_____
_____
_____
_____

B. What is the maximum downtime (of 1–3 in question A) during which the system, support function, operation, and so on can be unavailable before the following are affected:

1. Departmental work flow _____

2. Customers or clients _____

3. Revenue and/or cash flow _____

4. Accounts payable (vendor relations) _____

5. Company credentials _____

6. Company goodwill _____

7. Contract fulfillment _____

8. Long-term financial viability _____

9. Legal consequences _____

V. DISASTER/RECOVERY PROCEDURES

A. Do you already have a disaster/recovery plan for:

1. Your data processing system

    (a) Minicomputer?

    (b) PCs?

    (c) On-line terminals?

2. Other elements of your department's business services?

B. When was the last time the plan was ☐ updated/☐ tested?

_____

If such a plan exists, please make a copy and forward it to the disaster/recovery management planning team.

C. Do you have any form of manual system to fall back on in the event your department or the central data processing facility is affected by a disaster?

_____

D. For on-line terminals (systems), do you have any form of batch processing fallback?

_____

E. *Note:* Please indicate the most current schedule of *off-site* backups that could assist you in the event the database is lost.

☐ Daily ☐ Weekly ☐ Biweekly ☐ Monthly

Describe department conditions that would affect recovery of lost data sets at the latest level.

_____
_____
_____
_____

1. What would you require on-site for immediate recovery?

_____
_____
_____
_____

2. Please indicate your preference:

☐ The latest backup tape copies would be held off-site (at some cost in immediate dataset recovery).

_____
_____
_____
_____

☐ Additional backups would be taken (at your department's expense) off-site.

_____
_____
_____
_____

☐ Source documents would be stored off-site at your department's expense.

_____
_____
_____
_____

☐ All source documents would be logged by date and time of entry.

_____
_____

☐ Other suggestions to ensure recovery up to the time of an adverse incident?

F. Please list your data storage requirements.

G. Please list data set–dedicated volumes by serial numbers.

Note: If data sets cannot be identified by serial number, please list the data set naming conventions. Also list details on any tape data sets (including generation data groups) which are required.

VI. **OTHER CRITICAL RESOURCES REQUIRED**

A. Special forms (this should comply with the company's forms inventory manual):

1. Form types: _____ _____ _____ _____

2. Monthly volumes: _____ _____ _____

Note: Please indicate where forms are stored.

☐ In-house _____
_____
_____
_____
_____
_____

☐ Off-site _____
_____
_____
_____
_____
_____

☐ With the vendor _____
_____
_____
_____
_____
_____

B. Please list by job, job run instructions, operational documentation, and so on. *Note:* Use as many sheets as required.

C. List printing facilities (requirements).
   1. Letter perfect. _____
   2. Dot-matrix (near letter perfect, draft form)

   3. Laser printing facilities
      (a) Flashes overlays: _____   _____
      (b) Monthly volumes: _____   _____

D. List software.

E. Other. _____

VII. RECOVERY REQUIREMENTS: PERSONNEL

A. In the event of a disaster, can you provide personnel to assist in the backup and recovery procedures?
   ☐ Yes   ☐ No
   Please attach a list of personnel; include their special training or disciplines.

B. Can you make personnel available to assist in a test (or tests) of the backup and recovery plan?
   ☐ Yes   ☐ No
   Please attach a list of personnel; include their special training or disciplines.

VIII. SPECIAL INFORMATION

A. What is the normal processing cycle of critical support services (daily, weekly, and so on)?

Please list application and cycle. Use additional sheet(s) where required.

_____
_____
_____
_____
_____
_____
_____
_____
_____
_____

B. Is there any particularly critical period (i.e., is the system more critical at a particular time of the month, year, and so on)?

_____
_____
_____
_____
_____
_____
_____
_____
_____

IX. BACKUP PROCESSING

A. Please indicate by system the most critical system and time it requires to be processed at the backup site. (Please use a separate sheet if required.)

_____
_____
_____
_____
_____
_____
_____
_____
_____

B. Are you aware of the data backup procedures used by the data center? ☐ Yes ☐ No

C. Do you have regular copies of all critical data stored off-site?

D. Have you recently reviewed the data that was stored off-site?

Note: Since backups of disk volumes are kept off-site by the data processing center, the preceding items apply mainly to tape data sets.

X. **BACKUP PROCESSING SITE**

Additional costs may be incurred in maintaining adequate resources at an alternate processing site. Would the criticality and impact of the loss of these systems to your business activity justify a contribution to support the backup system?

XI. STRATEGIC PLANNING: CURRENT DEVELOPMENTS

Are there any current developments which might affect current backup processing requirements?

XII. SUGGESTIONS: GENERAL INFORMATION

Have you any suggestions, or are there any other items you think we should consider in the development of a disaster/recovery plan?

XIII. PCs, MINICOMPUTERS, AND OFF-LINE ACTIVITIES

Please describe your present off-line data processing activities and the type of work being performed.

PART 8

INTERFACE DISASTER/RECOVERY MINIPLAN MODEL: BRANCH OFFICE FOR A FINANCIAL INSTITUTION

OVERVIEW

In today's technological business society, the interface, interrelationship, and interaction of central operations and satellite facilities are as far away as the keyboard and monitor of the computer. Interference with computer performance can result in the cessation of all business activity from the central operation out. What happens if the satellite operations encounter a disaster? It could mean the loss of central information operations, resources, and communication. It could also be an actual disaster that strikes the district, divisional, or satellite facility. Despite all safeguards, disaster in today's business environment is an everyday reality.

The miniplan model that follows, together with the user and noncomputer operations questionnaire in Part 7, provides the guidelines for satellite facilities to develop a plan for disaster recovery and continuity of performance.

FINANCIAL INSTITUTION (BANK OR FINANCIAL SERVICES)

Interface Disaster/ Recovery Miniplan Model

The outline that appears on the following pages provides the development structure for a financial institution or financial services organization that operates one or more branch offices. The design, implementation, testing, and maintenance of a disaster/recovery plan is vital to the continuity of this financial institution and its financial services operations. Although many of the backup and recovery efforts are directed to noncomputer operations, it is important that consideration be given to those computer functions and communications that support branch activity and become an integral part of the plan.

This outline reflects the strategies and action plans that have to be developed by branch management based on four assumptions:

1. The data center has experienced a disaster and is partially or totally destroyed (long-term outage).
2. The data center has experienced a short-term interruption resulting from a disaster.
3. The branch office has experienced a disaster and is partially or totally destroyed.
4. The data center has experienced a short-term interruption resulting from a disaster, and the building is salvageable within a reasonable period of time.

The outline presented here should, in combination with the guidance and efforts of the organization's disaster/recovery plan and the various disaster/recovery teams, provide a mechanism for gathering the necessary data to build a comprehensive, workable plan.

Forward (for Use in the Plan)

The presence of a disaster/recovery plan for the [*name of organization*], [*name of branch*], located at _____
_____, presupposes the probability of an unexpected or undesirable event that could render the branch (division, subsidiary, and so on) functionally inoperative for some period of time. It assumes that the branch will be disrupted, and that the normal business activity located at this site will not be performed. This period of disruption will be of sufficient duration to incur potential losses. It is the intent of this plan to provide workable strategies and action plans so that the services of the branch can be restored in the shortest period of time, and that our primary objective is to bring the operation back to normal.

The plan reflects the commitment of branch management to provide the resources, planning, and know-how necessary to manage not only the risk of loss, but also the probable loss, and maintain the continuity of branch and customer services.

The purpose of this disaster/recovery plan is to reduce the number of decisions which must be made when and if an "adverse incident" occurs. The preplanned strategies and action plans will guide the staff of the branch as well as other disaster/recovery teams in response to any catastrophic event and the recovery of branch operations following the incident.

The assumptions that this plan can be based upon are numerous. They include, but are not limited to, disastrous events that can affect life, property, and continued business operations. The four assumptions previously outlined establish the basis for the "worst-case" condition and, therefore, provide the stimuli for preparing this plan. The responses, strategies, and action plans have been developed based on the worst-case condition; any lesser incident will require a subset of the initial high-impact plan. In most cases, the process would be considered complicated, as each incident (assumption) has a level of catastrophe. By defining the responses, strategies, and action plans for backup and recovery in a worst-case situation, every consideration is made in advance.

Objectives of the Branch Disaster/Recovery Plans

The objectives of the plan will be directed towards the results that can be anticipated and achieved with the plan's existence. The ability to quickly recover the branch operation and resume customer services from an adverse incident are our prime objectives.

The plan provides the following:

1. The specific goals to be achieved.
2. The methods (responses, strategies, and action plans) and the means to fulfill the goal.
3. The ability to measure or determine the degree by which the goals have been accomplished.

This is a predecision document. Since the nature and extent of disruption cannot be determined in advance, it is necessary, as we have previously stated, to develop the plan for a total-destruction condition.

These objectives are "user" oriented and establish the requirements to return the "branch" to its normal operation.

Goal or Objective (to be developed by each organization)
Example:

1. *Assumption*: The branch of _____ located at _____ has been totally destroyed. All records, supplies, equipment, and so on have been ruined or destroyed and cannot be recovered.
   Goal: The branch, because of its sensitivity to customer service, and the number of transactions performed at this branch, must be put back into operation at a convenient site within 72 hours.
   Method: Several alternatives exist. The specific alternative which must be initiated will most probably fall within the following alternatives.

   Alternative 1: Customers will be directed (temporarily) to perform their banking requirements at another branch of the bank. Banking hours will be extended there to accommodate the increased volume of people. Personnel not involved in the recovery of the affected site will be assigned to the alternate site, and shift rotation will be established to utilize personnel effectively.
   Alternative 2: The affected branch will establish its operations at the main office of the bank, and customers will be notified to use the most convenient branch.

   Alternative 3: The _____ bank has agreed to provide _____ hours after normal banking operations to allow the affected branch to use its facilities (if compatible) for our customer services.

2. *Assumption*: The branch at _____ has been temporarily disabled by fire, water, and so on, and an operating posture can be returned within a day, according to the disaster/recovery management team.
   Goal: Return the activity of the branch to normal operating condition within its own facility in 24 hours.

   Method: _____
   _____
   _____

   Alternative 1: _____
   _____
   _____

   Alternative 2: _____
   _____
   _____

   Alternative 3: _____
   _____
   _____

Activation of the Plan: Assumptions

1. *When will the action plans of this plan be invoked?* This disaster/recovery plan will be initiated upon the occurrence of an unexpected or undesirable event (adverse incident). It is the responsibility of the senior member of the organization on-site at the time of an incident to initiate the emergency response plan and notify the Disaster/Recovery Administrator of the incident once all actions have been taken to minimize the affects of the incident.
2. *How long will the disaster/recovery plan remain in effect?* Once the DRMT has declared a disaster and initiated the plan, the strategies, and the action plans, its assignments, responsibilities, and authorizations will remain in effect until all the conditions that resulted from the plan have been resolved and normalized, and the DRMT has notified all those teams involved that the state-of-emergency action no longer exists.
3. *Who is responsible for the recovery of the affected site?* This plan provides for the establishment of teams that are solely responsible for the implementation of all recovery processes of the affected site. The initiation of a disaster condition by the DRMT invokes the recovery team (organization) composed of, but not limited to, the Salvage and Insurance Teams, the Supplies Team,

the New Facilities Team, the New Equipment Team, and the New Hardware Team. These teams will remain mobilized until recovery is effected. Recovery is considered effected when the specific site is in an occupiable and workable condition and all services have been reestablished.

4. *What are the preplanning activities required to ensure a prompt and timely recovery?* Each branch office will prepare sufficient documentation to cover all the requirements and activities that are necessary to ensure prompt resumption of normal activities. The individual teams will prepare specific action plans that can direct backup and recovery procedures in order to achieve the goals of their specific branch plan. They will also coordinate their activities with the corporate disaster/recovery plan.

The specific teams will prepare working lists of all equipment, hardware, supplies, and so on that are required to effect backup and recovery of the affected site. Plans will be discussed with various vendors and purchase orders will be issued once the DRMT has initiated the state of emergency and has declared a disaster.

All major decisions necessary to recover the site (total destruction) will be made as part of this plan, and those individuals given the responsibility to perform these tasks will have the authority to invoke portions of, or the entire plan, based on the level of disaster.

Definition of Disaster Terms

Note: These are examples. Each organization will develop its own disaster definitions as established by the policy of the organization.

The [*branch*] of the _____ will initiate the branch disaster/recovery plan under either one (or more) of the following conditions:

- An unexpected or undesirable event that will disable, partially or completely, the branch office facilities of the [*bank*], whose site will be unoccupiable or inoperative for a period of 24 hours or more.
- An unexpected or undesirable event that will impair the branch of the [*bank*] from performing its normal business activity on a day-to-day operation.
- An unexpected or undesirable adverse incident that has resulted in injury (or death) to one or more persons who are employed (or are customers) of the [*bank*].
- For purposes of testing the plan to the satisfaction of management or insurance companies with the intent to prove that recovery is possible.

For the purposes of this plan, the following basic adverse incidents are covered (this list is to be developed by the branch disaster/recovery team):

- Power loss
- Heating or air-conditioning loss
- Fire
- Water (flood)
- Weather and natural phenomenon
- EPA disaster incident
- Sabotage
- Bomb threat

This plan also encompasses:

- Mandatory computer operations (partial or critical resumption at an alternate site)
- Necessary computer operations and services (partial or critical services resumption at the bank's main headquarters or appropriate branch)
- Desirable computer operations and services (full recovery of all services at an alternate or current site)

EMERGENCY RESPONSE PLAN

Incident Reporting

The senior manager, supervisor, or employee on the scene at the time of an adverse incident is responsible for the following (ranked in order of importance):

1. *Protecting the lives and welfare of all personnel and customers.* Evacuate all people from the premises and ensure that all people are accounted for who are on the premises at the time of an emergency.
2. *Mitigating damage.* If possible, after considering the safety of the people, every effort should be made to reduce the damaging effect of any incident.
3. *Notifying emergency services.* Notify the appropriate emergency service as soon as possible. Quick response, especially in the case of fire, can substantially reduce the damage. Where central alarm systems are installed, these should be activated immediately upon recognition of an incident.

   The following list of emergency phone numbers should be prepared and posted at convenient locations around the branch. (Add other numbers if appropriate.)
   a. Fire department
   b. Police department (if 911 is not used)
   c. Emergency medical services (if 911 is not used)
   d. Local hospital
   e. Poison information center
   f. Central alarm main office
   g. FBI
   h. Private guard service (for securing building)
   i. Power company
   j. Telephone company
   k. Heating and ventilating service and repairs
4. *Removing vital records.* The amount of reconstruction of data will be reduced if it is possible, during evacuation, to salvage whatever records you can.

5. *Securing the vault and/or building.* In the event of an emergency, and if it is *safe*, vault doors, cash drawers, and other areas where cash or valuables are kept should be locked (secured). However, the safety of people should never be compromised.

    If evacuation is required, owing to an incident not occurring at the branch office but in an adjacent area, time will permit securing of all areas and/or removal of all valuable data, documents, and items. Removal should not be required if the area occupied by the bank branch is not directly threatened.

6. *Notifying management.* The Disaster/Recovery Management Team should be notified of any adverse incident as soon as possible. Initiation of the pyramid calling tree to team members only on a need basis should be made only after the DRMT has had the opportunity to evaluate the situation and determine the level of disaster.

7. *Conducting a building sweep (accountability).* The senior manager (or designated alternate) should remain on or near the affected site to guide emergency services and advise of any special conditions that exist.

Evacuation Procedures

Training personnel as to the proper procedures for the orderly evacuation of the premises is vital to the safety of employees as well as customers. Individuals should be assigned specific duties concerning evacuation procedures.

Note: In high-rise buildings and high-traffic areas, specific designated assembly points should be selected so that each group can be accounted for.

Important: Consideration should be given to the assignment of each individual in the branch to perform (remove) one activity or vital document. These should be selected in the order of criticality (priority). (Removing a typewriter which is easily replaced is certainly a lower priority than removing the discs in the teller terminals.)

Personnel Safety

Personnel safety and the welfare of any individual on the premises is the most important part of the emergency response plan. It is the policy of the bank that no individual will be required to perform any service beyond the assurance of his or her own personal safety, and where possible, the safety of the people for whom they are responsible.

As part of this plan, the following assignment chart should be prepared specifically indicating the duties and tasks to be performed in the event of an emergency.

ASSIGNMENT CHART

Individual	Task to Perform	Item to Recover
1. _____	_____	_____
2. _____	_____	_____
3. _____	_____	_____

Emergency Response and Procedures: Examples

The emergency response and procedures require specific detailing by the Disaster/Recovery Administrator and the appropriate teams with regard to the way each of the items are to be accomplished. Following are two examples. Other emergency response and procedures include, but are not limited to:

- Power failure
- Weather and natural phenomenon
- Heating or air-conditioning failure
- EPA disaster incident
- Sabotage
- Bomb threat

Emergency Response and Procedures in the Event of a Fire: Example

(Detail the steps to be followed in the event of a fire.)

1. Evacuate, if necessary.
2. Report the fire.
3. Evacuate (remove) critical items, if possible.
4. Follow fire-fighting procedures.
5. Shut down power.
6. Shut down the air-conditioning.
7. Report the incident.
8. Initiation of the DRMT and appropriate teams.

Emergency Response and Procedures for a Water Emergency: Example

(Detail the steps to be followed in the event of any incident occurring from water.)

1. Determine the source of the water.
2. Power-down all equipment that could be affected by water.
3. Remove all vital records or any items that could be seriously damaged by water.
4. Inform the insurance company (to be done by the DRMT).
5. Report the incident (call emergency repair services).
6. Clean up the water. Call in service companies that specialize in cleaning and drying out equipment, and so on.
7. Salvage as much as possible to avoid further exposure to water damage.
8. Initiate appropriate disaster/recovery teams.

Establishing a Disaster/Recovery Operation and Control Center

In the event of a disaster which prohibits the occupancy of the primary site, it is necessary that the branch recovery team has a location where it can coordinate all the activities until a backup site and/or recovery can be put into operation.

The branch Disaster/Recovery Administrator must establish this control center and provide for all the necessary communications and personnel to perform the specific tasks required.

Table of Contents

The following table of contents review and brief description will allow the branch Disaster/Recovery Administrator to select those sections that are applicable to the specific branch operation. These guidelines will assist the satellite office in developing the branch plan.

1. *Selection of a backup site (preplan).* A number of alternatives have to be established that will ensure rapid transfer of bank branch operations. Owing to the erratic nature of the office real estate market, it is important that an update be made periodically of available sites.
2. *Movement to the backup site (preplan requirement initiated at the time of the disaster).* The plan must include all the requirements to establish the alternate site. Various branch and main office disaster/recovery teams will be assigned to call upon the appropriate team(s) to transport, prepare, and establish the alternate or backup site. The following teams, among others, will be involved in the backup operations:
   a. Transportation
   b. Supplies
   c. New Facilities
   d. Administration
   e. New Computer Hardware
   f. Auditing
   g. Public Relations
   h. User Liaison
   i. Vendor Liaison

Addenda

The items outlined in previous sections of the branch disaster/recovery plan composed information that remains fairly static. The addenda section of the plan will contain information that may have to be modified periodically to reflect the mobility of people, as well as items such as equipment, supplies, and forms, and the altering of procedures. This allows the plan administrator to easily maintain the plan as changes occur. The following items would be appropriate for inclusion in the plan as addenda.

1. *Pyramid calling list.* The calling tree (or pyramid list) lists all the individuals (team leaders) who are assigned specific tasks in the event of a disaster.
2. *Recovery team (staff).* Each of these team leaders will be responsible for preparing the team assignments and appointing specific individuals to carry out necessary action plans as well as alternates. There are various teams involved in establishing the new facility as well as recovering the backup site. The team members include, among others:
   a. Branch Disaster/Recovery Administrator
   b. Disaster/Recovery Management Team
   c. Communications Team leader
   d. Transportation Team leader
   e. User Liaison Team leader
   f. New Facilities Team leader
   g. Operations Team leader
   h. Business Continuity Team leader

Supplies, New Equipment, and Hardware

Prepare a complete list of all office equipment (desks, chairs, file cabinets, and computer and noncomputer office equipment)—every item that is required to operate the branch. This list should include the specific vendors (with alternates) from whom these items are readily available.

Forms and other specialty items require special attention since there may be a long lead time to obtain certain checks and computer forms. A complete forms book should be prepared with a sample of each item and its supply source. (See "Forms, Checks, and Special Documents," on page 270.)

Vendor Contact List

This section should contain a complete list of all vendors. Include anyone who provides a service. Group like items together, or make a separate sheet for each item.

VENDOR LIST

Name, Address, City, State, Zip	Item	Contact	Office Phone	Emergency Phone	Lead Time
____	____	____	____	____	____
____	____	____	____	____	____
____	____	____	____	____	____
____	____	____	____	____	____
____	____	____	____	____	____

Critical Services

List all the critical services that must be performed by the branch and the equipment, supplies, communications, and so on necessary to support these critical services. It is understood that these critical services do not necessarily constitute all the mandated services to maintain customer satisfaction, but it should be understood by bank management that if more services can be provided, they will; these, however, have priority.

The following are the priority list for the branch located at _____.

1. _____
2. _____
3. _____
4. _____
5. _____
6. _____
7. _____
8. _____
9. _____
10. _____

Forms, Checks, and Special Documents

This section should list all the specialized forms in use at the branch and their source.

1. The source of the form
2. The cost of the form
3. The minimum backup supply required for one week, thirty days
4. Lead time necessary to obtain forms
5. Alternate sources
6. Present rotating inventory
7. Location of warehousing

Contracts and Agreements

This section contains contracts which have been established with vendors for special equipment, supplies, new hardware, copy machines, phone service (if special system), and so on. All these contracts have been reviewed by the legal staff to ensure that the bank is protected and that the contracts incorporate specific commitments by vendors to support our organization at the time of an emergency.

Verification should be made to determine if contracts include:

- Statements of vendors to replace necessary equipment if a disaster strikes.
- Statement from vendors relative to providing specialized forms, checks, mailers, and so on.
- Agreements from any other user, bank, real estate firm, office recovery center, or any other firm that agrees to provide or perform services or offer assistance in time of distress.

Operations Policy and Procedures

It is impossible to describe all the operations (and services) performed at the branch. However, this section should contain an index to the branch operations, policy, and procedures, and the location of copies. Copies of all branch operation procedures should be kept off-site and updated when new services and procedures are introduced.

Branch Interface with Computer Operations

To ensure continuity of operations from the backup site and/or the new (or reconstructed) facility, assuming the data center is fully operational and that the only affected facility is the branch (user), the branch recovery plan should be included into the data center disaster/recovery plan. This would include only those activities and procedures that provide for the data processing (on-line) services necessary to perform branch operations.

For the most part, the same basic plan would be implemented at the time the data center experienced an outage and the branch would have to be put back on-line when the data processing operation went to backup operations or when the center recovered.

The branch should, at either the backup or the recovered site, be able to store daily transactions and forward these to the data center each day for updating files. Backup disks will be prepared by the branch to ensure that no loss of data will occur.

Staffing, Training, and Maintenance Program

Staffing

The branch (user department) must accept a full role in the development of its own disaster/recovery plan. In addition, the branch management should be prepared to cooperate with staff at various levels as the data center plan proceeds. Part of this activity will involve planning and carefully analyzing decisions as to what branch functions are deferrable, and for how long.

The branch disaster/recovery team, in conjunction with the organization DRMT, can coordinate an effective program for backup and recovery operations.

Training Requires Testing

One of the most important aspects of successful disaster/recovery planning is the continual testing and evaluation of the plan itself. Although it is difficult to perform such tests under day-to-day branch operations, many tests of small segments of the plan can be performed. It is important that branch management recognize that a plan that has not been tested cannot be assumed to work. Branch operations are, historically, a unique facility of the banking operation. Service to the community establishes a responsibility to that same community. Discontinuance of these services not only affects the reliability of the individual bank, but affects the stature of banking institutions as a whole. Suffice it to say that if the plan is not subjected to continual and rigorous management review as well as in-depth testing on a scheduled basis, it will fail when needed.

Test Plans

Devising test plans that adequately and reliably exercise the disaster/recovery plan requires considerable skill and great care. The objective is to provide tests that are entirely realistic while still conducting daily business but also economically feasible.

Conducting Tests

A good argument can be made that the only method to test a plan completely is to actually cease, or otherwise disrupt, operations at the branch for which the plan has been prepared. However, this is seldom practical and quite possibly could, in itself, create actual losses. It is generally only necessary to assume that operations at the branch are disrupted or otherwise not available.

For example, it is not essential to have an actual fire in order to test the emergency response plans and the evacuation procedures. What is needed is an understanding with the fire department and documentation of the specific test procedures to follow in simulating the fire and emergency condition.

Likewise, to test the procedure for moving to, and implementing backup operations at, another location, it is not mandatory to cease operations at the branch site. Rather, it is enough to gather those teams involved and other information required to begin operations at an alternate site. If nothing more occurs than the ability to establish communications on both a data processing and an operations level, a good portion of the test is performed satisfactorily. If the other teams in a "scenario" mode can establish contact with vendors and ensure that all supplies, equipment, and so on can be mobilized (but not actually delivered), another aspect of the test can be deemed acceptable.

Thus, the backup and recovery plan may be tested one strategy, action plan, or resource at a time so as not to disrupt normal business activity. Since in real efforts the backup and recovery will compete with each other for financial and personnel resources, it is wise to periodically perform a test that involves both areas at the same time.

Preplanning

Whether we deal with developing the plan itself, testing the plan, or maintaining the plan, the responsibilities, leadership, team members, management liaison requirements, disaster/recovery functions, and the necessary preplanning must clearly state the tasks and boundaries. The planning will cross department lines, but the activities to be handled for disaster/recovery must be strictly defined.

Plan Maintenance

There is an ongoing responsibility for any plan and more so for a disaster/recovery plan to ensure the plan's capability of providing guidance once a disaster occurs. This requires regular updating of the plan. A series of checklists, similar to those developed by the data processing disaster/recovery plan for the on-going maintenance items, should be developed.

Once the teams have been assigned, each should review its responsibilities and procedures from the point of view of its own "technical" understanding of the assignments and problems involved. Each team, in developing the mechanisms to perform its tasks, should be responsible for the procedures and, where modification is required, discussions with the Disaster/Recovery Administrator to make sure each action is feasible.

As we previously discussed, the ability to provide a backup location will require continuous updating. The volatility of the real estate market will continue to place a burden on the New Facilities Team. And, in developing the plan to ensure proper backup capability, continuous updating will be required.

Changes in operating procedures, as well as additions of new services which require support equipment and personnel, necessitates constant review. The branch Disaster/Recovery Coordinator should be made aware of, or provide his or her own system for updating plan and branch requirements.

PART 9

MODEL DISASTER/EMERGENCY RESPONSE PLAN

OVERVIEW

Every business, industry, institution, or government facility, regardless of size or location, is threatened daily by a variety of potentially devastating events that could impede the successful continuity of business performance. This entire book deals with the plans for recovery from such events. Part 9 deals with the immediate response necessary to mitigate the impact of a catastrophic event. In fact, a minor incident may result in a catastrophic conclusion if there is insufficient preplanning regarding how to handle disasters when they occur.

Whether the lack of emergency response planning is caused by the press of daily business activity or from apathy is unimportant. The fact remains that if a situation should arise that could substantially affect the health and welfare of people, and the destruction of property, and no plans exist to cope with the situation, the bottom line is total chaos.

The following is a plan that provides the model for your organization to formulate one or more plans to deal with unexpected or undesirable events.

The model was not intended to be a complete, detailed emergency response plan that would "fit" your organization's requirements. On the contrary, it is only a guide; there are too many complex conditions and facets to each organization's emergency response requirements. Instead, this model was prepared to define the key aspects to be considered in planning to cope with disaster.

MODEL DISASTER/EMERGENCY RESPONSE PLAN

Universal Emergency Number 911
(Fire, Law Enforcement, Medical)

Additional Phone Numbers Are on Page _____

Emergency Control Center (ECC) Operations

- ECC is in Room _____, Building _____
- ECC members are identified on page _____

Emergency Procedures

- See pages _____

Evacuation Routes

- See pages _____

Plant Shutdown Procedures

- See pages _____

Prepared by

_____
Manager, Facilities and Services

Approved by

_____
Director, Support Services

Emergency Call List
SDC Emergency Numbers

Security .... Ext.
Plant Engineering .... Ext.
Personnel .... Ext.
Emergency Control Center (Bldg. _____ Room _____) .... Ext.

Function	Name	Ext.
Emergency/Disaster Administrator		
Backup Administrator		
Security		
Facilities		
Plant Engineering		
Voice Communications		
Communications Engineering		
Data Communications		
Corporate Computer Facilities		
Research and Development		
Reprographics (Home phone numbers are on file with Security)		

Fire   911 or

Fire Department ....

Law Enforcement   911 or

Police Department ....
Sheriff Department ....
Highway Patrol ....
Federal Bureau of Investigation ....

Medical Service   911 or

Ambulance ....
Paramedic ....
Hospitals: (Name of Hospital) ....
            (Name of Hospital) ....
            (Name of Hospital) ....
Medical Clinics: Mark Hall Industrial (Oxnard) ....
Red Cross ....
County Emergency Services ....

Utilities/Services

Gas Company ....   County Communication Center ....
Electric Utility ....   Sanitation District ....
Water Company ....   OSHA (District Office) ....
Telephone (Local Service, AT&T, GTE, Sprint, MCI, etc.)   Civil Defense Office ....
U.S. Postal Services ....
After Business Hours:
List special phone numbers to correspond with the emergency service utility list

EMERGENCY RESPONSIBILITIES

The president or his or her designee will:
- Be notified of the emergency situation
- Receive status reports from the emergency plan administrator regarding the degree and extent of the disaster
- Declare existence of an emergency
- Authorize evacuation of the work place if necessary
- Approve information releases to the news media
- Review reports and recommendations and make necessary decisions

The emergency and disaster plan administrator or his or her designee will:
- Provide the president with status reports on damage and estimated recovery schedules
- Notify key personnel and emergency organizations as required
- Direct equipment and manpower to the disaster area if required
- Receive status from the emergency scene coordinators
- Prepare written report(s). Include information about
  - Community conditions
  - Plant conditions
  - Employee safety and welfare
  - Emergency team activity

The plant security member will:
- Request assistance from the local fire department, police department, hospitals, and so on
- Report to the emergency and disaster plan administrator
- Deploy security personnel to protect lives and property
- Notify key personnel and emergency organizations
- Assist fire and rescue teams
- Provide vehicular and pedestrian traffic control
- Assist in first-aid treatment
- Make announcements on the public address system
- Oversee the evacuation of buildings and grounds
- Secure vital records and documents

The personnel member will:
- Report to the emergency and disaster plan administrator
- Man emergency first-aid centers
- Direct operation of the first-aid teams
- Supervise removal of the injured from the disaster site
- Establish priority of medical attention to the injured and their removal to hospitals
- Notify OSHA and county coroner in the event of a fatality
- Recommend protective measures, medical assistance, and/or building evacuation

The plant engineering member will:
- Report to the emergency and disaster plan administrator
- Assist plant security in isolating the disaster area and enforcing evacuation measures
- Direct operations of damage control teams
- Order plant shutdown procedures as necessary
- Assess the extent of damage
- Effect temporary repairs
- Initiate cleanup and salvage activities

EMERGENCY PROCEDURES
Emergency Guidelines for Security, Plant Engineering, and Personnel

- Safeguard employees against injury
- Confine and terminate emergency situation
- Administer medical assistance to injured persons
- Prevent damage to products and equipment
- Restrict damage to facilities
- Return to partial or normal operations as soon as possible

Fire extinguishers are shown on pages _____.

Utilities shutoffs are shown on pages _____.

Evacuation routes are shown on pages _____.

This section briefly describes procedures for security, plant engineering, and personnel in the event of the following emergency/disaster situations:

- Accidents
    - Fire/explosion .... page _____
    - Chemical spill .... page _____
- Natural catastrophes
    - Earthquake .... page _____
    - Tornado .... page _____
    - Flood .... page _____
- Threatening situations
    - Civil disturbance .... page _____
    - Bomb threat .... page _____

FIRE/EXPLOSION

When fire/explosion is reported or observed, security will:
- Notify _____ Fire Department (phone: _____)
- Notify appropriate medical services (see page _____)
- Notify emergency disaster plan administrator (ext. _____ or _____)
- Notify emergency response members to report to the scene (see page _____)
- Meet the fire department and emergency crews and direct them to the scene
- Dispatch security personnel for assistance and crowd control
- On direction of plant engineering, effect building evacuation (see pages _____)
- Cordon off the emergency scene and restrict access to persons needed in area

Plant engineering will:
- Report to the emergency scene and take charge
- Order evacuation of all personnel from the affected area
- Direct all damage control operations
- Order shutdown of systems as required to isolate the emergency scene and prevent further damage

Personnel will:
- Report to the emergency scene and provide medical assistance
- Provide first-aid treatment for injured personnel and remove them to the first-aid centers
- Set up a unit at the first-aid center to provide additional first aid, and direct serious injuries to hospitals for treatment
- Report any injuries and extent of damage to the emergency disaster plan administrator

Model Disaster/Emergency Response Plan **277**

LOCATION OF FIRE EXTINGUISHERS — BLDG. "C"

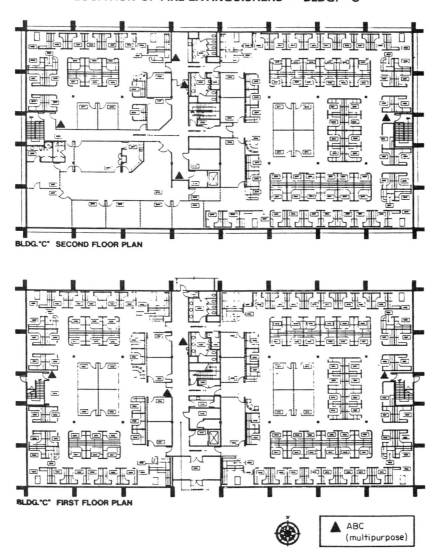

Note: Common floor plans can be used for multiple purposes, including location of fire extinguishers (shown here), smoke and/or heat detectors, security provisions, evacuation routes, etc.

LOCATION OF FIRE EXTINGUISHERS — BLDG. "B"

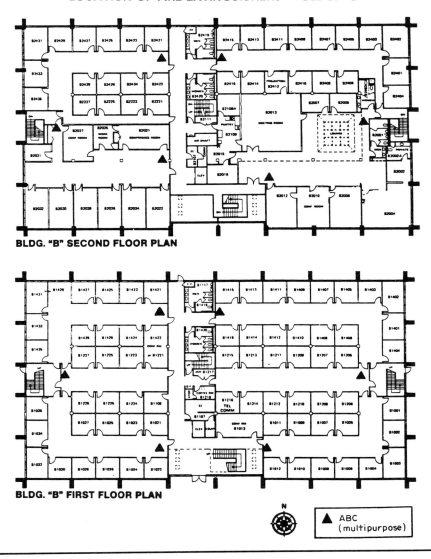

BLDG. "B" SECOND FLOOR PLAN

BLDG. "B" FIRST FLOOR PLAN

▲ ABC (multipurpose)

CHEMICAL SPILL

**When chemical spill occurs within the plant,
plant engineering will:**

- Confine or seal off the spill as much as possible
- Cordon off employees from the immediate area
- Implement any plant shutdown procedures required for safety
- Implement cleanup actions if chemicals pose no hazards or report incident to the fire department (phone: _____)
- Keep the emergency and disaster plan administrator or designee informed

Security will:

- Request medical assistance if necessary
- Set up first-aid centers to assist the injured
- Announce the evacuation of all affected areas

**When a chemical spill occurs outside the plant (e.g., overturned or burning chemical truck),
plant engineering will:**

- Take action to prevent fumes from entering the building (e.g., shut down air-conditioning) and notify the computer facility representative before the air-conditioning is shut down

Security will:

- Report the incident to the fire department (phone: _____)
- Report the incident to appropriate law enforcement agencies (e.g., highway patrol) and ask for advice for first-aid treatment

- Announce to employees to move to areas away from hazardous fires
- Assist in the evacuation or movement of employees

Personnel will:
- Request medical assistance from outside agencies if ill effects are reported (identify symptoms to medical authorities to get information on proper first-aid procedure)
- Set up a first-aid center in a chemical-free area
- Assist in removal of seriously ill persons to the treatment centers

TORNADO

**After danger has subsided,
plant engineering will:**
- Inspect damaged areas
- Order shutdown of systems as required for safety
- Report damage assessment to the emergency and disaster plan administrator
- Direct all damage control operations

Plant security will:
- Oversee the evacuation of employees from damaged areas
- Assist trapped or injured employees
- Notify the fire department in case of fire threat or for rescue assistance (phone: _____)
- Request help from outside medical services if injuries require treatment
- Secure classified material

Personnel will:
- Set up first-aid treatment centers for injured employees
- Assist medical agencies when they arrive on the scene

EARTHQUAKE

**When earthquake danger has subsided,
security will:**
- Establish an alternate emergency control center if the central control center becomes unusable
- Establish communications link with the local civil defense office or the office of emergency services and maintain it as required
- Enlist management personnel to control access and egress to/from the buildings and site
- Assist in evacuation as required
- Attempt to obtain professional medical assistance if necessary

Plant engineering will:
- Order shutdown of all facilities systems immediately to prevent further damage to the buildings or site
- Direct damage control operations; bring all fires under control and extinguish
- Conduct a search of all buildings for survivors and injured personnel; direct first-aid teams to these locations by the safest routes
- Inspect building structure(s) and assess the extent of damage
- Inspect all building systems for damage; restore to operation systems providing life support and communication
- Define safe areas within the building(s) to house personnel unable to leave the site

Personnel will:
- Establish an alternate emergency first-aid center if the normal first-aid center becomes unusable, and inform emergency control center
- Provide first-aid treatment at the scene and remove injured personnel to the emergency first-aid center
- Establish a unit at the emergency first-aid center and provide additional first-aid treatment
- Report the number and extent of injuries to security personnel for relaying information to local authorities
- Request from security personnel updated data on the availability of transportation and hospitals for serious injuries

FLOOD

**For a flood resulting from a fractured pipe,
plant engineering will:**
- Shut valves to stop water flow
- Shut down electrical circuits in areas of heavy moisture

- Drain water from flooded areas
- Dry equipment, materials, and furniture to prevent or limit damage
- Report status to the emergency and disaster plan administrator

Security will:
- Notify the fire department if pumping of water is required
- Assist employees displaced by flooding

**For a flood resulting from an overflow,
plant engineering will:**
- Note conditions and notify the emergency and disaster plan administrator when water level threatens road to and from facilities or the plant itself

Emergency and disaster plan administrator will:
- Recommend to the president that employees be released early from work if flooding threatens to close access roads
- Direct appropriate actions to prevent water from entering the buildings

CIVIL DISTURBANCE

**When notified of a disturbance (labor unrest, protest actions, picketing, and so on),
security will:**
- Determine the nature of the disturbance while avoiding confrontation
- Meet with spokesperson if appropriate
- Monitor events closely
- Make timely reports to the disaster and emergency plan administrator
- Apprise local law enforcement agencies of the event
- Post more security personnel at entrances
- Keep employees away from the area

**When a disturbance threatens to become violent,
security will:**
- Close and block entrances
- Notify local law enforcement agencies if problems arise for employees coming or going from the facilities; take steps to protect employees
- Assign emergency personnel equipped with communication equipment on the roof of the building
- Unobtrusively take film records of the incident
- Provide barricades, screens, and other equipment required
- Call in additional civil or off-duty security/law enforcement personnel if necessary
- Oversee the evacuation of personnel if the disturbance occurs within the facility

The ranking corporate officer or designee will:
- Decide whether to order a plant shutdown or early release of employees
- Issue press releases in conformance with company policy

The emergency and disaster plan administrator will:
- Inform department managers on how to handle problems and about what can and cannot be said to employees

BOMB THREAT

**Upon notification of a bomb threat,
the security manager will:**
- Notify the president (ext. _____) and/or the emergency disaster plan administrator (ext. _____)
- Notify the police (phone: _____), fire department (phone: _____), and FBI (phone: _____)
- Notify _____ Sheriff Bomb Squad (phone: _____)

The president or emergency disaster plan administrator will:
- Apprise the security manager of an authorized evacuation
- Switch the tape cartridge "message for actual evacuation" into the PA system of the affected building(s); the tape cartridge is stored at the security office, building _____, room _____
- Add supplementary information after the taped message is completed

**During evacuation,
security will:**
- Search all areas to ensure total personnel evacuation (giving special attention to the physically handicapped)
- Assist police in its search

- Make emergency equipment available (including emergency lights, ropes, first-aid kit, two-way radios, and sound-powered megaphones)
- Man exit doors, corridors, stairs, gates, and so on for control and orderly movement
- Direct personnel to a safe distance beyond the building or fence perimeters
- Notify Industrial Security Division (phone: _____)
- On locating explosive device, notify the security manager; report progressive development until the danger is removed (phone: _____)

**When the explosive device is located,
the security manager will:**
- Control subsequent investigations requiring cooperation with the FBI and other law enforcement agencies; employees will answer questions as factually as possible

The police department or sheriff's bomb squad will:
- Remove or disarm any suspected explosive device(s)

A corporate officer will:
- Decide when personnel may return to the building(s)

The corporate relations office will:
- Release any information to the news and television media about the bomb threat

EVACUATION ROUTES

Evacuation may be required for several types of emergencies:
- Fire/explosion
- Chemical spill
- Bomb threat
- Sprinkler alarm

Evacuation will be authorized by the president or emergency plan administrator; evacuation will be implemented by the security manager

During evacuation, security will:
- Search areas to ensure total evacuation
- Assist police or others in their search
- Assist handicapped or injured persons
- Man doors, corridors, stairs, and gates to ensure orderly evacuation
- Direct personnel to safe distances

Evacuation routes are as follows:

_____
_____
_____
_____
_____

PLACE YOUR DIAGRAM HERE

PLANT SHUTDOWN PROCEDURES

During an emergency response, implementation of plant shutdown procedures may be required to:
- Reduce hazards to employees
- Minimize loss of essential materials or equipment
- Restrict damage to the facilities

One table should show emergency shutdown actions to be taken by employees associated with the appropriate trades:
- Mechanical
- Electrical
- General maintenance
- Construction

Another table should show actions to be taken for the curtailment of electrical usage; these actions are primarily the responsibility of plant security and plant engineering

Pages _____ show the locations of gas, water, and electricity shutoffs

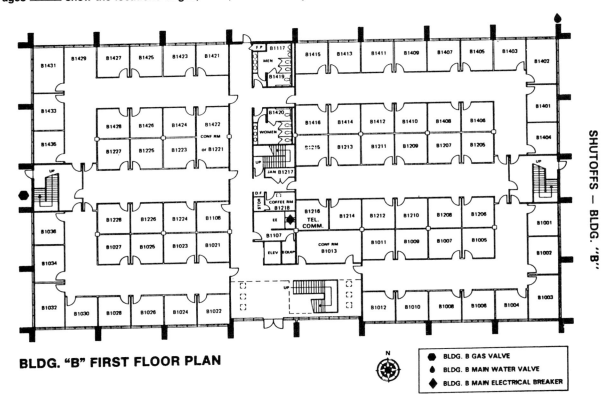

BLDG. "B" FIRST FLOOR PLAN

- ● BLDG. B GAS VALVE
- ● BLDG. B MAIN WATER VALVE
- ◆ BLDG. B MAIN ELECTRICAL BREAKER

SHUTOFFS — BLDG. "B"

Model Disaster/Emergency Response Plan **283**

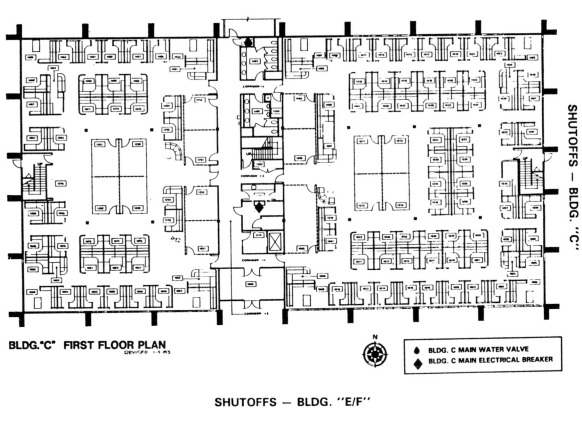

BLDG. "C" FIRST FLOOR PLAN

- ● BLDG. C MAIN WATER VALVE
- ◆ BLDG. C MAIN ELECTRICAL BREAKER

SHUTOFFS — BLDG. "E/F"

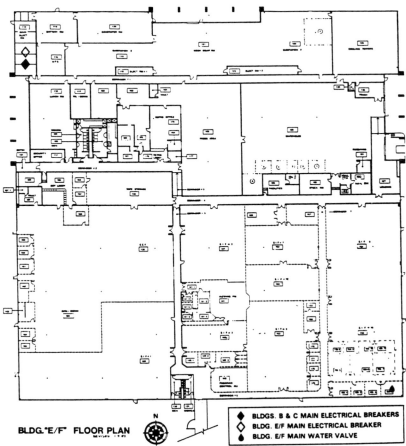

BLDG. "E/F" FLOOR PLAN

- ◆ BLDGS. B & C MAIN ELECTRICAL BREAKERS
- ◇ BLDG. E/F MAIN ELECTRICAL BREAKER
- ● BLDG. E/F MAIN WATER VALVE

PART 10

SAMPLE INTERACTIVE SESSIONS

OVERVIEW

The most productive method to obtain information is to have participants of the preplanning seminar volunteer it. In performing these seminars worldwide, the authors have found that on a one-on-one basis, most companies are reluctant to disclose information about their organization. Yet, in an open classroom environment with the use of interactive sessions, we are able to obtain a wealth of information about the organization. In fact, participants are anxious to show how they can contribute to the success of their project as well as those of other participants.

The few sample interactive sessions, and those listed in the organization and participant profile, provide the seminar leader with a step-by-step process for accumulating data as well as training personnel in disaster/recovery planning. Use these models to develop interactive sessions more applicable in content to your organization.

Protecting an organization from every risk that could result in business discontinuity is virtually impossible. First, it is not practical; second, it is not cost-effective; third, there are more threats than we are able to identify or protect ourselves against. Baseline security* provides the planner with a methodology to implement security provisions without having to spend extensive hours identifying risks and planning countermeasures.

INTERACTIVE SESSION: BASELINE SECURITY TECHNIQUES
Exercise: Organizational Status Test
Overview

The baseline security concept is new. It assists your organization in establishing the policies, procedures, methods, and strategies to prevent disasters. The objective of this exercise is to determine how well your organization ranks in prevention/security and disaster/recovery planning.

To aid you in responding to the status checklist, we have abstracted the definitions for each of the baseline security items for you to refer to while proceeding with this exercise. We have numbered the yes/no boxes so that you can determine a rough percentage of the baseline items presently within your data center.

Important: Having an unenforced policy, or one that was instituted and has eventually disappeared, means that the control is no longer in place. Remember that the level of criticality of a particular control and its implementation is more important than just having a high number of minimal controls. The percentage is only a guideline. Having incorporated 50 percent or 75 percent of these controls, not having them enforced, or having controls that have little effect on the overall activity does not mean you are safe.

For example, having a no smoking/eating control is far less important than having an adequate access control. Therefore, if these two were your only controls, you could say that you have covered 50 percent of the requirement. Yet, we would consider the access control to have a ranking of 10 on a scale of 10; the no smoking/eating control would be considered a 3 or 4. If all the other controls were in place, the no smoking/eating control would be a self-imposed restriction.

1. *Control title*: Assets accountability assignment
   Description: Specific data producers, computer users, and computer center staff are assigned explicit ownership or custodial accountability and usage rights for all data, data handling, and processing capability, controls, and computer programs. This can be done by establishing policy—*establishing* meaning ownership, usage, and custodianship—and requiring that forms be completed and logs be made designating and recording such accountability for data and programs. Copies should be placed in all locations and for specified times. For example, a set of booklets for each data activity area would state ownership, usage, custodianship, and control requirements. Another method is to make this information part of the organization's policy manual.
2. *Control title*: Data accountability assignment to users
   Description: Users are formally assigned the responsibility for the accuracy, safekeeping, and dissemination of the data they handle. If the data processing department does not handle data properly, then it is up to the users to require corrections. Organiza-

---

*This concept appeared in the United States Department of Justice, Bureau of Justice Statistics document, "Computer Crime—Computer Security Techniques." The complete document as revised by the authors is available through DIA*log Management, Inc.

tionally, users provide a data processing department with the resources to assist them with their functions. In terms of controls, users should be able to tell the data processing department what is required in terms of data accuracy, relevance, timeliness, handling procedures, and so on.

3. *Control title*: Separation and accountability of data processing functions

    Description: Holding managers accountable for security in the areas they manage requires that these areas be clearly and explicitly defined so that there is no overlap or gaps in managerial control of data processing functions. Data processing functions should be broken down into as many discrete, self-contained activities as is practical and cost-effective under the circumstances. Besides being a good general management principle to maintain high performance, this control also provides the necessary explicit structure for assignment of controls, responsibility for them, accountability, and a means of measuring the completeness and consistency of meeting all vulnerabilities adequately. Separate, well-defined data processing functions also facilitate the separation of duties among managers, as is required in the separation of duties of employees. This reduces the level of trust needed for each manager. The functions of authorization, custody of assets, and accountability should be separated to the extent possible.

4. *Control title*: Computer security management committee

    Description: A high-level management committee is organized to develop security policy and oversee all security of information-handling activities. The committee is made up of management representatives from each part of the organization concerned with information processing. The committee is responsible for coordinating computer security, reviewing the state of security, ensuring the visibility of management's support of computer security throughout the organization, approving computer security reviews, receiving and accepting computer security review reports, and ensuring proper control interfaces among organization functions. It should act, in some respects, similarly to a board of director's audit committee. Computer security reviews and recommendations for major controls should be made to, and approved by, this committee. The committee ensures that privacy and security are part of the overall information-handling plan. The steering committee may be part of a larger activity within the organization to carry out the function of information resource management. For example, in a research and development organization, an oversight council made up of representatives from organizations that send and receive databases from the research and development organizations was established. They were charged with oversight responsibilities for the conduct and control of the research and development organization relative to the exchange of databases. Especially important were questions of individual privacy concerning the content of the databases.

5. *Control title*: Remote terminal user's agreement

    Description: All remote users are required to sign a user's agreement before they are permitted to use system resources. The agreement covers who will pay for system-related expenses, identifies physical location and relocation of terminals, establishes maintenance and service of equipment, assigns training of users, states hours of usage, instructs on further dissemination of information obtained from the system, details proper usage of the system, assigns physical security of terminals and other equipment, states the service provider's rights to deny service and to inspect equipment, establishes insurance coverage and liability for losses and renegotiation of the agreement, and other related matters.

6. *Control title*: Confirmation of receipt of documents

    Description: The confirmation process consists of verifying the receipt of documents. Confirmations of delivery can be made by obtaining master files of names of input/output documents and their addressees and performing a selection of a sample of addressees by running the master file on a computer separate from the production computer or at least at a time different from normal production work. Confirmation notices and copies of the documents are then sent to the addressees to confirm that the documents are correct and that the documents were received as expected. Confirmation of smaller volumes of documents can be easily done on a manual basis. Receipt forms are used by recipients of particularly sensitive documents and returned to the sender to confirm correct report distribution and encourage accountability.

7. *Control title*: Discarded document destruction

    Description: Input/output documents, including any human-readable documents or nonerasable computer media (carbon paper, punch cards and tape, one-time-use printer ribbons), should be reviewed for potential loss sensitivity and appropriately destroyed when no longer needed. Appropriate protection of material awaiting final disposition should be used. Logging of all actions to ensure an audit trail and adherence to rules is essential. Strict assignments of tasks and accountability are essential. Documents such as obsolete system development materials, test data and manuals, and obsolete criminal histories should be considered.

8. *Control title*: Proprietary notice printed on documents

    Description: Sensitive and valuable documents have a classification (e.g., "sensitive," "private," "proprietary," "confidential," "for authorized parties only") or an explicit warning indicating that the information is the property of a certain organization, that it should be handled according to special criteria, that it is not to be used for certain purposes, and so on. One organization printed "confidential" in the middle of the page; although this made reading a bit more difficult, it prevented people from cropping the record and photocopying it to remove any indication that it was confidential. Other approaches include having the computer print appropriate words on only sensitive output or having preprinted paper with appropriate designations (classifications). (This has the advantage of warning display terminal users that the information should be specially treated.) Policies and procedures establishing and defining the classification of information must also be written and established as company policy.

9. *Control title*: Courier trustworthiness and identification
    Description: Couriers are frequently used to distribute computer output reports to computer users. Couriers must be especially trustworthy, have a background investigation similar to that for computer operators, and be bonded. A new courier should be personally introduced to all those persons to whom he or she will be delivering computer output and from whom he or she will be receiving materials for delivery. Couriers should be required to use signed receipts for all transported reports. Couriers should be required to keep all reports in their personal possession in properly locked or controlled containers. *All users should be informed immediately upon the termination of any couriers delivering or picking up reports.* Couriers should carry special identification to show that they are authorized to function in claimed capacities. *Telephone calls in advance of delivery of highly sensitive reports should be made to recipients of those reports.*

10. *Control title*: Keeping security reports confidential
    Description: Computer security requires the use and filing of numerous reports, including results of security reviews, audits, exception reports, documentation of loss incidence, documentation of controls, control installation and maintenance, and personnel information. These reports are extremely sensitive and should be protected to the same degree as the highest level of information classification within the organization. A clean-desk policy should be maintained in the security and audit offices. All security documents should be physically locked in sturdy cabinets. Computer-readable files should be secured separately from other physically stored files and should have high-level access protection when stored in a computer.

11. *Control title*: Suppression of incomplete or obsolete data
    Description: Dissemination and use of incomplete and obsolete data are prevented or restricted by directive of the organization. This directive must be implemented by receivers of data that are to be processed, converted, or stored by reasonableness checks within application systems and by output control and dissemination activities. For example, in criminal justice information systems, access to nonconviction and arrest data that are one year old or more and do not contain a disposition are restricted to certain types of requesters. The same concept (i.e., if a record is incomplete or outdated, it should not be disseminated) can be applied to applications other than criminal histories. Such data may also be selectively restricted by requester type.

12. *Control title*: Completion of external input data
    Description: If missing essential data are still missing beyond a time limit, take steps to obtain the appropriate data. Within the criminal justice environment, a request for disposition information is issued when a particular record has remained incomplete beyond a time limit.

13. *Control title*: Personal data input/output inspection
    Description: An organization that receives databases from or disseminates them to outside sources should have an input/output control group. This group checks the databases it has received and disseminated. It checks for the inclusion of improper data fields, such as individual names and social security numbers. Also, more sophisticated checking of the relational aspects of the data field is done to determine whether individuals can be identified by combining information from multiple fields. The group screens all files to be received and investigates anomalies. A log is kept of all activity.

14. *Control title*: Human subjects review
    Description: An independent review board (a human subjects review board) reviews all proposals in an organization concerning treatment of subjects in studies. The board is made up of members of the parent organizations: some from the department in question, and some from outside the department. The charter of the board is to determine whether the subjects of a study will be put "at-risk" or "at a disadvantage" because of participation in the study. The manner in which individual privacy (data confidentiality) is handled is a key issue. The board reviews the original plans of the project, the mode of operation, and the justification of any risks to ensure that the potential benefits of the activity outweigh the potential costs. The board also has the responsibility of evaluating staff decisions. The reason for this evaluation is that not all problems can be anticipated by the board. Three areas of qualifications are examined: (a) sensitivity to issues of privacy; (b) personal values; and (c) general competence and ability to cope with unforeseen problems. All decisions are documented.

15. *Control title*: Separation of personal identification data

    Description: For databases that identify individuals as well as contain sensitive information about individuals, the database is separated into a file of personal identifiers and a file of data with an index linking the identifiers with the data.

16. *Control title*: Sufficient personal identifiers for a database search

    Description: To reduce the probability that an erroneous match between personal data and identification will occur, a sufficient set of personal identifiers is required before searches are permitted. Using techniques for the location of a personal record involves the ranking of several matches or near matches on several fields, such as name, date of birth, race, and sex. Because the erroneous identification, such as a criminal history or other record for an individual, may involve potential harm to the individual, the probability of a correct match should be very high. One installation identifies a sufficient set as complete name including known aliases (or maiden name if applicable), race, sex, and date of birth.

17. *Control title*: Low building profile

    Description: Buildings that house computer systems and the computer facilities should be unobtrusive and give minimum indication of their purpose. There should be no obvious signs identifying computing activities outside or inside the buildings. The buildings should look unimpressive and ordinary relative to nearby buildings. Building lobby directories and company telephone books should not identify the locations of computer activities except for offices and reception areas that serve outsiders (users, vendors, and so on) and are located separately from operational areas. Physical access barriers (locked doors, required escorts, and so on), including access control signs, guards, and receptionists, should be reasonably visible, however.

18. *Control title*: Physical security perimeter

    Description: The physical perimeter within which security is to be maintained and outside of which little or no control is maintained should be clearly established. All vital functions should be identified and included within the security perimeter. Physical access control and prevention of damage immediately outside security perimeters should be carefully considered. For example, physical barriers should extend to the base floor and to the base ceiling around sensitive areas. Areas beneath false floors and above false ceilings must be controlled consistent with the control of working areas between them. Important equipment, such as electrical power switching and communication equipment and circuits, must be made secure and included within the security perimeter. Employees and on-site vendors should be made aware of perimeters on a least-privilege basis. The perimeter should be easily discernible, simple, uncluttered, and sufficiently secure relative to the value of the assets inside the perimeter. Drawings and specifications of the perimeter should be available and used for planning any facilities changes. Additional barriers between areas with different security requirements within the exterior barrier also should be established.

19. *Control title*: Placement of equipment and supplies

    Description: Equipment, such as telephone switching panels and cables; utilities; power and air-conditioning plants; computer devices; and supplies such as paper, cards, chemicals, water, tapes, and disks, should be placed or stored to ensure their protection from damage and minimize the adverse effects they may have on other items. Dust, vibration, chemical effects, fire hazards, and electrical interference are produced by some equipment and supplies, and they should be kept separate from equipment and supplies affected by these phenomena. Items requiring special safeguards should be isolated to reduce the extent of required safeguard coverage. In multifloor buildings, vertical as well as horizontal proximity should be considered.

20. *Control title*: Security for sensitive areas during unattended periods

    Description: Sensitive areas during unattended time should be made physically secure with locked doors, significant barriers, and automatic detection devices for movement or natural disaster losses. Periodic inspection by guards and closed-circuit TV monitoring are also important. In addition, sensitive areas not generally visible to others should never be occupied by a lone employee for safety and prevention of malicious acts. Some computer-related work areas, such as the computer room, are occupied by employees at all times. Other areas and some computer rooms are left unattended for varying periods of time from several hours per day to only one or two days (such as holidays) each year. Safeguarding when employees are present and not present represents significantly different security requirements.

21. *Control title*: Areas where smoking and eating are prohibited

    Description: Smoking and eating are not permitted in computer equipment areas. Prevention requires signs, written policy, enforcement, and penalties rigorously applied. In addition, personal grooming to eliminate long hair and loose clothing should be voluntarily practiced to avoid personal injury and interference with the moving parts of peripheral equipment.

22. *Control title*: Alternative power supply

    Description: A power supply independent of the public utility source for uninterrupted service is provided by batteries charged from public utility power providing a few minutes of independent power or by an independent power source, such as a diesel generator, for longer durations. An alternative source of energy, such as a diesel generator without batteries but with adequate power quality regulators, can be used when uninterrupted service is not important, but long durations of outage are harmful. This control is needed only where power is sufficiently unreliable relative to the seriousness of computer failure or unavailability. The location, environment control, and access security are important to ensure the integrity of the alternative power equipment and fuel. Periodic full tests are important for maintenance. Some organizations use the independent sources as the primary supply and the public utility as a backup. One organization has located a new computer center at a site between two public electric power grids and obtains power alternatively from both to reduce the likelihood of public power failure.

23. *Control title*: Delivery and loading dock access

    Description: The loading dock area is made secure with the use of a window and an intermediate holding room. The window is used by truck drivers when they wish to speak to someone from the facility, have receiving papers signed, and gain authorization for access to the intermediate holding room. An employee from the inside can release the lock on a door opening onto the loading dock from the holding room. The truck driver can then unload supplies or other items onto the dock and into the holding room without having access to any other areas of the building. When the delivered material is entirely within the holding room, and when the delivery person has gone, the outside door can again be locked by the employee at the receiving window. Then an inside door leading to the holding room can be unlocked and opened for the movement of the material to its proper storage/use location.

24. *Control title*: Emergency preparedness

    Description: Emergency procedures should be documented and periodically reviewed with occupants of areas requiring emergency action. Adequate automatic fire and water detection and suppression capabilities are assumed to be present. Reduction of human injury is the first priority, followed by saving other important assets. Emergency drills that enact the documented procedures should be periodically held. It should be assumed that occupants of an area in which an emergency occurs do not have time to read emergency procedures documents before action. Procedures should include activation of manual alarms and power shutoff switches, evacuation routes, reporting of conditions, safe areas for regrouping, accounting for all occupants, use of equipment such as fire extinguishers to aid safe evacuation, and actions following complete evacuation. A hierarchy of emergency commands should be established with backup assignments. Emergency drills should be organized to minimize loss of critical activities such as computer operation. Close supervision of drills by managers who are aware of practice or real emergencies is necessary. Large, clearly visible signs providing basic directions are required. For example, switches should be clearly identified with signs that can be read from likely positions of occupants. First-aid kits should be available in regrouping areas. Emergency food, water, tools, waste disposal, waterproof equipment covers, and communication and sleeping supplies should be available for prolonged emergencies. All civil ordinances and insurance policy requirements must be met.

25. *Control title*: Minimize traffic and access to work areas

    Description: Employee and vendor work areas and visitor facilities should be located to minimize unnecessary access. Persons should not have to pass through sensitive areas to reach their workstations. Sensitive functions should be placed in low traffic areas. Traffic points through security perimeters should be minimized. Employee convenience facilities such as lavatories, lounges, lockers, and food and drink dispensers should be located to minimize traffic through barriers and sensitive areas. Toilets outside of security perimeters, such as in the lobby and receiving areas, are essential. Areas with many workstations should be separated from areas with few workstations. For example, computer peripheral equipment requiring human operation should be in rooms separate from computer equipment requiring little human attention.

    Access authorization should be granted on a privileged basis. Three access levels can be granted: *general*, *limited*, and *by exception*. General access is granted to those whose workstations are in a restricted area. In a computer room, this includes computer operators, maintenance staff, and first-level supervisors. Limited access is granted for specified periods of time to those responsible for performing specified preplanned assignments, such as auditors, security personnel, and repair or construction crews. Finally, exceptions can be made in emergencies as long as those having access are escorted and after which extraordinary measures are taken to ensure integrity of the area. *Application programmers no longer need access to computer rooms except on an emergency basis. Systems programmers need access on a limited basis. Visitors should be restricted entirely from computer rooms unless by exception and accompanied by a high-level manager who explicitly accepts responsibility and is personally accountable.* Other sensitive areas, such as programmers' offices, job setup areas, and data entry work areas, should be similarly restricted to authorized access. Signs identifying limited access areas should be posted, and rules should be strictly enforced.

26. *Control title*: Physical access barriers

    Description: Physical access through a security perimeter from a less sensitive area to a more sensitive area or between areas where different privileges apply must be limited to as few openings as possible. The remaining barrier between openings should be made of sufficiently sturdy materials to resist entry. Openings should have entrance controls consisting of one or more of the following methods:
    — Sign in/out log
    — Challenge of unauthorized entry by authorized persons
    — Challenge access by posted signs
    — Mechanically or electrically locked doors
    — Guards (local or remote using closed-circuit TV)
    — Mantrap (double) or turnstile doors

    In computer centers, limited access should be maintained for all areas except public entry lobbies, lavatories, lounges, food areas, and all areas outside the outermost security perimeter. There should be a central administration of access throughout a computer center. Procedures must be documented and include exception condition procedures. Emergency exit doors must be provided for safety and for compliance with ordinances and insurance requirements.

27. *Control title*: Remote terminal physical security

    Description: Physical access barriers, accountability for use, and resistance to visual and electromagnetic monitoring of terminals and local communication loops are maintained and periodically reviewed, consistent with security of the computer system being used from the terminal. Terminals are frequently owned or are under the control of computer users and often do not come under the jurisdiction of computer centers supplying services. Therefore, this control is directed to users or indirectly to computer center employee functions as liaison to terminal users and has the authority to disallow system access from any terminal where acceptable controls are not in place. Signed agreements are used to enforce the requirements. Resistance to visual or electromagnetic monitoring can include line-of-sight barriers to prevent reading of displays from a distance and placing terminals sufficiently removed from a security perimeter so that electromagnetic emanations would be costly to monitor. Securing work papers and terminal media should also be ensured. Locks on terminals, clearing of work areas after use, and bolting of terminals to fixed objects might be considered.

28. *Control title*: Universal use of badges

    Description: To control access to sensitive data processing facilities, all persons are required to wear badges. Different color badges, including photos in some cases, are used for employees, visitors, vendor representatives, and those employees requiring temporary badges (used when an employee has forgotten or lost his or her badge). All persons are required to wear their badges in conspicuous places on their person; visitors and in some cases everybody could be required to leave an item of identification such as a driver's license at the front desk when they are issued a badge. The decision to require badges depends on business practices, numbers of people, amount of traffic, and other access controls in use. For two or three people in an area with little traffic, the need for badges in that area may be precluded. However, minimization of exceptions may warrant their use. Positive badge administration is essential. Disciplinary action should result from infractions of the rules.

29. *Control title*: Programming library access control

    Description: Computer program libraries containing listings of programs under development and in production and associated documentation must be protected from unauthorized access. In larger organizations, a full-time or part-time librarian may be used to control access, logging in and logging out all documents. The program library should be physically separated by barriers from other activities. Documents should be distributed only to authorized users. It may be necessary to enforce strict access control to programmers' offices as a means of protecting programs and documentation. Programmers should have lockable file cabinets in which they can store materials currently in use. A clean-desk policy at the end of each working day may be justified as an extreme measure. Program and documentation control is particularly important when using or developing licensed software packages because of the strict contractual limitations and liabilities. Only production media should be outside the library; all tapes, packs, or documents should be returned upon completion of jobs. Backup tapes/packs should be removed from the premises immediately upon completion regardless of when they are run.

30. *Control title*: Separation of equipment

    Description: Different types of computer equipment (central processors, disk drives, tape drives, communications equipment, printers, power supplies, tape libraries, terminals, consoles, check sorters) require different environments for optimum operation and different numbers and types of people in attendance. Therefore, they should be placed in different rooms with appropriate separation walls, distances, and accesses. For example, printers and check sorters create dust and vibration from paper movement and should be separate from disk and tape drives that are sensitive to air quality and vibration. Central processors are normally unattended and should be in a low-traffic environment.

31. *Control title*: Electrical equipment protection

    Description: Every item of computing equipment that is separately powered should have a separate circuit breaker in the electrical supply for that equipment. Alternatively, equipment may be supplied with other protective mechanisms from power failures or other electrical anomalies. Circuit breakers should be clearly labeled for manual activation. The locations of all circuit breakers should be documented and available in disaster/recovery plans.

32. *Control title*: Electrical power shutdown and recovery
    Description: Emergency master power-off switches (manual, semiautomatic, and/or fully automatic) should be located next to each emergency exit door. The switches should be clearly identified, and easily read signs giving instructions for the use of switches should be posted. Activation of alarms and/or power and environmental equipment should be automatic. Detection devices should be placed in strategic locations. Activation of any of these switches should be followed with reports documenting the circumstances and persons responsible for their use. Alternative power supplies should be available when data processing needs justify continuous operation, and they should be tested on a periodic basis. The power supply should be used during the test for a sufficiently long period of time to ensure sustained operation under emergency conditions. Fuel supplies for alternative power should be periodically measured, and the quality of the fuel tested. Pumps, switches, and valves for switching from alternative fuel tanks should also be periodically tested. One computer installation had an uninterruptible power supply: a diesel generator and two independent, separately located oil tanks. Either tank could independently supply the entire diesel generator. Each tank was filled by a different oil company. Two diesel generators and engines were also installed for backup purposes.

33. *Control title*: Inspection of incoming/outgoing materials
    Description: Certain materials and containers are inspected, and entry or departure is restricted. Within constraints of all applicable laws and personal privacy, guards would prevent movement of materials and inspect contents of closed containers into and out of sensitive areas. Materials may include tapes, disks, listings, equipment, recorders, food and beverages, chemicals, and such containers as lunch boxes and briefcases. Some unneeded materials could be stored outside for later retrieval by owners. Authorization forms may be used to control movement. Spot checks and posted signs rather than continuous inspection may be sufficient.

34. *Control title*: Isolation of sensitive computer production jobs
    Description: Some production systems, such as those producing negotiable instruments or processing personal information (as in organized crime intelligence files, customer financial records, security transactions, medical and treatment records, and so on) are sufficiently sensitive to potential loss to require special handling. Such systems should be run on dedicated computers or only share computer systems with harmless or other trusted applications. For example, data communications access might be shut down during such a job. Some sensitive systems may be run at times when general activity is minimal, such as on Sundays, and run by an operations team especially held accountable for the operation. Extraordinary physical and computer security measures may be taken during the job run. Special marking may be done of all materials used.

35. *Control title*: Protection of data used in system testing
    Description: Application and test programmers usually need test data to develop, debug, and test programs under development. In some cases, small amounts of fictitious test data can be generated independent of users and production data. However, many application programs require significant amounts of test data that are exact copies of a full range of production data. Test data are frequently obtained as samples or entire files of production input data currently being used or recently used for the application being replaced or as output from other preprocessing computer programs. There is sometimes significant exposure by providing real, current production data to programmers. Often data can be obtained from obsolete production input files, *but in some cases even these data may be confidential*. Customers for whom production programs are being developed should be made aware of the exposure problem, and advised and assisted on how to produce test data in the least confidential but expedient manner. Sensitive test data should be treated with the same care as equivalent production data. In any case, development and test programmers should not be given access to real production files in a production computer system except in the case of an emergency, and then under highly controlled conditions.

36. *Control title*: Magnetic tape erasure
    Description: Computer centers should have magnetic tape erasure devices, commonly referred to as "degaussers," for the erasure of the contents of magnetic tapes. Such devices should be kept under strict control of the computer centers. Preferably, the device should be kept in a locked cabinet and authorized for use by selected individuals. The device should also be kept a significant distance away from magnetic tape storage areas. An erasure should be offered to computer users, and an option for tape erasure should be made available on magnetic tape disposition forms providing a date upon which erasure should be performed. All magnetic tapes used for temporary storage (scratch tapes) should also be routinely erased before reuse. Dual control or separation of functions

should be established to ensure that tapes containing valuable information are not mistakenly erased without authorization. Two sign-offs should be part of the procedures, and a log should be kept of the tapes erased.

37. *Control title*: Data classification

    Description: Data may be classified at different security levels to produce cost savings and effectiveness of applying controls consistent with the various levels of data sensitivity. Some organizations maintain the same level of security for all data, believing that making exceptions is too costly. Other organizations may have only small amounts of data of a highly sensitive nature and find that applying special controls to the small amount of data is cost-effective. When data are classified, they may be identified in two or more levels, often referred to as nonrecords, vital records, general information, confidential information, secret information, and other higher levels of classification named according to the functional use of the data, such as trade secret data, unreported financial performance, and so on.

38. *Control title*: Cryptographic protection

    Description: A high level of data communications and storage protection can be obtained by using the data encryption standard (DES). However, effective encryption key management is essential. Frequently, applications do not require this level of encryption, and much simpler forms of encryption may be used. Data compression is a particularly simple form of encryption that also increases the efficiency of data storage. Data compression can be achieved by eliminating redundant information (spacing, and so on) and by encoding data fields. *The cryptanalysis work factor should be determined and compared to the value of compromising the data being protected.* Financial institutions and military contractors processing sensitive corporate data should consider cryptographic transmission.

39. *Control title*: Correction and maintenance of production system

    Description: In spite of implementation and strict enforcement of security controls and good maintenance of application and systems programs, emergencies arise that require violation or overriding of many of these controls and practices. Occasionally, production programs will fail during production runs on the computer. This may happen on the second and third shifts during periods of heavy production computer activity. If a failure occurs in a critical application production run, it is frequently necessary to call upon knowledgeable programmers to discover the problem, make a change in the production computer program, make changes in input data, or make decisions about alternative solutions (e.g., reruns using previous versions of the production program). When such emergency events occur, all necessary and expedient measures must be taken, including physical access of programmers to computer and production areas, access by such programmers to data files and production areas, access by such programmers to data files and production programs, correction of production programs, and ad hoc instructions to operations staff. During any of these activities, it is important to record all the events as they occur or shortly thereafter. Following the termination of the emergency, programmers should be required to make the necessary and ordinary permanent changes that may have been made on a temporary basis during the emergency and document the emergency actions. This usually requires updating and testing production programs and the normal process of introducing tested updated programs for production use. After an emergency and before permanent corrections have been made, the production application program should be treated in a suspicious mode of operation requiring increased levels of observance by users, production staff, managers, and possibly data processing auditors. These extra efforts should continue until confidence has been built up in the production activities through acceptable experience.

40. *Control title*: Limited use of system utility programs

    Description: Most computer installations have one or more system utility programs capable of overriding all or most computer system and application controls. In some computer installations, one such computer program is called *Superzap*. In one large computer installation, five such utility programs existed. These programs should be controlled by password or kept physically removed from the computer system and the program library and physically controlled so that they are available only to a limited number of trusted, authorized users. Occasionally, if the programs are made available on-line, they can be protected by special passwords required for their use. Changing the name or password frequently is another way to better safeguard these on-line programs.

41. *Control title*: Production program authorized version validation

    Description: The authorized versions or copies of production programs, according to identifiers, are checked with a list of authorized copies and changes made to the production programs to determine that the version of a production program to be run is authorized. Updating the list is part of the ordinary maintenance process of production programs. Separate test and production program libraries are maintained.

42. *Control title*: Automation of computer operations

    Description: Computer operations should be made as automatic as possible, using such capabilities as production, program and test program libraries, automatic tape library management, reduction of job control by punch cards, and computer operator activity logging.

43. *Control title*: Computer user trouble calls logging

    Description: All calls from users and staff regarding problems with a computer and communications system are logged detailing the caller's name, the time and date, and the nature of the problem. A brief disposition report is then prepared for each problem report. A manager reviews each of the problem disposition reports to determine that the problem has been satisfactorily resolved and also to determine that there are no adverse impacts of the solutions (e.g., a correction of the operating system may have some side effect with a security or privacy implication). The reviewing manager also determines whether or not the responding operating person taking care of the problem was within bounds of authority. Simple requests for information are not considered problems within this procedure.

44. *Control title*: Computer program quality assurance

    Description: A testing or quality control group should independently test and examine computer programs and related documentation to ensure integrity of program products before production use. This activity is best authorized by software development management or by the quality assurance or test department. Excessively formal program development standards should be avoided. Basic life-cycle procedures should be established before more elaborate practices are required. However, compliance with the established standards and procedures should be strongly enforced. Users should be required to sign off to ensure that applications going into production meet standards and needs. This aspect of control should be generated by the user.

45. *Control title*: Computer programs change logs

    Description: All changes to computer programs are logged in a permanent written document. The log can be used as a means of ensuring formal approval of changes.

46. *Control title*: Exception reporting

    Description: Exception reporting on a timely basis should be built into the computer operating system, utility programs, and application systems to report on any deviation from normal activity that may indicate errors or unauthorized acts. For example, if a user defines a data file that allows public access, a message will be printed out warning the user, and possibly the operations staff, that the file is not protected. Exception reporting should occur when a specific control is violated, or the exception report may constitute a warning of a possible undesirable event. Exception reports should be recorded in a recoverable form within the system and, when necessary for timely action, displayed to the computer operator, or, in the case of on-line terminal use, displayed to the terminal user.

47. *Control title*: Independent control of audit tools

    Description: Audit programs, documentation, and test materials are kept in secure areas by the internal auditors. Audit programs do not remain in the data center tape library. The audit programs are not kept on disk or in any other way kept on the system where they might be subject to tampering.

48. *Control title*: Independent computer use by auditors

    Description: Audit independence can be considerably enhanced by using a computer not associated with the data processing activities being audited. Otherwise, if the same computer is being used, the computer should be used in isolation from all other activities. Where data tapes are being audited, they may be taken to a service bureau to perform audit activities.

49. *Control title*: Tape management avoiding external labels

    Description: A tape management system can be used to keep track of all tapes by using a serial number appearing on the tape reel. Serial numbers may contain storage rack location information as well as a serial number. Operators handling the tapes do not know the contents of the tapes because the identity of the data set owner, creation and update dates, data set names, and like information are recorded only on internal (machine-readable) labels. The software package for managing tapes contains an index of serial numbers and the corresponding label information. An up-to-date copy of the index relating serial numbers and tape information is maintained at off-site storage location(s).

50. *Control title*: Separation of test and production systems

    Description: When an organization is large enough to have need for more than one computer, there is a distinct advantage to limiting the development and test to one computer system and production work to another computer system. Further separation of activities can also be achieved by using multiple production systems and even multiple test systems where each application is run on a separate computer system. Likewise, each group of programmers could do testing on separate computer systems. The cost benefits of large-size and high-memory capacity would be lost, but applications could be more nearly matched to the appropriate size of computer and memory. Compilers may be moved to the test system.

51. *Control title*: Minimizing numbers of copies of sensitive data files and reports

    Description: The number of copies of sensitive tape, disk, or paper files should be minimized. Destruction dates should be specified and destruction instructions followed. It may be advisable to destroy most paper copies of files on the basis that the information can be retrieved and reprinted from computer media when necessary. This is based on the concept that files stored in computer systems and computer media are generally often more secure than on paper. Normal backup procedures often require that several copies of computer media files be made and stored at different sites. However, some files may be so sensitive that numerous copies in different locations may contribute seriously to their exposure. As many as 20 to 30 copies of computer-stored files may be produced in a single year in a large computer installation. The organization primarily accountable for highly sensitive information should have control and logs of all copies and their locations. Adequate backup must be balanced with the exposure danger of multiple copies and backup procedures.

Note: Establishing classification of documents, records, and so on is important. Dissemination should be made on a need-to-know basis.

52. *Control title*: Data file and program backup

    Description: The current form of every data file that may be needed in the future should be copied at the time of its creation, and the copy should be stored at a remote, safe location for operational recovery purposes. It may be advisable to store several copies: one immediately available in the computer center, another available some short distance away, and a third archived at some remote distance for longer-term (archival) storage. Periodically updated data files should be cycled from the immediate site to the local site to the remote site by data file generations (father, grandfather, and so on). In addition, copies of the computer programs necessary to process the backed-up data files, documentation of the programs, computer operation instructions, and a supply of special printed forms necessary for production running of the programs should also be stored at a remote, safe location. This hierarchical arrangement of backup data files provides for convenient restarting of production runs in case of damaged or missing files. More serious problems that could result in loss of local backup data files can be resolved by using copies of remote backup data files. When a backup file is returned to the computer center for use, there must be assurances that it also is backed up safely with another copy.

53. *Control title*: Secrecy of data file and program name

    Description: Names for data files and computer programs are necessary for computer program development and documentation. They are also necessary for job setup and, in some cases, for computer operation. However, file and program names need not be known by those people who are in a transaction relationship with the computer system and not concerned with programming computer applications. Therefore, a different set of terminology and naming of entities should be developed for documentation of users manuals and for transaction activities.

54. *Control title*: Input data validation

    Description: Validation of all input to a computer system should be performed in both applications and computer operating systems to assist in the assurance of correct and appropriate data. Validation should include examination for out-of-range values of data, invalid characters in data fields, exceeding upper and lower limits of data volume, and unauthorized or inconsistent control data. Program errors dependent on the content or meaning of the data can also be checked. For example, inconsistent criminal justice disposition data relative to previously entered dispositions can be flagged for manual checking and correction. The same procedure can be applied to financial validation, statistical data, and so on.

55. *Control title*: Limit transaction privileges from terminals

    Description: In addition to controlling resources (files, off-line data storage volumes, and so on), the transactions that a particular user is permitted to initiate are limited. What the system commands that a user can use, or is informed of, is controlled by the user's job duties. Thus, the systems level and application commands, such as reporting who is currently logged into the system, are restricted on a need-to-know basis. Logs may be kept for all attempts to use an authorized system command; this can be used to determine who needs training or perhaps disciplinary action. In larger systems, logging is done automatically, and after a number of unsuccessful attempts, the system is closed down to any particular terminal.

    Note: A major cause of such incidents is the sign-on of restricted terminals that remain active even though there is no use for extended periods or operators are absent from their stations. Terminals should be shut down after a reasonable period of nonuse, so that the user is required to sign on again.

56. *Control title*: Computer terminals access and use restrictions

    Description: Access to the use of all terminals owned or under the control of the organization should be restricted to authorized users. This can be done by physically securing rooms in which terminals are located and, where justified, by using metal key or electronic key locks to activate terminals. Terminals within security perimeters that are used frequently may be turned on at the beginning of the work day and left unlocked throughout the business day, then locked again at the end of the business day. Those terminals that are used only occasionally may be left locked except during use at any time of day. It may also be advisable to use various commercial locking devices to prevent terminals from being removed from assigned areas.

57. *Control title*: Contingency recovery equipment replacement

    Description: Commitments should be obtained in writing for computer equipment and supplies within a specified period of time following a contingency loss. Some vendors will commit to replacement of their products within a reasonable period of time and will specify that period of time as a commitment. For example, in one computer installation the vendor agreed to replace a central processor within five days and a second processor, if necessary, within ten days. The paper forms supplier agreed to deliver a two-week supply of all special forms in the same time frame. In contrast, other vendors would not guarantee replacement times but would only indicate that best efforts would be provided. This usually means that the next available equipment within the vendor company inventory or equipment (possibly) in transit would be provided with a priority over other normal product deliveries. Emergency ordering procedures should be established as part of a contingency recovery plan.

58. *Control title*: Disaster/recovery planning (business interruption preparedness planning)

    Description: Every computer center must have a written/tested disaster/recovery plan and a recovery management team. Primary and backup managers must be assigned specific responsibilities for each aspect of recovery from all types of partial or complete disasters. Each aspect of the disaster/recovery plan should have assigned to it a specific individual responsible for its execution; it should also be based on critical application downtime maximums. Separate individuals should be assigned to coordination, systems support, hardware recovery, facilities, administration, scheduling, communications, documentation and supplies, backup data files and security recovery funding, insurance, personnel, historical recording of events, and public affairs. Priority processing needs of all *time-dependent* applications to be recovered after a disaster must be identified.

    This requires that all computer users specify the importance of their computer applications, processing requirements and alternative means of processing, resource requirements, and consequences of failure to process. Top management as well as data processing management is responsible for meeting the critical needs of computer users in the best interests of the organization. Priorities will assist in the scheduling of processing at backup facilities or at the home site when it is restored. A designated person (the Disaster/Recovery Administrator) should serve as liaison to users informing them of special needs and the status of the processing of their work.

    A detailed history of the recovery process must be documented, and recovery activity must be verbally reported during the recovery process. After recovery, the historical documentation should be analyzed to determine how to better handle future contingencies, insurance claims recovery, and any litigation that may follow a disaster. *Every job function* and available personnel at the time of a disaster should be analyzed relative to its performance during and prior to a disaster. Measures of criticality and priority of functions should be determined and documented in the plan.

59. *Control title*: Financial loss contingency and recovery funding

    Description: Self-insured organizations, such as government agencies, should be assured of readily available emergency funds for contingencies and recovery. Specialized data processing insurance is available and should be considered when insurance covering other types of losses in a business may not apply. Financial risk protection should cover asset losses, business interruption, and extra expenses resulting from contingency recovery. Organizations not self-insured should bond all employees against fraud in high-risk areas of data processing activities. Blanket bonds will normally cover this activity.

60. *Control title*: Computer systems activity records

    Description: Most computer systems produce a number of system activity logs, journals, and exception reports. Such recordings should be periodically and selectively examined both manually and through automated means looking for key indications of possible unauthorized activities. Such recordings on tape and disk, and sometimes paper listings, should be archived for a reasonable period of time, and records should be kept to ensure that no reports are missing. For example, printed console logs should be on continuous forms. Any breaks in the forms should require signatures indicating integrity of operation and no missing pages. In one computer installation the console logs are examined on a sample basis monthly. All logs should be dated and timed with an indication of operational personnel on duty at the time the logs were produced. It may be necessary to keep manually written logs of some computer operation activities to compare with or complete the automatic logging of system activity. This is vital to the detection of theft-of-services and implanting of damaging programs.

61. *Control title*: Monitoring computer use

    Description: On a random or periodic selective basis, communications between the host computer and remote terminals are monitored. File names and contents are examined. Such monitoring must be limited to computer activity that is established for business purposes only to avoid privacy invasion. The usage is logged and analyzed to determine that the user is only doing actions that have been explicitly authorized.

62. *Control title*: Employee's identification on work products

    Description: All computer operators and other employees should have standard identifications in the form of official names, numbers, or passwords. The identification is to be entered into all records, data input, and activity logs and journals to identify workers associated with all work products. Identification can be accomplished by manual signatures or by keying of identification into equipment keyboards. Data entry clerks should be required to initial all forms or batch control forms used for data entry and enter their identifications into computer input data. Computer operators should sign computer console printer listings or enter their codes through console keyboards indicating the starting and ending of their work periods.

63. *Control title*: Data processing auditor

    Description: Organizations with internal audit resources should establish data processing audit expertise within the internal audit function. In small organizations, general auditors can acquire data processing knowledge and skills. In large organizations, full-time data processing audit specialists should be established to carry out data processing audits and assist general auditors in financial audits.

64. *Control title*: Computer security officer

    Description: An organization with sufficient computer security resources should have an individual identified as a computer security officer. In small organizations, the individual appointed may share this responsibility with other duties. In larger organizations, one or more full-time employees should be assigned computer security administration responsibilities. The computer security officer should ideally report to the protection or security department covering the entire organization. This provides proper scope of responsibility for information and its movement throughout the organization. For practical purposes, the computer security officer often functions within the computer department. Job descriptions are highly variable; examples may be obtained from many organizations with established computer security officers.

65. *Control title*: Cooperation of computer security officers

    Description: Maintaining an effective computer security function can be enhanced by exchange of information with computer security functions in other outside organizations. Local computer security organizations can be developed within a city, a part of a city, or regionally. Monthly or other periodic meetings of computer security officers can be held to exchange useful information and experience. Security seminars and workshops offered by private firms or the government provide excellent forums to evaluate security and controls instituted and used in other companies or agencies. A hotline communication capability can be established for exchange of information on an emergency basis to provide warning of possible mishaps or losses. It is important to limit the details of information exchanged to ensure that confidential controls information is not disseminated to unauthorized parties.

66. *Control title*: Responsibilities for application program controls

    Description: The inclusion of controls in application programs should be explicitly ensured and documented starting with design requirements and continuing through specifications development, production, and maintenance stages. The responsibility for adequacy and types of controls should be shared among data processing auditors, systems analysts, computer programmers, users, and data owners. Explicit documentation of controls is essential to ensure completion of their implementation, testing, and development of operational procedures to carry out the intent of the controls and to ensure their integrity during change and maintenance.

67. *Control title*: Participation of computer users at critical development times

    Description: Computer users, including those providing input data and those using computer output reports, should supply explicit control requirements to systems analysts and programmers who are designing and developing application systems. Users should also be required to explicitly agree that necessary controls have been implemented and continue to function during production use of the system and programming maintenance.

68. *Control title*: Requirements and specification participation by data processing auditors

    Description: Data processing auditors should participate in the development of requirements for important applications systems to ensure that the audit requirements in applications systems are adequate and that sufficient controls have been specified. Data processing auditors should be required to sign off on all formalized application system requirements and specifications.

69. *Control title*: Vendor-supplied program integrity

    Description: To the greatest extent possible and practical, vendor-supplied computer programs should be used without modification. Many new vendor-supplied computer programs have been developed with controls and integrity built into them. Documentation on vendor software controls should be reviewed to ensure the understanding and methodology or implementation. Any modifications to these programs will possibly compromise the built-in capabilities. Desired changes to the programs should be obtained from the vendor as standard program updates.

70. *Control title*: Technical review of operating system changes

Description: Whenever changes are to be made to the computer operating system programs, the changes must be reviewed. The intent is to make sure the new changes are valuable and will not compromise controls and integrity, have an unanticipated impact on some other part of the system, or interfere excessively with vendor updates.

71. *Control title*: Compliance with laws and regulations

    Description: A statement regarding the new or modified system's compliance with relevant laws and regulations must be provided in requirements and specifications. Direct quotes from laws and regulations regarding data processing security and privacy applying within a legal jurisdiction, or those that may apply, should be included.

72. *Control title*: Telephone access: universal selection

    Description: Limiting access to a computer and data files can be an important means of security. There are several means for accomplishing this. It may be possible and important to eliminate dial-up access to a computer. A computer interfaced to the dial-up public telephone network is exposed to access from any telephone in the world. There may be a trade-off in computer security by giving up or limiting the benefits of dial-up access. This can be accomplished by using only point-to-point wire, leased-line telephone access, or modems that will dial back to the user requesting access to the computer, thus prohibiting direct access. An alternative is to provide dial-up access to a small computer for development or other time-sharing purposes while reserving another computer for more sensitive production activity that is not interfaced to dial-up telephones. A control computer providing access to two or more other computers can also be used as a means of protecting them from dial-up access. An alternative method of restricting access is to provide for dial-up access at limited periods of time of day. During periods of dial-up access, particularly sensitive files or applications would not be resident in the computer system or secondary storage. A variation is to remove all sensitive files from secondary storage except at the explicit times of use of these files. A partial degree of protection for dial-up access systems is to maintain strict need-to-know availability of the telephone numbers and log-in protocol for accessing the computer system. Most dial-up time-sharing computer services have similar access protocols; therefore, a unique, very different initial access exchange of identifying information may be useful to limit access. The telephone numbers should be unlisted, different in pattern of digits, and have different prefixes from voice telephone numbers for the organizations that are publicly listed. A callback to verify the source of telephone access is also popular.

73. *Control title*: Terminal identifiers

    Description: Automatic terminal identification circuits can be installed in or associated with terminals for identification in host computers. Terminal identifiers are used to indicate whether a particular terminal is permitted to initiate or receive certain transactions. This access control requires that remote terminals be physically secured and that only certain known individuals be able to access remote terminals. Cryptographic devices can be used as terminal identifiers. Certain record change requests must be handled by means other than the use of these remote terminals, such as through the mail to a central facility; in this way records integrity can be preserved. Unauthorized intentional or accidental use of applications programs is prevented. A log records all unauthorized attempts to use applications programs.

74. *Control title*: Passwords for computer terminal access

    Description: Secret passwords are commonly used for access to computer systems through terminals. However, there is wide variation in the procedures for password administration. Passwords are normally accompanied by a protocol of exchange of recognition between the user and the computer, including the input from the user of a project or account number and a password. Normally, one or more users are working with the computer under a single project or account number. Occasionally, only one password is used for a group of people as well. *However, each user should have his or her own secret password.* In some cases, each user may select his or her own password; it is then known only to him or her and stored in the computer system. Others select their passwords but must receive approval of them from the computer security coordinator to ensure that they are appropriate and not easily guessed. Some organizations use computer programs to produce appropriate, easily remembered, but somewhat random passwords. In other cases, passwords are chosen by a computer security administrator and assigned to users. And finally, passwords can be generated automatically by the computer system and assigned to users. Another variation is the assignment of a password to a user with instructions that he or she is to use his or her password for initial access, at which time the user must then change his or her password in the computer system. The user should be prevented from using the initial password again. Frequently, privileged passwords are identified in the computer system so that systems programmers and others requiring password access are allowed a wider range of system usage and use special commands to carry out their work. It is generally accepted that passwords should be changed among a group of computer users who might share their passwords every time an individual leaves the immediate group by terminating his or her employment or by being given new assignments. Privileged passwords should be changed more frequently than others. Passwords should also be changed whenever there is any indication of possible system abuse or compromise. If passwords are manually conveyed to users, it should be done in confidential, sealed envelopes personally delivered by a trusted employee or orally in face-to-face conversation in confidential surroundings. A receipt should be received from the user indicating that he or she has received and accepted a new password and agrees to keep it confidential. These receipts should be kept on file by the computer security administrator. It is best to keep no paper record of passwords, and the master password file in the computer system should be encrypted or otherwise protected. If a password is forgotten by the user, it should be removed from the computer system and a new password should be assigned. The user should destroy any written record of the password once memorized, and severe penalties should be enforced for writing or revealing the password. An alternative is to keep a record of passwords locked in a safe place, such as a vault. This can be done by the project leader for each group of users and is more desirable than having a centralized record of project numbers.

75. *Control title*: Passwords generated and printed by the computer in sealed envelopes

    Description: User passwords are provided by a computerized random number/letter generator and printed directly through sealed envelopes, using the same carbon paper in envelope techniques that are used for many direct deposit receipts. These sealed envelopes are delivered directly to the user without the password ever having been seen by humans. Because the user expects a new password at a certain time, a missing envelope will be noticed and the previously generated password will be canceled and reissued. Similarly, if an envelope is opened or has evidence of tampering the password is canceled and reissued. Receipts are returned to ensure delivery.

76. *Control title*: Dynamic password change control by the user

    Description: Users are allowed to change their passwords any time once they have logged in to the system. A parameter can be set at log-in time or at any time during a logged-in session that prevents changing a password. This would be useful when an individual logs in to the system, but gets up and leaves the terminal for a short period of time, and does not want anyone to come along and change the password while he or she is away. The user must enter a new password twice to prevent an incorrect password entry caused by a typing error. If the second password is not the same, the user must begin again.

77. *Control title*: Terminal log-in protocol

    Description: The protocol for logging into a computer system from a computer should be designed to reduce unauthorized access. The terminal response to a log-in should provide a minimum of information to avoid providing an unauthorized user with any assistance. No system identifying information should be provided until the full user identification process has been successfully completed. There should be no feedback aids to an unauthorized user at any time during the log-in process that would provide clues to correct or incorrect input. Incorrect input should result in no assistance, and the system should disconnect. When the user identification and password are being typed in, there should be no intermediate feedback from the system that indicates whether the system has accepted any partially completed identification input. This requires a user to enter the complete set of identification and password information before there is any indication of whether or not this information is correct. Identification information should consist of the user name or other nonsecret identification, such as an account number, followed by input of the secret password. Display terminals should provide display suppression while the password is being typed in to avoid its observation by another person. Printer terminals should provide nonprinting character mode or provide underprinting and overprinting of the spaces where the password is printed on the page. Additional, personal questions may be posed by the computer system to be answered by the terminal user to further ensure correct identity. No more than three attempts at entry of an unacceptable identification or password should be allowed. Three unsuccessful attempts should cause a telephone line disconnect. Time delay after an incorrect identification or password input of several seconds should occur to increase the work factor of automated exhaustive search for passwords. Also, a limited amount of time should be allowed for entry of a password before a telephone disconnect is performed. A variation of a password should be provided as a duress alarm. For example, if an individual is being forced to enter his or her password at a terminal, he or she might interchange the last two characters, which would result in an immediate alarm at the host computer system that an entry is being attempted under duress. Any log-in that deviates from normal or accepted ranges of activity should be noted in an exception report at the host computer console in a timely manner for immediate action by a computer operator. All log-ins, whether authorized or unauthorized, should be journaled for later audit trail analysis. A means of allowing an unauthorized terminal user to gain authorized access to the system under totally monitored conditions should be provided to assist in locating sources of unauthorized attempts. Unauthorized users can be provided enough benign services to keep them at the terminal long enough for other detection activity to take place. Each time authorized users log into the system successfully, they should be provided with information concerning the date and time of the last time they logged into the system. Other information about their last sessions may also be summarized. Users can be made aware of any possible unauthorized use of their password in this manner.

78. *Control title*: Computer system password file encryption

    Description: The password file in the computer system contains master copies of passwords to verify correct identification and password input from terminal log-ins. This data file is one of the most sensitive in the entire computer system and, therefore, must be properly protected. Passwords in the file should be individually encrypted using a one-way encryption algorithm (i.e., the password can be encrypted, but there is no reasonable means of decryption that would be computationally feasible given the current state of the art in switching speeds and cryptanalysis). When a password is entered from a computer terminal, it is immediately encrypted using the same algorithm and compared with the encrypted form of the master password for matching. In this manner, clear text passwords reside within the computer system for the shortest possible amount of time.

BASELINE SECURITY
ORGANIZATIONAL STATUS TEST

Use this checklist to see how well your organization measures up to baseline security.

		Yes	No
Prevent Asset Responsibility Loss			
Assets Accountability Assignment	1.	☐	☐
Data Accountability Assignment to Users	2.	☐	☐
Separation and Accountability of Data Processing Functions	3.	☐	☐
Computer Security Management Committee	4.	☐	☐
Remote Terminal User's Agreement	5.	☐	☐
Prevent Disclosure, Taking, or Unauthorized Use of Documents			
Confirmation of Receipt of Documents	6.	☐	☐
Discarded Document Destruction	7.	☐	☐
Proprietary Notice Printed on Documents	8.	☐	☐
Courier Trustworthiness and Identification	9.	☐	☐
Keeping Security Reports Confidential	10.	☐	☐
Prevent Modification, Disclosure, or Unauthorized Use of Obsolete or Incomplete Input/Output Data			
Suppression of Incomplete or Obsolete Data	11.	☐	☐
Completion of External Input Data	12.	☐	☐
Prevent Disclosure or Unauthorized Use of Personal Information			
Personal Data Input/Output Inspection	13.	☐	☐
Human Subjects Review	14.	☐	☐
Separation of Personal Identification Data	15.	☐	☐
Sufficient Personal Identifiers for a Database Search	16.	☐	☐
Avoid Destruction of Assets and Interruption of Business			
Low Building Profile	17.	☐	☐
Physical Security Perimeter	18.	☐	☐
Placement of Equipment and Supplies	19.	☐	☐
Security for Sensitive Areas During Unattended Periods	20.	☐	☐
Areas Where Smoking and Eating Are Prohibited	21.	☐	☐
Alternative Power Supply	22.	☐	☐
Delivery and Loading Dock Access	23.	☐	☐
Prevent Human Injuries and Other Damages From Contingencies			
Emergency Preparedness	24.	☐	☐
Prevent Unauthorized Access to Sensitive Areas			
Minimize Traffic and Access to Work Areas	25.	☐	☐
Physical Access Barriers	26.	☐	☐
Remote Terminal Physical Security	27.	☐	☐
Universal Use of Badges	28.	☐	☐
Programming Library Access Control	29.	☐	☐
Prevent Damage to Equipment			
Separation of Equipment	30.	☐	☐
Electrical Equipment Protection	31.	☐	☐
Electrical Power Shutdown and Recovery	32.	☐	☐
Prevent Unauthorized Taking and Facility Damage			
Inspection of Incoming/Outgoing Material	33.	☐	☐
Prevent Compromise of Data			
Isolation of Sensitive Computer Production Jobs	34.	☐	☐
Protection of Data Used in System Testing	35.	☐	☐
Magnetic Tape Erasure	36.	☐	☐
Data Classification	37.	☐	☐
Cryptographic Protection	38.	☐	☐

		Yes	No
Prevent Unauthorized Program or Data Modification			
Correction and Maintenance of Production System	39.	☐	☐
Limited Use of System Utility Programs	40.	☐	☐
Production Program Authorized Version Validation	41.	☐	☐
Automation of Computer Operations	42.	☐	☐
Detect Computer, Application and Communications Systems, and Operations Failure			
Computer User Trouble Calls Logging	43.	☐	☐
Computer Programs Quality Assurance	44.	☐	☐
Computer Programs Change Logs	45.	☐	☐
Exception Reporting	46.	☐	☐
Prevent Interference with Auditing			
Independent Control of Audit Tools	47.	☐	☐
Independent Computer Use by Auditors	48.	☐	☐
Prevent Loss, Modification, Disclosure, or Destruction of Data Assets			
Tape Management Avoiding External Labels	49.	☐	☐
Separation of Test and Production Systems	50.	☐	☐
Minimizing Numbers of Copies of Sensitive Data Files and Reports	51.	☐	☐
Data File and Program Backup	52.	☐	☐
Secrecy of Data File and Program Name	53.	☐	☐
Input Data Validation	54.	☐	☐
Limit Transaction Privileges From Terminals	55.	☐	☐
Computer Terminals Access and Use Restrictions	56.	☐	☐
Recover From Business Interruption			
Contingency Recovery Equipment Replacement	57.	☐	☐
Disaster/Recovery Planning (Business Interruption Preparedness Planning)	58.	☐	☐
Financial Loss Contingency and Recovery Finding	59.	☐	☐
Detect Unauthorized System Use			
Computer Systems Activity Records	60.	☐	☐
Monitoring Computer Use	61.	☐	☐
Detect Unauthorized Activities of Employees			
Employee's Identification on Work Products	62.	☐	☐
Prevent Inadequacy of System Controls			
Data Processing Auditors	63.	☐	☐
Computer Security Officer	64.	☐	☐
Cooperation of Computer Security Officers	65.	☐	☐
Responsibilities for Application Program Controls	66.	☐	☐
Participation of Computer Users at Critical Development Times	67.	☐	☐
Requirements and Specification Participation by Data Processing Auditors	68.	☐	☐
Vendor-Supplied Program Integrity	69.	☐	☐
Technical Review of Operating System Changes	70.	☐	☐
Avoid Violations of Laws and Regulations			
Compliance With Laws and Regulations	71.	☐	☐
Prevent Unauthorized Computer Access			
Telephone Access: Universal Selection	72.	☐	☐
Terminal Identifiers	73.	☐	☐
Passwords for Computer Terminal Access	74.	☐	☐
Passwords Generated and Printed by the Computer in Sealed Envelopes	75.	☐	☐
Dynamic Password Change Control by the User	76.	☐	☐
Terminal Log-in Protocol	77.	☐	☐
Computer System Password File Encryption	78.	☐	☐

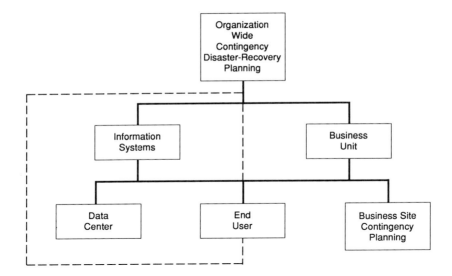

INTERACTIVE SESSION: SCOPE OF PLAN

Overview

The recognition for formalized disaster/recovery planning beyond the scope of data processing, and even disaster/recovery planning for data processing, has been very slow in coming. As companies have grown to become more dependent on their data processing operations, and data processing is completely dependent on the activities and input of the users, organizations are beginning to comprehend just how dependent and vulnerable they are and will continue to be.

Organizations still believe that the requirements for disaster/recovery planning are limited to data processing, since disaster/recovery planning has become synonymous with this activity. And although there is an understanding that data processing is sensitive to conditions beyond the ability of data processing management to protect the entire environment, the concept of disaster/recovery planning as it relates to business activities outside of data processing is totally foreign to management.

It is only recently that planning of this nature has been given some consideration in small- to medium-sized organizations, who by virtue of their size alone are more vulnerable and more susceptible to failure in the event they lose their data center or major user activity.

Restoration of Business Functions as Important as Data Processing Recovery

Let's take a look at the problems inherent in maintaining business performance, above and beyond the ability to recover data processing operations. If we use the following list, we can visualize the problem of minimizing the scope of disaster recovery.

1. Identify the business functions of the organization (other than data processing) that must be recovered in order to sustain critical services.
2. Determine if provision has been made to sustain the senior management functions. Have all the legal requirements in maintaining the corporation under all conditions been met?
3. Users have defined their disaster/recovery requirements for their function.
4. Funds and other resources have been budgeted and assigned for the developing, implementing, training, testing, and maintaining the plan.

Insurance Only Provides Dollars . . . Not the Ability to Continue Business Services

The cost of business interruption is high on two counts:

1. It costs a lot of money.
2. It does not provide for backup and recovery requirements.

The development of a disaster/recovery plan for data processing and the rest of the organization is a reasonable investment to ensure business survival. There are other considerations to which the organization has a responsibility that should be contained inside the plan:

- The organization has an ethical responsibility for the safety and welfare of individuals.
- The organization has a social responsibility to the community.
- The organization has a business responsibility to customers and vendors.

- The organization has a legal responsibility to customers, shareholders, and employees.

How to Determine Whether You Can Sustain Business Performance if the User Community Has a Disaster

If you are looking for a real eye-opener, conduct a risk assessment of your user departments and determine how long you would stay in business if one or more users lost its ability to operate its activities. Estimate your cost of downtime. *Do not include data processing operation.* For the purposes of this interactive session, consider a disaster that will adversely affect (destroy or disable) the premises of the organization's primary business location.

The disaster scenario represented by a worst-case situation is, or should be, the motivating factor to look beyond the data center for the development of a *business* disaster/recovery plan in addition to the data processing plan.

Identifying the Scope of the Plan

Modern business organizations are no longer simple entities. The inner workings, the actions, interactions, and interfacing of departments, divisions, subsidiaries, and so forth are highly complex, performed with specialized technology and dependent upon accurate, timely, and audited information. Consider what functions in your organization are vulnerable to natural or human-engineered disasters.

Exercise 1: Reviewing the Organization and Determining the Facility Scope of the Disaster/Recovery Plan

On Form SOP-1, evaluate the scope of your disaster/recovery plan. This is broken down into two segments. The first part of defining scope is to define all the facilities the organization presently occupies. In the second part list in order (priority) the succession of the plan development, and what functions of the organization will be involved in the plan development (Exercise 2). When identifying the critical facilities, the following should be considered:

1. Will the loss of this facility affect the normal performance of the business?
2. Is the senior management of the organization housed in this facility?
3. How many business functions are housed in this facility?
4. Is the data processing operation housed in this facility?
5. How many people are housed in this facility?
6. Can the rest of the company maintain business performance if this facility did not recover for one week, two weeks, one month, and so forth?
7. Which of these functions are computer-dependent or noncomputer operations?

Instructions for Use of Form SOP-1

Column 1: Facility, Location

Identify each of the locations of the company by listing the facility name and location. Use the space below to insert any other pertinent information.

Column 2: Occupants

Identify the department, division, or function housed at the location.

Column 3: Total Personnel

Indicate the total number of people in each of the departments, divisions, or functions listed in column 2.

Columns 4 and 5: Owned or Leased

Put a check in the box indicating whether the facility is owned (O) or leased (L).

Columns 6 and 7: Noncritical or Critical Activity

Indicate by a check mark whether the specific function is noncritical (NC) or critical (C).

Form SOP-1

Facility, Location	Occupants	Total Personnel	O	L	C	NC
A.						
B.						
C.						

Form SOP-1

Facility, Location	Occupants	Total Personnel	O	L	C	NC
D.						
E.						
F.						

Form SOP-1

Facility, Location	Occupants	Total Personnel	O	L	C	NC
G.						
H.						
I.						

Exercise 2: Identification of Critical Lines of Business and Supporting Functions

Overview

The success of the data processing or organizational disaster/recovery planning program is based on the determination of the critical product and/or services, including "upstream" support or staff function. If the company's overall disaster/recovery plan is to provide the continuity of business services, it is important to address every aspect of critical functions. Management must carefully review the list of critical product and/or services to ensure that only the really *critical* functions are included. The ensuing plan will be dramatically diluted in effort, create a far more complex plan than may be needed, add substantially to the backup and recovery costs, and create a risk of recovery failure if the scope of the plan goes beyond the really critical functions.

The basic assumption in approaching disaster/recovery planning for the organization is that a plan for the information processing support services is clearly the first objective of the program. The selection of the critical functions must parallel the critical applications. This exercise is used to determine the critical lines of business and supporting functions and double checks as to whether all the critical applications have in fact been listed. Again, this is a management and user responsibility.

Critical Business Functions

Outside the realm of disaster/recovery planning for data processing operations, where critical applications have been selected and approved, the selection of critical business function is almost determined by default. Although the use of critical data processing applications is an excellent guideline in pinpointing critical business functions, critical application selection alone is not the only criterion. If critical applications are selected by the data processing department and approved by management, a vital step has been excluded from the process: *consultation and input from the user. It is important to note in doing this exercise that all critical business functions are not necessarily supported by computer processing*, and if this is the case in your organization, these critical support services have a tendency to be overlooked.

Critical applications/business functions can be identified as those activities which, in the event of a loss of services, have a direct dollar loss. Further, critical business functions and critical applications can have an indirect loss which can be as detrimental to the organization as a financial loss. Some of these indirect losses are as follows:

1. The inability to make strategic, tactical, or operation decisions owing to a lack of information
2. Business discontinuity
3. Loss of cash flow control
4. Contractual agreement violation
5. Legal/regulatory compliance
6. Increased operational expenses not accounted for and included in insurance coverage
7. Loss of customer base, or at a minimum, poor customer relations
8. Reduction in sales and marketing efforts
9. Loss of highly qualified personnel
10. Loss of public confidence
11. Improper or erroneous financial statements
12. Loss of business opportunities
13. Loss of vendor confidence, credit, and so forth
14. Loss of logistical and operational control
15. Erratic purchasing policies and decisions
16. Loss of shareholder confidence
17. Poor production, distribution, and marketing decisions
18. Increased conflict with personnel and unions
19. Failure of branch, division, or agency support and communication
20. Loss of ongoing research and development control

Regardless of the loss, each of the adverse incidents imparts an impact on the organization at the same rate. Impacts are based on *time-critical* or *time-sensitive* business services. In the airline reservation activity, losses start the moment service is interrupted. Even though a business function/service is considered critical to the organization, you may not consider its inclusion in the disaster/recovery plan if its interruption would not cause losses until an extended period of time has elapsed (see "Interactive Session: Maximum Allowable Downtime and Cost of Downtime").

In determining the scope of the plan, the criticality of business services and functions must be evaluated based on the impact of the business interruption. Data processing is the most obvious of the critical support services. Other likely support functions that are needed in order to resume critical business functions include transportation, communications, and purchasing. Corporate staff involved in public relations and the security department have special responsibilities during an emergency and should develop disaster/recovery plans as well as be part of other departments' disaster/recovery plans.

The only method to determine the scope of the disaster/recovery plan is to perform a management-level, mandated criticality assessment. Senior management at corporate, departmental, or divisional level must create the priority; final approval of the plan must be made by senior corporate management.

Sample List of Users/Services

The following are some of the critical business functions that would create business paralysis or business amnesia should the present data processing support function fail.

1. Finance/general ledger
2. Payroll
3. Sales
4. Marketing
5. Manufacturing
6. Material inventory (raw materials)
7. Inventory picking
8. Logistics/shipping/receiving
9. Accounts payable
10. Accounts receivable
11. Research and development
12. Engineering
13. Quality control
14. Risk management/actuarial
15. Data center/information processing
16. Corporate headquarters
17. Senior management (CEO, president, vice presidents)

Form SOP-2

	Business and Supporting Function	Capacity Level of Business and Supporting Functions					
		Loc.	VH	H	M	L	VL	NC
1	Finance/General Ledger							
2	Payroll							
3	Sales							
4	Marketing							
5	Manufacturing (DP Process Controlled)							
6	Material Inventory (Raw Materials)							
7	Material Inventory (Finished Goods)							
8	Inventory Picking							
9	Logistics/Shipping/Receiving							
10	Accounts Payable							
11	Accounts Receivable							
12	Research and Development							
13	Engineering (Cadcom)							
14	Quality Control							
15	Data Center/Information Processing							
16	Risk Management/Actuarial							
17	Marketing Research							
18	Application Development							
19	Branch Office (Financial)							
20								
21								
22								
23								
24								
25								
26								
27								

VH = Very High L = Low
H = High VL = Very Low
M = Medium NC = Noncritical

Form SOP-2

	Business and Supporting Function	Capacity Level of Business and Supporting Functions					
		Loc.	VH	H	M	L	VL	NC
1								
2								
3								
4								
5								
6								
7								
8								
9								
10								
11								
12								
13								
14								
15								
16								
17								
18								
19								
20								
21								
22								
23								
24								
25								
26								
27								

VH = Very High   L = Low
H = High   VL = Very Low
M = Medium   NC = Noncritical

Use this chart to develop your own critical business functions.

Form SOP-2A Final Plan Selection Priority Chart

Priority	Business Activity	Location	Start Date	Responsibility
1	Data Center			

Exercise 3: Objectives of Plan by Indirect Losses

Objective: Plan Development

	Indirect Loss Potential	Critical Unit Business Service	Function Selection
1	Ensure ability to make strategic decisions.		
2	Ensure ability to make tactical decisions.		
3	Ensure ability to make operational decisions.		
4	Avoid business discontinuity. (Select those business services that will create the most severe conditions if interrupted.)		
5	Avoid loss of cash-flow control		

Objective: Plan Development

	Indirect Loss Potential	Critical Unit Business Service	Function Selection
6	Avoid contractual agreement violations.		
7	Avoid legal/regulatory violations.		
8	Maintain continuity to assure customer satisfaction.		
9	Maintain vendor relationship.		
10	Assure stability of qualified personnel base.		

Objective: Plan Development

	Indirect Loss Potential	Critical Unit Business Service	Function Selection
11	Provide sufficient support to enssure ability to take advantage of all business opportunities.		
12	Avoid breakdown of logistical and operational control.		
13	Aviod erratic purchasing decisions.		
14	Maintain continuity of research and development.		

Objective: Plan Development

	Indirect Loss Potential	Critical Unit Business Service	Function Selection

Blank chart for you to develop your own.

Objective: _____ Plan Development

	Indirect Loss Potential	Critical Unit Business Service	Function Selection

Objective: _____ Plan Development

	Indirect Loss Potential	Critical Unit Business Service	Function Selection

Objective:		Plan Development	
	Indirect Loss Potential	Critical Unit Business Service	Function Selection

INTERACTIVE SESSION: MAXIMUM ALLOWABLE DOWNTIME AND COST OF DOWNTIME

Overview

The performance, workings, operations, and functions of a business, financial institution, or governmental agency in today's accelerated environment is a highly complex structure. Unlike their predecessors, information-hungry, information-dependent, information-deluged organizations are operating with specialized technology and are totally reliant on accurate, complete, and timely information. The dispersal of on-line reliant services, the growth of interaction between customer and supplier, and the spread of extremely complex computer and communications technology has substantially reduced an organization's ability to recover from an adverse incident within a reasonable time frame. Thus, there is a growing concern about this additional vulnerability that exists within the organization as natural and human-engineered disasters grow in numbers.

Just surviving a disaster is not the answer. There is a need for broader and more fundamental planning at the management level. *First*, disaster/recovery planning must be recognized as a business planning process, not a technical plan dedicated to data processing operations. *Second*, it is *management's* responsibility to determine what critical *business* services and resources must be recovered in order to maintain the continuity of business performance. Although data processing is understandably a critical resource, it does not constitute the only scarce or difficult-to-replace resource in an organization. *Third*, there must exist specific management policy on the maximum allowable downtime to recover operational and functional resources in order to sustain the continuity of critical business, financial, or governmental services.

The various interactive sessions in this seminar manual should provide the keys for identifying critical business functions and critical data processing applications. The objective of this exercise is to view how long your organization can maintain business activities at the time of an adverse incident, as well as determine the "cost" (in terms of dollars and other business losses) of downtime.

Our objective is to highlight first, how a business deteriorates without critical support functions, and second, to determine the maximum allowable downtime in order to ensure that critical functions should be restored in order to maintain the continuity of the business services.

Viewing Cost

Loss of critical support services (computer or noncomputer activities) can create a domino effect on business performance. In responding to the activities of this interactive session, visualize how your organization's business activity would degenerate (deteriorate) and what the "domino effect" would be throughout the organization, ultimately affecting external business activities and relationships. A number of business factors must be considered beyond the actual financial losses. These could be, but are not limited to:

- Failure to fulfill contractual obligations
- Increased expense in performing normal business functions
- Poor business decisions (or the lack of decisions) owing to the inability to obtain proper information
- Cessation of research and development
- Degeneration of vendor relationship
- Loss of credit standings
- Loss of customers
- Loss of confidence and organizational credentials
- Potential legal implications and/or penalties

Some losses may occur minutes after a disaster; others may take hours or days. Contingency actions may alleviate momentary degradation of services, but without the ability to recover critical support services within an acceptable (identifiable) time frame, chaos may be the order of the day.

There seems to be an acceptance by management that plans are needed to prevent a major catastrophe within the organization in the event of a disaster, evidenced by having a group participate in this program. The ability of such a small group to influence management concerning the ongoing support necessary to sustain a program such as disaster/recovery planning is usually difficult. Preparing scenarios such as the maximum allowable downtime and the cost of downtime gives credence to the reality of an organization's vulnerability.

Modern Technology Is Its Own Disaster

All of us recognize the sensitivity of today's growing dependence on computer information support. Beyond this, our expansion of these systems into every facet of operation and business performance increases the *manmade* vulnerabilities, errors, malicious attacks, and fraud. Add to this the growing list of external hazards of chemical spills, radiation leaks, falling aircraft, accidental or malicious destruction of communications, the increased incidence of earthquakes in known zones (and in areas earthquakes have not been normally active), toxic fumes, regular and forest fires, floods, tornadoes, killer storms, and so forth. Regardless of the concentration of activities within a single location, or dispersal of operations over a wide geographic area, the risk of serious business disruption does not seem to diminish.

Exercise 1: Maximum Allowable Downtime

Using Matrix 1, chart the maximum allowable downtime (estimate) for each of the business functions detailed. A blank matrix is provided so that you can create a business function chart to more accurately represent your organization. Here, we are not looking for losses; we are more interested in determining how long each of these functions can be sustained without support services, and when computer backup must be provided for them. There are two objectives to this exercise:

1. Define those critical business services and when they must return to operation.
2. Assist in determining the strategies to be employed based on their restoration of support requirements.

DIA*log 3-DAY DISASTER/RECOVERY PLAN PREPARATION WORKSHOP
- © DIA*log 1987

MATRIX 1

INTERACTIVE SESSION
MAXIMUM ALLOWABLE DOWNTIME

CHECK APPROPRIATE BOX WHEN SERVICES MUST BE RESTORED

USER FUNCTION APPLICATION GROUP	DAYS — 1st WEEK							DAYS — 2nd WEEK							3rd WEEK	4th WEEK	5th WEEK	HOURS				NA
	1	2	3	4	5	6	7	1	2	3	4	5	6	7				1	2	3	4	
1. FINANCIAL SERVICES																						
a. Accounts Receivable																						
b. Accounts Payable																						
c. General Ledger																						
d. Payroll																						
e. Commissions																						
f. Premiums/Incoming																						
g. Debt Service																						
h. Interest Payments																						
2. CONTRACT FULFILLMENT																						
a. Order Processing																						
b. Purchasing																						
c. Logistics																						
d. Inventory Control																						
e. Manufacturing																						
f. Assembly																						
3. SALES																						
a. Inquiries																						
b. Ordering Processing																						
c. Estimating																						
d. Shipping Schedules																						
4. RESEARCH & DEVEL.																						
5. APPLICATION DEVELOPMENT																						
6. FUNDS TRANFER																						
7. MARKETING																						
a. Research																						
b. New Product																						
c. New Market																						
8. RESERVATIONS SYSTEM																						
9.																						
10.																						
11.																						
12.																						

DIA*log 3-DAY DISASTER/RECOVERY PLAN PREPARATION WORKSHOP

• © DIA*log 1987

INTERACTIVE SESSION
MAXIMUM ALLOWABLE DOWNTIME

MATRIX 1

CHECK APPROPRIATE BOX WHEN SERVICES MUST BE RESTORED

USER FUNCTION APPLICATION GROUP	DAYS — 1st WEEK	DAYS — 2nd WEEK	3rd WEEK	4th WEEK	5th WEEK	HOURS	NA
	1 2 3 4 5 6 7	1 2 3 4 5 6 7				1 2 3 4	

Exercise 2: Cost of Downtime

Form 1 is used to list, by time, the deterioration of critical support services and its impact on the organization. Assume that the entire data center has been destroyed (worst-case situation) and that the primary users are restricted from entering the damaged facility. Assume also that no backup provisions have been made, and that data stored off-site (together with programs) can only recover the system to midnight of the previous day.

Time After Disaster

We have listed time periods when major impacts usually affect an organization. If these do not comply with your findings, use the blank area next to these times to create your chart.

User Function

Select a specific user activity. If you feel that the loss of support services has an immediate impact on the entire organization, change the user to the entire organization.

Effect

Stop the deterioration effects on services.

Suggested Strategy

This will be discussed at the seminar.

FORM 1

PAGE ___ of ___

INTERACTIVE SESSION
COST OF DOWNTIME
IMPACT & DETERIORATION OF BUSINESS SERVICES

COMPLETE AT LEAST 2 USER FUNCTIONS • 3 FORMS SUPPLIED

TIME AFTER DISASTER	YOUR TIME	USER FUNCTION	EFFECT	SUGGESTED STRATEGY
1 hr				
2 hrs				
3 hrs				
6 hrs				
12 hrs				
24 hrs				
1 day				
2 days				
3 days				
4 days				
5 days				
6 days				
7 days				
1 week				
2 weeks				
3 weeks				
4 weeks				
5 weeks				
6 weeks				
7 weeks				
8 weeks				
9 weeks				
10 weeks				

FORM 2

INTERACTIVE SESSION
COST OF DOWNTIME
ACTUAL COST IN DOLLARS

TIME AFTER DISASTER	USER 1	USER 2	USER 3	TOTAL
1 hr				
2 hrs				
3 hrs				
6 hrs				
12 hrs				
24 hrs				
1 day				
2 days				
3 days				
4 days				
5 days				
6 days				
7 days				
1 week				
2 weeks				
3 weeks				
4 weeks				
5 weeks				
6 weeks				
7 weeks				
8 weeks				
9 weeks				
10 weeks				

© DIA*LOG 1987

INTERACTIVE SESSION: EFFECTS AND IMPACT ON CRITICAL SUCCESS FACTORS—ORGANIZATION GOALS AND OBJECTIVES

Overview

Companies run successfully when they plan. Well-managed and disciplined organizations have specific critical success factors—goals and objectives. These are usually prepared as a "road map" for directing the organization's present and future activities. The coalition of departmental performance provides the planned profit, efficiency, and operational goals of the company. Unexpected, undesirable, adverse incidents represent a threat to the organization's ability to perform normally and to the achievement of its goals and objectives.

Further, these catastrophic events have the ability to affect the very survival of the organization, at worst, or substantially reduce the profitability of the company and delay future growth for an indefinite period of time.

This interactive session is designed to highlight the key issues and economic impact that *will* result from the interruption or the destruction of the computer support services or key user activity. This interactive session, like all the others performed during the disaster/recovery preplanning seminar, is designed to create a greater awareness of the critical issues that will affect the normal performance of business.

When responding to the impact chart on page 325, view the activities of your organization on a real-time basis. By creating a high level of awareness of the effects and impact, the importance of implementing preventive and protective

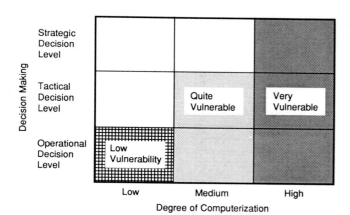

Figure 1 The degree of vulnerability to computer power loss in relation to the level of computerization penetration. (Figure 1 and chart on pages 324, 325, and 327 reproduced through the courtesy of Management Advisory Publications; *How to Prepare an EDP Contingency Plan for Business Continuity*, Javier F. Kuong and Gerald I. Isaacson, Wellesley Hills, MA, 1986.)

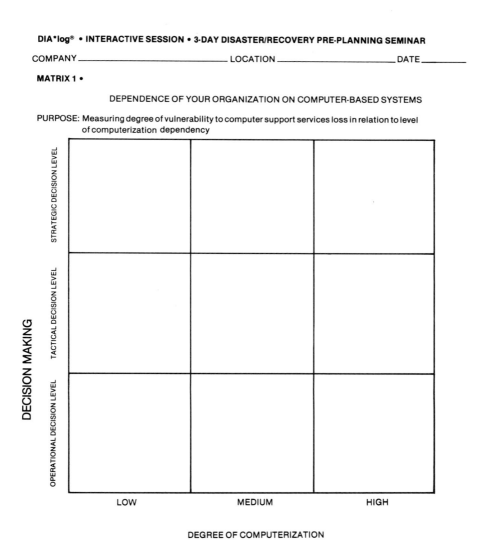

Figure 2 The three levels of decision making and the impact of computer systems on each level.

Decision Characteristic	Decision Making Level	Relative impact of computer systems on desision level
• Decisions are highly judgmental. • Non-repetitive. • A human actually makes decisions based upon information system. • Computerized decisions are usually a small precentage of tasks and decisions made via computers at this level.	S T R A T E G I C	• Impact on organizations is usually limited as managers can continue to operate without computers. • The major impact is one of lost convenience. • If managers are very atuned to the use of computers and rely very heavily on them to operate, the degree of dependence can be more serious.
• Decisions may have a great deal of judgment, but some of them could be automated or heavily aided by a computer system. In certain types of applications, such as manufacturing planning and control, the dependency on the computer may be a great deal more serious than for strategic systems. • On-line/real-time systems which drive business operations (airline systems).	T A C T I C A L	• If many of the decisions are automated, then a loss of the information processing facility would inflict serious damage and potential business continuity could be seriously jeopardized, especially when the systems are of the advanced/real-time interactive type.
• Decisions are of a routine nature and highly repetitive. • The computer is used as means of keeping clerical costs down and to handle large volumes of data on timely basis. • In an emergency, business procedures could be handled by manual means, even if mthis slows the pace of business.	O P E R A T I O N A L	• The degree of importance of the computer on business continuity depends upon the extent of the automation and whether the volume of transactions handling and speed with which the information is needed is paramount to running the business. If the extent of computerization is drastic and not easily revertible to manual procedures, then the impact is likely to be more serious.

measures to negate or mitigate the affects of an adverse incident becomes more apparent. It is important to recognize that what you and your peers might consider technical issues, realistically derive from normal business and management considerations. Further, once the effects of adverse incidents are recognized as issues beyond a technical concern, this aspect of planning should be mandated by management, properly staffed and funded.

The loss of operating effectiveness, resulting from the loss of computer support services or key user(s), is based on time; time without support services is based on dependency (on computer and/or users); and degradation of operational effectiveness relates specifically to length of time without support. The graph on page 325 shows the various deteriorating stages that you or any organization can go through before losing full operational readiness in relation to the time period after the adverse incident has occurred. When doing the impact assessment chart, keep this chart as well as Matrix 1 handy. Make sure that you have correlated the degree of vulnerability with the potential degradation of business activities (include the impact on critical success factors from Form I-1). The underlying issue is that the loss of critical support services and/or key user departments that are unable to resume support could create total and irrevocable business failure.

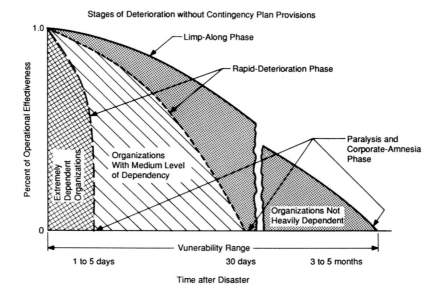

Stages of Deterioration without Contingency Plan Provisions

DIA*log 3-DAY DISASTER/RECOVERY PRE-PLANNING SEMINAR

COMPANY _____ LOCATION _____ DATE _____

FORM I-1 AFFECTS & IMPACT ON CRITICAL SUCCESS FACTORS AND ORGANIZATION GOAL AND OBJECTIVES

• ADD ADDITIONAL ITEMS THAT MIGHT PERTAIN TO YOUR ORGANIZATION

PAGE 1 of 2

N/A		INABILITY TO SUPPORT CRITICAL SUCCESS FACTORS OR MEET ORGANIZATIONAL GOAL AND OBJECTIVES	IMPACT DATA PROCESSING			USER: IMPACT		
			Major	Medium	Minor	Major	Medium	Minor
	1	Accuracy and Integrity of Information						
	2	Timeliness of Information						
	3	Maintenance of Business Continuity						
	4	Security, Fraud, Vandalism, Malicious Damage						
	5	Welfare of Personnel						
	6	Compliance with Legal/Regulatory Requirements						
	7	Legal Implications						
	8	Contractual Fulfillment Obligations						
	9	Operational Effectiveness/Efficiency						
	10	Public Relations • Public/Shareholder/Customer Relations						
	11	Vendor Relations						
	12	Credit Rating/Relations Financial Community						
	13	Credibility • Corporate Stature						
	14	Cash Flow						
	15	Secondary Cash Flow Affects						
	16	Profitability • Short-Term						
	17	Profitability • Long-Term						
	18	Account Receivable						
	19	Operating Logistics • Supplies/Materials/Shipping Warehouse/Inventory Control						
	20	Customer Service Dispatching						
	21	Administrative Services						
	22	Payroll						
	23	Financial Management						
	24	Financial Statements/Budgetary Requirement						

© DIA*log Management, 1987

N/A			IMPACT DATA PROCESSING			IMPACT USER:		
			Major	Medium	Minor	Major	Medium	Minor
	25	Financial Statements/External Audit						
	26	Auditability and Audit Trails						
	27	Competitive Position						
	28	Compromise of Data/Data Manipulation						
	29	Privacy/Confidentiality • Exposure						
	30	Illegal Access						
	31	Generally Accepted Accounting Procedures						
	32	Management Decision						
	33	Loss of External Communication						
	34	Errors and Omissions						
	35	User Discontinuity						
	36	Labor Relations						
	37	Affects on Third Party						
	38	Policy • Standards • Procedures						
	39							
	40							
	41							
	42							
	43							
	44							
	45							
	46							
	47							
	48							
	49							
	50							

MATRIX 2 — INTERACTIVE SESSION • 3-DAY DISASTER/RECOVERY PRE-PLANNING SEMINAR
DOCUMENTING THREATS • HAZARDS • VULNERABILITIES
PURPOSE: Determining Probable Causes and Impacts on Business Critical Success Factors

COMPANY _____ DATE _____
PAGE ____ OF ____

CAUSE ▼ / IMPACT ▶	FINANCIAL LOSS	OPERATIONS STOPPAGE	CUSTOMER RELATIONS	LEGAL/REGULATORY IMPLICATIONS	DATA INTEGRITY	CASH FLOW MAINTENANCE	VENDOR RELATIONS PURCHASING FUNCTIONS	CONTRACT FULFILLMENT	BUSINESS CEASES	COMPETITIVE ADVANTAGE	PRIVACY/CONFIDENTIALITY LOSS	PERSONNEL & UNION RELATIONS	SHAREHOLDER & PUBLIC RELATIONS	BRANCH • AGENCY • INTERNATIONAL COMMUNICATIONS	LOGISTICS & OPERATIONS CONTROL	BASELINE SECURITY
EXAMPLE: TOTAL DESTRUCTION OF DATA CENTER	✓	✓	✓	✓	✓	✓	✓	✓	✓	✓	✓	✓	✓	✓	✓	✓

© DIA*log Management, 1987

Index

Actual cash value, 30
Administrative team, 34–35, 40
Air-conditioning, 132
    failure of, as data center disaster, 23–24
All employee training, 162
Alternatives analysis, 6–8
Alternatives identification, 26–29
Annual loss expectancy (ALE), 12
    calculation of, 8, 25–26, 29
    derivation of, for each risk, 26
    development of, 121
Application development standards, 134–135
Application recovery plan, 42, 46–48
    in contingency plan implementation, 4, 57
    for data center production applications, 42–43
    work sheet for, 48–51
Application recovery priorities:
    essential applications, 3
    priority scheme, 3–4
    ranking criteria for creating applications recovery, 3–4
    recovery thresholds, 3
    service-level needs, 4
Applications team, 33, 37
Application system documentation, 231–232
Area layout, 168
Assets, theft of, 25
Assumed downtime maximum, 123
Auditing staff, role of, 135–136
Audit team, 41
Authorized entrants, 169
Awareness training, 60

Backup site activation teams, 33, 36, 37
Backup site strategy, 139
Backup site test, 163
Banking, regulation of, 8
Baseline security:
    interactive session on, 285–299
    organizational status test on, 300–302
Broadcast, employee notification by, 150
Building layout, 167
Business interruption preparedness planning, 90
Business interruption recovery plans:
    components of, 240–241
    four-day workshop on, 245–246

Business interruption recovery plans (*Cont.*):
    one-day workshop on, 243–244
    personnel participation, 241
Business vulnerability analysis:
    dependency on electronic data processing (EDP), 1
    developing qualitative and quantitative risk model, 2–3
    identification of EDP-dependent business processes, 2
    legal requirements, 2
    management's viewpoint in, 1–2
    restoration policy for critical applications, 3

Calling tree:
    for all employees, 140–141
    in contingency plan implementation, 57
    for disaster/recovery teams, 140
    pyramid, 231
Card-key access, 28
Card key holders, 169
Change management, 64–67
Checklists, 165
Cold site, 7
Cold site contract, 9
Combination locks, 28
Command (control) center, 35–37
Commercial cold site, 7, 8
Commercial hot site, 7, 8
Communications team, 33, 37
Company medical facilities hospitals, 233
Computer dependency, 9, 10
Computer equipment failure:
    countermeasures against, 29
    risks of, 24
Computer and noncomputer disaster/recovery planning questionnaire, 247–263
Computer security risks, 9
Computer virus, 26
Computer worm, 26
Confidential information, theft of, 25
Confidentiality statement, 203
Constant voltage transformer, 29
Contingency plan implementation:
    file backup, 57
        backup strategy, 59

Contingency plan implementation (*Cont.*):
        features of off-site storage location, 59–60
        hard copy, 59
        microcomputers, 59
        number of generations saved, 59
        reasons for, 57, 59
        recovery up to minute, 60
        retention policy, 59
        scratch tape management, 59
        tape rotation strategy, 60
    preparedness standards, policies, and procedures
        application recovery plans, 4, 57
        calling tree, 57
        disaster kit, 56
        employee commitment letters, 57
        forms book, 56
        purchased software and software escrow program, 56–57
        security of contingency plan, 57
        vendor commitment, 56
        vendor list, 56
Contingency planning, 9
Control center, 149–150
Cooperative agreement, 7, 8, 9
Cooperative cold site, 7, 8
Cooperative hot site, 7, 8
Corporate controller's activity, 46–48
Corporate data center file retention philosophy, 43–55
    cost, 44
    criticality, 44
    file activity and dump frequency, 44
    file management, 43
    interdependencies, 43
    legal requirements, 43
    summary of standard retentions, 44–45
Cost-benefit analysis, 29
Cost justification, in choosing backup strategy, 8
Cost-of-downtime evaluation, 13
Cost savings opportunity analysis, 30–31
CPR and first-aid training, 162
Credit cards, 148
Crime, classifying, 26
Critical success factors, interactive session on effects and impact on, 322–327

Damage assessment, 37–38

Data center:
    access to, 128
    layout for, 166–167
Data center production applications, applications recovery plan:
    requirements for, 42–43
Data center risk analysis:
    computer equipment failure risks, 24
    environmental control risks, 22
        air-conditioning failure, 23–24
        power failure, 22–23
    logical security risks, 25
        hackers and malicious damage, 25
        loss expectancies, 25–26
        theft of assets, 25
        theft of confidential information, 25
        theft of secret information, 25
    network failure risks, 24
        equipment failure, 25
        line failures, 25
    physical security risks, 19–20
        cold and hot weather, 22
        earthquakes, 22
        fire, 20
        flood, 21
        labor disputes, 22
        lightning, 22
        rioting, 22
        terrorism, 22
        theft, 22
        vandalism, 22
        wind, 22
    quantitative risk model, 26
    threats analysis, 19
Data collection:
    application system recovery, 231–233
        application system documentation, 231–232
        backup facilities, 232
        communications, 232
        computer activities, 232
        database management support, 232
        documentation, 232–233
        emergency communications, 233
        emergency shelters, 233
        first aid and emergency medical services, 233
        mutual assistance, 233–234
        technical support documentation, 232
        vendors, contracts, contacts, 232
        vital records reclamation team action, 233
    communications, 229
    data center operating procedures and policies, 229
    data processing facility layouts, 225–226
    data processing planning coordination, 224
    disaster/recovery information- gathering task force, 223
    emergency action plan responsibilities, 230
    emergency response plan, 230–231
    facility layout, 227

Data collection (*Cont.*):
    hardware/auxiliary equipment, 226–229
        contracts and agreements, 228
        emergency services, 228
        forms, 227
        hardware, 227
        office equipment, 228
        office recovery sites, 228
        off-site storage, 228–229
        services, 228
        software, 227
        supplies, 227–228
    information collection, 221
    insurance notification, 234
    insurance reporting and loss recovery, 234
    management's participation, 223
    normal communications, 229–230
    organizational charts, 226
    personnel assignments and participation, 221–222
    personnel training, 235
    plan maintenance, 235
    plan preparation workshop, 235
    plan test monitoring, 235
    pyramid calling tree, 231
    questionnaires, checklists, guidelines, and forms, 223–224
    recovery/repair and restoration, 234
    risk assessment, 226
    security systems, 229
    senior management team, 222–223
    task force assignments, 224
        recommended courses of action, 224
        task requirements, 225
    task force disaster/recovery coordinator, 225
    task forces and disaster/recovery action teams, 222
    testing plan, 235
    user coordination/cooperation, 234–235
    user operating procedures, 229
    workshop disaster/recovery administrator, 225
Datacomm configuration, 139–141
Datacomm security, 133–134
Data diddling, 26
Data Encryption Standards Devices, 133–134
Data leakage, 26
Data preparation team, 34, 39
Data processing facility layouts, 225–226
Data processing planning coordination, 224
Decision criteria, 136
Dedicated disaster/recovery centers, as backup alternative, 13–14
Dial-up lines, versus leased lines, 4–5
Diesel generator, 29
Disaster, probabilities of, 27, 122
Disaster countermeasures selection:
    cost benefit analysis, 29
    countermeasures against computer equipment failure, 29
    countermeasures against fire, 26–27

Disaster countermeasures selection (*Cont.*):
    countermeasures against intrusion, 28
    countermeasures against power problems, 28–29
    countermeasures against water damage, 28
Disaster kit, 142–143
    in contingency plan implementation, 56
Disaster-notification fee, 14
Disaster/recovery action teams, 151–159
Disaster/recovery administrator, duties and responsibilities of, 118–119
Disaster/recovery emergency response plan, 273
    emergency call list, 274
    emergency procedures, 276
        bomb threat, 280–281
        chemical spill, 278–279
        civil disturbance, 280
        earthquake, 279
        evacuation routes, 281
        fire/explosion, 276–278
        flood, 279–280
        plant shutdown procedures, 282–283
        tornado, 279
    emergency responsibilities, 275
Disaster/recovery information-gathering task force, 223
    disaster preparedness, 136
        backup site strategy, 139
        datacomm configuration, 139–141
        disaster kit, 142–143
        disaster/recovery action teams, 151–159
        employee commitment letters, 145
        file backup, 136–137
        forms books, 143–144
        insurance, 150–151
        logistical considerations, 147–150
        mainframe configurations, 139
        monthly supply list, 143
        off-site storage, 138–139
        priorities for limited processing environment, 145–147
        software escrow, 145
        vendor information, 142
        vital records retention, 144
Disaster/recovery management team:
    charter, 119
    members, 119
    risk assessment, 120–124
    scope, 119
Disaster/recovery plan, 9
    application recovery plan worksheet, 175
    area layout, 169
    authorized entrants, 169
    authorizing management data, 170
    backup facility requirement and general information, 184
    backup site agreement, 172
    backup site alternatives, 171–172
    backup site datacomm configuration, 172

Disaster/recovery plan (*Cont.*):
  backup site mainframe configuration, 172
  building layout, 167
  card key holders, 169
  certification checklists, 177
  characterization of test, 179
  confidential employee questionnaire, 174–175
  copies of policies and annual premiums, 175–176
  current operations schedule, 175
  damaged items inventory, 176
  data center layout, 166–167
  data center sign-in sheet, 170
  disaster kit, 174
  distribution and security, 165
    copies, 165
    list of recipients, 165
    physical security of plan, 166
    planned enhancements, 166
    screening information, 165
    security and confidentiality agreement, 166, 181
    serialization, 165
    termination or transfer of copy holder, 165
  emergency call list, 168–169
  escrow agreement, 174
  evacuation route, 170
  hardware replacement, 176
  home phone numbers, 173
  information security policy, 115
  instructor's guide for seminar on, 195–202
  interactive session on scope of, 302–316
  maintenance agreements, 170
  management's letter of commitment, 115
  monthly checklists, 182
  monthly supply list, 174
  off-site agreement, 170–171
  personal liability of officers, 115–116
  phases of, 159–161
  plan postscript, 115
  planned enhancements, 181, 183
  postscripts to, 166
  preface for, 97–107
  preplanning and assumptions, 116–117
    disaster/recovery planning and data collection team, 117–119
    other manuals, 124–125
    plan initialization, 125
    risk assessment, 120–124
    role of disaster/recovery management team, 125
  preplanning seminar/workshop, 186–189
  prevention/security, 125–126
    air-conditioning, 132
    application development standards, 134–135
    application purchase, 135
    application support standards, 135
    datacomm security, 133–134
    document security, 132
    emergency response plan, 130–131

Disaster/recovery plan, prevention/security (*Cont.*):
    fire protection, 128–130
    housekeeping, 131
    isolation transformer, 132
    maintenance agreements, 132
    medical security, 128
    physical security, 126–128
    power, 131–132
    role of auditing staff, 135–136
    role of user departments, 136
    software security, 132–133
  primary site datacomm configuration, 172
  primary site mainframe configuration, 172
  probable backup site operations schedule, 175
  problem log, 179
  procedures for reporting and obtaining damaged/lost clothing and property, 184
  purchased software, 174
  purpose of plan, 114–115
  required ancillary equipment, 172
  review of, 161
  security and confidentiality agreement, 181
  table of contents for, 90–97, 108–114
  tape library report, 171
  team plans, 177
  testing schedule, 178–179
  threat evaluation form, 177–178
  update/control, 164
    approval, 164
    checklists, 165
    formal review, 164
    update log, 164, 179–181
  vendor information, 171
  vendor lists, 173
  vendor support letters, 173
  vital records retention, 174
  workshop for planning:
    cost and expense considerations of disaster/recovery strategies, 242
    creating environment for, 191–193
    facilities, office equipment, supplies, forms, and communications, 242
    functional plan components, 242
    functions, operations, and services criticality, 242
    other considerations, 242
    other resource requirements, 242
    records management, 242
    risk assessment, 242
Disaster/recovery planning, 9
Disaster/recovery planning and data collection team, 117–119
Disaster/recovery planning seminar, 190
Disaster/recovery planning team, 32, 34, 35
Disaster/recovery plan preparation workshop, 190–191
Disaster scenarios, 122–123

Distributed network, as backup alternative, 17–18
DMS-based system, 44
DMSII-database backups, 44
Document security, 132
Documents off-site, 138
Downtime:
  costs of, 123
  interactive session on maximum allowable and costs of, 317–322

Earthquakes, as data center disaster, 22
EDP-dependent business processes, identification of, and business vulnerability analysis, 2
EDP recovery requirements:
  computer system sizing, 4
  end-user computing, 5
  implications of future applications, 5
  network configuration, 4–5
  software configuration, 4
EDP restart/recovery testing, 61
EDP teams recovery procedures:
  interim processing, 41
  mobilization, 38
  restart processing, 38, 41
  restoration, 41–42
Electronic data processing (EDP), dependency on, and business vulnerability analysis, 1
Emergency action plan responsibilities, 230
Emergency call list, 168–169
Emergency communications, 233
Emergency medical services, 233
Emergency power off and lighting, 27
Emergency response/action coordinator, 131
Emergency response plan, 130–131, 230–231
Emergency response team, 32, 35
Emergency shelters, 233
Employee commitment letters, 57, 145
Employee fidelity insurance, 151
Employee notification:
  by broadcast, 150
  and emergency phone numbers, 127, 140
Empty shell, 7, 14–15
End-user computing, 5
End users recovery procedures:
  outage period processing, 45, 51
  restart processing, 51, 55
Environmental control risks, 22–24
Equipment failure, 25
Expanded scope of program, 240

Facilities evaluation, interactive session on, 68–87
Facilities team, 34, 39
File backup, 136–137
Fire:
  countermeasures against, 26–27
  as data center disaster, 20
Fire alarms, 129
Fire drills, 130

Fire extinguishers, 27, 129
Fire hoses, 129
Fire protection, 128–130
First aid, 233
Flood, as data center disaster, 21
Foreign Corrupt Practices Act (1977), 2
Formal review, 164
Forms books, 143–144
Fortress strategy, 6, 8
   with full redundancy, 16–17

Getting the job done, 185–193
Guards, 126

Hackers and malicious damage, 25
Halon flooding, 27
Hardware/auxiliary equipment, 226–227
Hardware/software team, 33, 36
Hot site, 7, 13–14
Hot site contract, 9
Housekeeping, 131
Housing, 148

Industry, maximum downtime allowed by, 2
Insurance, 150–151
   employee fidelity, 151
   as not an alternative for advance planning, 19
Insurance coverage, 29
   cost savings opportunity analysis, 30–31
   coverage analysis:
      types of insurance, 30
      valuation method, 30
Insurance notification, 234
Insurance reporting and loss recovery, 234
Interface disaster/recovery miniplan model for financial institution branch office, 265–267
   emergency response plan, 267
      addenda, 269
      assumptions in activation of plan, 266–267
      branch interface with computer operations, 270
      contracts, and agreements, 270
      critical services, 269–270
      definition of disaster terms, 267
      emergency response and procedures, 268
      establishing operation and control center, 268–269
      evacuation procedures, 268
      forms, checks, and special documents, 270
      incident reporting, 267–268
      objectives of, 266
      operations policy and procedures, 270
      personnel safety, 268
      staffing, training, and maintenance program, 270–271
      supplies, new equipment, and hardware, 269

Interface disaster/recovery miniplan model for financial institution branch office, emergency response plan (*Cont.*):
      table of contents, 269
      vendor contact list, 269
Interim processing, 41, 55
Intrusion, countermeasures against, 28
Intrusion alarms, 127
Intrusion detectors, 28
Isolation transformer, 29, 132

Labor disputes, as data center disaster, 22
Layouts, 126
Leased lines, 134
   versus dial-up lines, 4–5
Lightning, as data center disaster, 22
Limited processing environment, priorities for, 145–147
Line failures, 25
Locks, 28, 126
Logical security risks, 25–26
Logic bombs, 26

Mainframe configurations, 139
Mainframe failure, 24
Maintenance agreements, 29, 132
Man trap, 28
Manual backup, as backup alternative, 18
Manual processing, 6, 8
Medical alert, 127
Medical security, 128
Microcomputers, as backup alternative, 17
Microfiche reports, 137
Motor generator, 29
Multidrop circuits, 25
Multiple data centers, as backup alternative, 16
Mutual aid agreements, 15
Mutual assistance, 233–234

Network failure risks, 24–25
New hardware team, 34, 38
No backup strategy, 13
Null strategy, 6

Office space, 149
Off-site rotation schedule of current tapes, 138
Off-site storage, 138–139
   desirable features of, 59–60
Off-site storage team, 33, 36
Off-site tests, 63
Ongoing maintenance, 64–67
Ongoing testing, 64
On-site tests, 61, 163
Operations team, 33, 37
Organizational charts, 226
Organizational status test, 285–302
Organization and participant profile, 203–219
Outage period processing, 45, 51

Pack family usage rules, 136
Partner agreements, 15
Performance and procedures profile:
   addressing the corporatewide problem, 237–238
   executive summary, 237
   functional requirements review, 241
   initiating corporatewide disaster/recovery plan, 239
   management considerations in prevention and security planning, 240
   organizational activity for initial response, 241–242
   outline of work of project initiation performance, 238
   preliminary activities, 238–239
Period-end file retentions, 137
Period-end tapes, off-site archive of, 138
Peripheral failures, 24
Personnel training, 60
PERT chart, 41
Physical security, 126–128
   risks of, 19–20
Piggybacking, 26
Point-to-point circuits, 25
Portable sites, as backup alternative, 18
Positive approach, need for, in backup planning, 19
Power, 131–132
   countermeasures against problems of, 28–29
   failure of, as data center disaster, 22–23
Preparedness testing, 61, 62
Primary site recovery teams, 34, 38, 39, 41
Priority scheme, 3–4
Private cold site, 7, 8
Private hot site, 7, 8
Problem log, 163
Public relations team, 40
Purchased software and software escrow program in contingency plan implementation, 56–57
Pyramid calling tree, 231

Qualitative and quantitative risk model, 2–3, 26

Reciprocal agreements, as backup alternatives, 15
Reconstruction, 136
Records management, 242
Recovery management cycle, 35
Recovery management plan:
   command center, 35–37
   damage assessment, 37–38
   notifications, 38
   organization of recovery teams, 32, 33
      administrative and special teams, 34–35
      backup site activation teams, 33, 36, 37
      disaster/recovery management team, 33, 36

Recovery management plan, organization of recovery teams, (Cont.):
    disaster/recovery planning team, 32, 34, 35
    emergency response team, 32, 35
    primary site recovery teams, 34, 38, 39
    user teams, 34, 39
  recovery management cycle, 35
Recovery/repair and restoration, 234
Recovery strategy, 6
  alternative analysis, 6–8
  backup alternatives, 9–10
    alternate processing strategy, 10
    disaster/recovery planning, 9
  contracts and agreements, 9
    cold site contract, 9
    cooperative agreement, 9
    hot site contract, 9
  cost justification, 8
  strategy selection, 8
Recovery team training, 162
Recovery thresholds, 3
Redundancy, 29
Reliable equipment, purchase of, 29
Replacement cost, 30
Restart processing, 38, 41, 51, 55
Restoration, 3, 41–42
Results analysis, 64
Rioting, as data center disaster, 22
Risk assessment, 120–124, 226, 242
Risk management, 10
  defining application-recovery requirements, 11
    communications, 11–12
    compatibility, 11
    contracts, 13
    cost considerations, 12–13
    environment, 11
    evaluating alternatives, 13–19
    location, 12
    physical capacity, 11
    security, 12
    testing, 12

Salami techniques, 26
Salary checks, 148
Salvage team, 34, 38, 41
Scavenging, 26
Scenarios planning, 61
Scratch floppy disk management, 139
Scratch tape management, 59, 138–139
Secret information, theft of, 25
Security and confidentiality agreement, 67
Senior management team, 222
Service bureaus, 7, 8
  as backup alternative, 13
Shared contingency facilities, membership in, 15–16
Site selection, 28
Software escrow, 145
Software security, 132–133
Special teams, 34–35
Sprinklers, 27
Superzapping, 26
Supplies team, 34, 40
Surprise test, 63, 163
Surveillance cameras, 126–127

Tape library management system, 44, 138
Tape rotation strategy, 60
Task force disaster/recovery coordinator, 225
Task forces and disaster/recovery action teams, 222
Technical training, 60
Terminal floppy dumps, 137
Terminals, physical security of, 133
Terrorism, as data center disaster, 22
Terrorist organizations, 23
Testing (see Validation and testing)
Test schedule, 163
Theft, as data center disaster, 22
Threats analysis, 19
Time brokers, 7, 8, 15
Time-division multiplexers, 25
Training for disaster recovery, 161
  all employee training, 162
  CPR and first-aid training, 162
  recovery team training, 162
Transportation, local, 149
Transportation team, 34, 40
Trap doors, 26

Trojan horse, 26

Unauthorized use of computer time, 26
Uninterruptible power system, 29
Unoccupied, unrented, surplus office buildings as backup alternative, 18–19
Update log, 164
User coordination/cooperation, 234–235
User departments, role of, 136
User liaison team, 34, 39, 41, 55
User operations teams, 34, 39
User teams, 34, 39

Validation and testing, 162, 163
  backup site test, 163
  EDP restart/recovery testing, 61, 63
  end-user restart/recovery testing, 63
  problem log, 163
  off-site tests, 63
  on-site tests, 61, 163
  preparedness testing, 61, 62
  scenarios planning, 61
  surprise test, 63, 163
  test schedule, 163
  validation, 163
Vandalism, as data center disaster, 22
Vendor commitment, in contingency plan implementation, 56
Vendor facility, as backup alternative, 17
Vendor information, 142
Vendor list, in contingency plan implementation, 56
Vendor product list, 142
Vendor support letters, 142
Vital records reclamation team action, 233
Vital records retention, 144

Water damage, countermeasures against, 28
Weather, as data center disaster, 22
Wind, as data center disaster, 22
Wire tapping, 26
Workshop disaster/recovery administrator, 225
Workshop task force coordinator, 222–223
Worst-case assumption, 124

ABOUT THE AUTHOR AND TECHNICAL EDITOR

ALVIN ARNELL, the creator of the planning methodology presented in this Handbook, is President of DIA*log Management, Inc., a Long Island–based corporation devoted to assisting organizations worldwide in the development of prevention, security, emergency response, crisis management, and contingency-disaster/recovery plans for corporations, business functions, and data-processing operations.

Prior to establishing DIA*log Managment, Inc. in 1980, Mr. Arnell's experience and expertise in corporate planning led him to develop emergency response training programs and install prevention and security measures in compliance with OSHA requirements within the paper converting and other specialized industries.

Mr. Arnell's 43-year career in senior management in a broad range of companies in business, automation engineering, industry, commerce, and consulting, has exposed participating companies in the DIA*log program, lectures, and presentations to a unique, in-depth appraisal and perspective of contingency-disaster/recovery planning as it applies to their specific data-processing operations, business continuity, departmental functions, environment, and corporate culture.

In assisting over 165 organizations in the development of their contingency-disaster/recovery plans, Mr. Arnell and his associates are responsible for further enhancing information processing security by aiding organizations in the implementation of internal investigative auditing and preparation and enforcement of computer security standards, policies, and procedures. He has also served as an internal consultant to many of the corporations participating in the Seminar/Workshop Program.

Mr. Arnell has published many articles relating to the various subjects that comprise the total contingency-disaster/recovery planning process, and has been a frequent speaker at educational and professional group meetings throughout the world.

Mr Arnell is the author of *Standard Graphical Symbols for Science and Engineering*, published by the McGraw-Hill Book Company, 1962.

DONALD DAVIS is Executive Vice President of DIA*log Management, Inc. Prior to joining DIA*log, Mr. Davis held a number of senior-level positions during a 31-year career with Burroughs Corporation (UNISYS). In these positions he had wide-ranging responsibilities in field engineering, quality assurance, data center management, hardware and software application, new systems development and installation, and customer and corporate technical support and education.

Mr. Davis has traveled extensively for Burroughs to present contingengy-disaster/recovery planning education and training programs. He was responsible for the initation of disaster/recovery planning and the establishment of extensive standards and procedures for the protection and security of Burroughs data centers worldwide.

Subsequent to his retirement from Burroughs, Mr. Davis joined DIA*log to apply his expertise to assisting seminar/workshop participant corporations in developing contingency-disaster/recovery plans.

DIA*log DISASTER/RECOVERY PLANNING SEMINAR/WORKSHOP
MANUALS & SUPPLEMENTAL DOCUMENTS

As a purchaser of the *Handbook of Effective Disaster/Recovery Planning*, the author and DIA*log Management, Inc. have made special arrangements with McGraw-Hill Publishing Company, to make their entire Seminar/Workshop Program available to your organization.

The following manuals and documents compose the entire Seminar/Workshop Program. These are available from DIA*log Management, Inc. to supplement this *Handbook Of Effective Disaster/Recovery Planning*, and will assist planners in developing their in-house Seminar/Workshop Program on Disaster/Recovery Planning. All of the documents offered are used as training materials in the performance of the Disaster/Recovery Planning Seminar/Workshop Programs presented by DIA*log, worldwide.

3-Day Disaster/Recovery Preplanning Seminar Materials

1. *Effective Disaster/Recovery Reference Manual*
   57 sections, over 650 pages, provide the disaster/recovery planner with a wealth of educational materials.
2. *Baseline Security Techniques Manual*
   An invaluable tool in establishing a superior security and controls program throughout the organization.
3. *Disaster/Recovery Planning Interactive Session Manual*
   All 25 interactive sessions described in this Handbook are contained in this manual, together with all of the instructions, forms, and details.
4. *Security and Facilities Evaluation Checklists Manual*
Over 350 pages of checklists and questionnaires relating to disaster/recovery planning.

4-Day Disaster/Recovery Plan Preparation Workshop

5. *Disaster/Recovery Master Plan*
   This manual can be used by planners in developing their organization's or data center's disaster/recovery plan.

Supplemental Documents and Seminar Workshop Materials

A complete listing is available on request.

Organizations purchasing the complete DIA*log Program, will continually receive up-to-date revisions and additional documents as they are prepared, at no additional cost for a period of one year.

Order/Inquiry Form—Description of Other Documents Supplied

**FOR MORE INFORMATION ABOUT THE DIA*log MANAGEMENT
DISASTER/RECOVERY SEMINAR/WORKSHOP PROGRAM AND MATERIALS . . .**

Please send us the following information:

Name _____ Title _____

Company _____

Address _____

City _____ State _____ Zip _____

Telephone No. Area Code ( ) _____

Fax No.   Area Code ( ) _____

[ ] Yes, I am interested in receiving complete details about the DIA*log In-House Disaster/Recovery Planning Seminar/Workshop Program.

[ ] Please send me literature concerning the purchase of the entire DIA*log Disaster/Recovery Seminar/Workshop Program described above, so that I can use it to develop a comparable program for my organization.

[ ] Please contact me as soon as possible, as we are interested in having DIA*log Instructor/Consultants present the Seminar/Workshop Program methodology for the development of our Disaster/Recovery Plan.

• Do you presently have a Disaster/Recovery Plan? [ ] Yes   [ ] No
• Are you planning to initiate a disaster/recovery planning program in the near future? [ ] Yes   [ ] No   When _____
• Will you be responsible for this program once it is initiated? [ ] Yes   [ ] No
• Are you interested in Disaster/Recovery Planning for:
  [ ] Data Center only
  [ ] Corporate planning

Please Send Your Inquiries to:

Alvin Arnell, President
DIA*log MANAGEMENT, INC.
1945 BYRON ROAD, MERRICK, NY 11566
Tel. (516) 867-3333 or fax inquiry (516) 623-8955

Order/Inquiry Form—Description of Other Documents Supplied